D1765794

CHEMISTRY OF SULPHUR DIOXIDE IN FOODS

CHEMISTRY OF SULPHUR DIOXIDE IN FOODS

B. L. WEDZICHA

Procter Department of Food Science,
University of Leeds, UK

ELSEVIER APPLIED SCIENCE PUBLISHERS
LONDON and NEW YORK

ELSEVIER APPLIED SCIENCE PUBLISHERS LTD
Ripple Road, Barking, Essex, England

Sole Distributor in the USA and Canada
ELSEVIER SCIENCE PUBLISHING CO., INC.
52 Vanderbilt Avenue, New York, NY 10017, USA

British Library Cataloguing in Publication Data

Wedzicha, B. L.
Chemistry of sulphur dioxide in foods.
1. Food—Sulphur dioxide content
I. Title
641.1′7   TX553.S/

ISBN 0-85334-267-9

WITH 27 TABLES AND 56 ILLUSTRATIONS

© ELSEVIER APPLIED SCIENCE PUBLISHERS LTD 1984

Printed in Great Britain by Galliard (Printers) Ltd, Great Yarmouth

To Glenis and Alexandra

Preface

The important and long-standing use of sulphur dioxide as a food additive, its widespread presence in the atmosphere and its technological importance in the chemical and other industries, such as paper making, has led to a proliferation of scientific literature from a wide range of disciplines on the chemical behaviour of sulphur(IV) oxospecies. Despite this, it is only in the last 10–15 years that some of the more fundamental aspects of sulphur(IV) oxoanion chemistry have been elucidated: for example, unequivocal evidence for the structure of hydrogen sulphite ion in solution was published as recently as 1979. During the last decade a detailed explanation of what is perhaps the best known reaction of hydrogen sulphite ion, that with carbonyl groups, has emerged. There is a continuing debate concerning the mechanism of autoxidation of sulphite ion. However, the understanding of such basic questions as the nature of the bonds between sulphur and oxygen is still incomplete. Perhaps the fascination with reactions of sulphur(IV) oxospecies is not only their exceptional diversity but also the completely unexpected reactivity which has frequently punctuated research in this area. An important new reaction discovered some 15 years ago is the sulphite catalysed deamination of cytosine responsible for sulphite induced mutagenic effects. More recently the well known cleavage of thiamine by sulphite has been shown to proceed by way of a surprising intermediate and there is also interest in the ability of sulphite ion to promote oxidation of unsaturated natural products. Despite this activity, there is still comparatively little known concerning the interactions between sulphur(IV) oxospecies and food components. The reactions of sulphur dioxide and sulphites in foods is a relevant issue to pursue since foods are frequently exposed to this additive at elevated temperatures and for long periods of time during storage when high concentrations may also be present. The efficacy of the additive depends on its wide reactivity with food components producing a spectrum of function which is greater than that of any other food additive, but which also leads to extensive conversion to reaction products. These have been only partly identified. The biochemical function of the additive in controlling microbial action and its effects on food enzymes is an area of considerable interest. The accumulation of knowledge over the last two decades in the area of interactions between sulphur(IV) oxospecies and components of biological systems, and in the basic chemistry of the oxospecies, renders timely an appraisal of the present understanding of reactions relevant to food systems.

This book assumes no knowledge of the chemistry of sulphur(IV) oxospecies, but it is taken for granted that the reader will have a grounding in basic chemistry and is familiar with nomenclature in biochemistry. Its aim is to enable the non-specialist chemist, biochemist or technologist to gain an appreciation of the factors responsible for the reactivity of sulphur(IV) oxospecies in a variety of situations and to illustrate these reactions by reference to food components and food systems. Thus, the reader is introduced to the structure and solution properties of sulphur(IV) oxospecies, to generalisations concerning their reactivity and to a number of inorganic and organic reactions which form the 'core' of knowledge for application in later chapters. The conventional methods of analysis of sulphur(IV) oxospecies in foods are both time consuming and subject to error. There is also a continuing requirement in the food industry for rapid methods of analysis which are automatic or instrumental. Much attention is therefore given to a consideration of both established and new methods for the analysis of this additive and also to the separation and identification of products arising from the reactions of sulphur(IV) oxospecies with food components. The reactions with food components have been elucidated in studies on model systems not necessarily intended to simulate foods. The known activity of sulphur(IV) oxospecies with respect to non-enzymic browning, enzyme action and microbial growth is first considered from the point of view of the basic chemistry and biochemistry of the processes and these findings are then discussed, in Chapter 6, in the context of food systems as a whole. During the last 15 years there has also been a considerable increase in knowledge concerning the toxicology of sulphur(IV) oxospecies. The current state is reviewed in the final chapter.

The subject matter of this book is therefore relevant to the practical worker in the food industry or in food research, those working in analysis as well as to those concerned with environmental aspects of sulphur dioxide.

I would like to express my thanks to Dr S. M. Hammond, Mr E. J. Redfern and Prof. David S. Robinson for their comments on parts of the manuscript, to Dr A. Möller of the Icelandic Fisheries Laboratories for her comments on the sulphiting of prawns, to Mrs N. Green for typing the references, to Mrs E. Romanec for assistance with some of the tables and particularly to my wife for typing the main part of the manuscript. I acknowledge a communication from Mr E. F. Eaton. I am also indebted to the reading room staff at the British Library Lending Division, Boston Spa, for their help in tracing many obscure references, and to the societies, publishers and authors who have given permission to copy figures and tables from published works.

B. L. WEDZICHA

Contents

1

Principles, properties and reactions

1.1 NOMENCLATURE

This book will be concerned primarily with reactions of the anions HSO_3^-, $S_2O_5^{2-}$ and SO_3^{2-}. The IUPAC name of the sulphur(IV) oxoacid, $H_2S_2O_5$, is disulphurous acid, and its salts, disulphites, the latter replacing the often used name, metabisulphite. Partially dissociated acids will be referred to with the prefix hydrogen- replacing that of bi-. For details of rules of nomenclature, the reader is referred to the collection of references to IUPAC definitive recommendations prepared by Fernelius and coauthors (1973).

1.2 NUCLEAR PROPERTIES OF SULPHUR

The chemical reactivity of sulphur is almost exclusively a function of its electron distributions, but nuclear effects, arising from mass, may appear and may be used to advantage as second order effects. The use of radioactively labelled sulphur compounds is becoming increasingly important for mechanistic studies and the nuclear magnetic resonance of sulphur, without isotope enrichment, may be reliably observed. The important nuclear properties of sulphur will be reviewed here.

The most direct consequence of nuclear composition is varying atomic mass. Ten isotopic forms of sulphur are known, of which four are stable and are found in nature as ^{32}S (95·1 %), ^{33}S (0·74 %), ^{34}S (4·2 %) and ^{36}S

(0·016 %), although the actual isotopic composition of a specific compound of sulphur depends upon the mechanisms of its formation (Nickless, 1968). The recommended relative atomic mass of the element (^{12}C reference) is 32·064 $\pm$ 0·003, which embraces its natural variability.

Known unstable isotopes of sulphur are those with relative atomic masses of 29, 30, 31, 35, 37 and 38, but the only one sufficiently stable for use in tracer studies is ^{35}S. It is readily produced in the following reaction:

$$^{34}_{16}S + ^{1}_{0}n \rightarrow ^{35}_{16}S + \gamma$$

in which the target for neutron irradiation may be sulphur of natural isotopic composition. However, the specific activity of the product (200 MBq g^{-1} target material at saturation for a neutron flux of $10^{16} n \, m^{-2} \, s^{-1}$) is limited by its decomposition under irradiation and the production of large amounts of ^{32}P. The derived product is also chemically identical to the target material and, therefore, cannot be separated from it. An alternative process for the production of ^{35}S is neutron irradiation of ^{35}Cl:

$$^{35}_{17}Cl + ^{1}_{0}n \rightarrow ^{35}_{16}S + ^{1}_{1}p$$

which could theoretically lead to the production of 'carrier free' isotope (maximum possible specific activity of $5·6 \times 10^{13}$ Bq milliatom^{-1}) although the maximum practical specific activity of organosulphur compounds is sometimes limited by radiation decomposition. A case of some interest is that of ^{35}S-thiosemicarbazide, which rapidly becomes dark green or almost black on self-irradiation, but the change is not accompanied by the presence of chromatographically detectable impurities. Self-irradiation may also interfere with the progress of chemical reactions when intermediates are susceptible to radiation damage. An example is the preparation of mercaptoethanol by the reaction of ethylene oxide with hydrogen sulphide. This reaction, though slow, gives an excellent yield of mercaptoethanol with inactive reactants but, when ^{35}S-labelled hydrogen sulphide, at a specific activity of 370 MBq mmol^{-1}, is used under the same conditions, no product is obtained (Wilson, 1966). Thus, although ^{35}S offers considerably higher sensitivity as a radioactive tracer than, say, ^{14}C (maximum specific activity of 2·4 GBq milliatom^{-1}), possible radiation effects at high specific activities may become troublesome. The isotope decays with a half-life of 87·2 days by emission of a β-particle at maximum and mean energies of 0·167 and 0·049 MeV, respectively, to yield ^{35}Cl as the product. Since the β-emission from sulphur is similar to that from ^{14}C, liquid scintillation counting methods and quench correction data for carbon are equally applicable to sulphur. The

International Atomic Energy Agency toxicity rating for ^{35}S is that of medium–low toxicity with an effective biological half-life, in simple inorganic form, of 44·3 days. The whole body burden is 15 MBq. Data for neutron activation analysis of sulphur have also been published (Koch, 1960).

The only isotopes of sulphur whose nuclei possess a non-zero magnetic moment are ^{33}S and ^{35}S, both of which have a spin of 3/2 and, therefore, an associated quadrupole moment. With regard to usefulness in nuclear magnetic resonance, only the stable isotope will be considered here. The relevant magnetic parameters for this nucleus are (Harris, 1978):

Magnetic moment   = 0·8296 (nuclear magnetons)
Magnetogyric ratio = 2·0517 × 10^7 rad $T^{-1}s^{-1}$
Quadrupole moment = 6·4 × 10^{30} m^2

Observation of sulphur resonances is difficult in compounds with natural isotopic composition because the abundance of ^{33}S is low, and its sensitivity relative to hydrogen nuclei at constant frequency is 0·384 (Schultz et al., 1971). Since most NMR work is now carried out at constant field, a crude guide to the ease of obtaining a spectrum from ^{33}S in sulphur with natural isotopic composition is the receptivity of the nucleus relative to the proton at constant field, which has a value of 1·71 × 10^{-5} in solution (Harris, 1978). The second difficulty is imposed by the large quadrupole moment of ^{33}S which leads to rapid relaxation of the nucleus and a broadening of the NMR resonance, possibly to such an extent that it becomes undetectable. Since the relaxation mechanism involves coupling through the electric field gradient at the nucleus, broad resonances will occur when a large electric field gradient is present, arising from an unsymmetrical distribution of charge around the nucleus. Thus, spherical or near spherical charge distributions will favour narrow resonances which may be expected for species such as CS_2, SO_4^{2-}, RSO_3^- and R_2SO_2. This effect is well known for ^{14}N in which tetrahedral environments (e.g. NH_4^+, $(CH_3)_4N^+$) show sharp resonances whilst broad resonances are found for less symmetrical molecules such as $(CH_3)_2NH_2^+$ (Ogg and Ray, 1957).

The practical observation of ^{33}S spectra requires a sensitivity enhancement of at least 100 for carbon disulphide, which contains a particularly high mass fraction of sulphur (Schultz et al., 1971), and spectra have been successfully recorded using both continuous wave (Dharmatti and Weaver, 1951; Lee, 1968; Retcofsky and Friedel, 1970; Retcofsky and Friedel, 1972) and pulsed (Schultz et al., 1971; Lutz et al., 1973; Vold et al., 1978; Kroneck et al., 1980; Faure et al., 1981) NMR techniques.

The observed range of chemical shifts is greater than 600 ppm and, as expected, sulphur attached to electronegative groups is deshielded, the resonance of sulphate ion, sulphones and sulphonic acids appearing at the highest frequencies. Line widths are also in accordance with expectation. Where a species exists in one or more ionic forms, a sharp resonance is promoted by rapid exchange with the medium, for example the narrow line widths of sulphonic acids are attributed to this effect. Neat methylsulphonic acid, on the other hand, exhibits no detectable resonance under the conditions used for its observation in solution (Faure *et al.*, 1981).

The narrow resonances of sulphones and sulphonic acids allow good precision for the determination of chemical shifts, although the variability of resonances of structurally widely differing sulphonic acids is not significant and is unlikely to distinguish sulphonic acid salts present in food systems. However, this technique permits the observation of sulphite derived products of the sulphonate type in food free from interference from other forms of sulphur, which possess large chemical shifts relative to sulphate ion.

1.3   ELECTRON PARTICIPATION IN BONDING TO SULPHUR

Sulphur is a group VI element with the ground state electronic configuration $[Ne]3s^23p^4$ and should, therefore, be capable of participating in bonding as a divalent atom. Although divalent sulphur has been observed with two, three and four bonds to the element, leading to compounds of angular, pyramidal and square planar geometry (Cotton and Wilkinson, 1966), the most striking feature of sulphur is its ability to exist in a 'hypervalent' state, exhibiting oxidation states as high as +6. Group VI of the periodic table includes as its first element oxygen, with a ground state electronic configuration $1s^22s^22p^4$, but large differences are evident between the chemistry of oxygen and that of the remaining elements of the group. The differences are partly attributable to the higher electronegativity of oxygen and the limitation of coordination number to four and valency to two. The main theme of this book is related to reactions of oxoanions of sulphur. These compounds occur widely and contain sulphur in average oxidation states +2 to +6. The bonding involved in their formation will now be reviewed briefly.

Although the σ-framework for bonding in tetrahedral molecules such as SO_4^{2-} may be accounted for in terms of sp^3 hybridisation, the shortness of the S—O bonds at 0·148 nm (Nord, 1973) compared with an estimated length of 0·169 nm (Pauling, 1952) implies the involvement of multiple bonding. Cruickshank (1961) has pointed out the angular suitability of d-orbitals on sulphur for π-bonding, the most important orbitals being $d_{x^2-y^2}$ and d_{z^2} which may effectively overlap with oxygen $p\pi$-orbitals in tetrahedral oxoanions. By assuming full participation by these d-orbitals and averaging their contributions over all four oxygen atoms, each d-orbital contributes a bond order of one-quarter to each S—O bond, the total π-bond contribution of two such orbitals being a bond order of 0·5. Assuming a linear relationship between π-bond order and bond length, Cruickshank (1961) found excellent agreement between estimated π-bond orders and experimental bond lengths for $C_2H_5SO_4K$, $K_2S_2O_7$, $K(SO_3)_2NH$, $K_2(SO_3)_2CH_2$, SO_3NH_3, $(NH_4)_2SO_3N_2O_2$, KSO_3NH_2, $SO_2(NH_2)_2$ and $(SO_3)_\infty$. Despite this apparently consistent explanation, there has been much continuing debate regarding the extent of d-orbital participation in the proposed $d\pi$–$p\pi$-bonding, theoretical calculations failing to provide an unambiguous answer to the bonding potential. The most useful experimental evidence which has entered this debate arises from studies of $L_{\text{II,III}}$ X-ray fluorescence spectra of sulphur-containing compounds. These spectra result from transitions between the valence orbitals and the $2p$-level for second row elements. Thus, since it is postulated that $3s$-, $3p$- and $3d$-orbitals are involved in bond formation in sulphur oxoanions, the technique allows the acquisition of evidence of $3d$-character in the bonding (Urch, 1969, 1970).

An example of the interpretation of such spectroscopic data for SO_4^{2-}, SO_3^{2-} and SO_2 shows some 30 % of the electronic population of the $3s$-, $3p$- and $3d$-orbitals as existing in $3d$ (Taniguchi and Henke, 1976), although a lower value, in the region of 10 % for SO_2, has been estimated by Noodleman and Mitchell (1978). Despite the fact that the participation of d-orbitals in sulphur bonding has not been universally accepted, it forms a convenient representation of the 'hypervalent' bonding in sulphur. The inability of oxygen to exhibit a valency greater than two is a result of the unavailability of d-orbitals of sufficiently low energy.

Sulphur $L_{\text{II,III}}$ X-ray emission spectra are sensitive to the oxidation state of the element (Fischer and Baun, 1965) and, in principle, the technique is potentially useful in establishing the state of the element in any chemical form, provided that caution is taken to avoid decomposition of the sample under electronic bombardment or irradiation. Merritt and Agazzi (1966), for example, suggest that the spectra of some sulphur-containing compounds, including simple salts such as lithium sulphate, may be altered significantly when the source of excitation is changed from electron beam to X-ray irradiation, and the difficulties posed by natural products are likely to be correspondingly greater.

1.4 STRUCTURE OF OXOSULPHUR COMPOUNDS

The structural parameters of some oxosulphur compounds in oxidation state $+4$ and of sulphate ion are shown in Fig. 1.1. Although the composition of the hydrogen sulphite and disulphite ions has been known for nearly 140 years, it is only relatively recently that the structures of the two ions have been confirmed. The two possible arrangements of atoms in disulphite ion are shown below:

$$\begin{bmatrix} \text{O} & \quad & \text{O} \\ | & \quad & | \\ \text{S}-\text{O}-\text{S} \\ | & \quad & | \\ \text{O} & \quad & \text{O} \end{bmatrix}^{2-} \qquad \begin{bmatrix} \text{O} & \text{O} \\ | & | \\ \text{O}-\text{S}-\text{S} \\ | & | \\ \text{O} & \text{O} \end{bmatrix}^{2-}$$

$$\text{I} \qquad\qquad\qquad \text{II}$$

The original assignment of structure **I** to this ion by Simon and coworkers (1956) using Raman spectroscopy was on the basis of identification of the C_{2v} symmetry necessary for that species. However, it was later shown that the result involved incorrect assignment of a peak at $655\,\text{cm}^{-1}$ to an S—O—S vibration. Since the Raman spectra of polythionates also contain this vibration and crystallographic data are consistent with a structure having C_1 symmetry with an S—S bond, structure **II** is the more likely one (Golding, 1960). The molecular dimensions shown in Fig. 1.1 were obtained from X-ray analysis of $K_2S_2O_5$ (Lindqvist and Mörtsell, 1957). In dilute aqueous solutions disulphite is hydrolysed to hydrogen sulphite ion but, at concentrations suitable for the crystallisation of these salts, it is expected that the position of equilibrium will favour the formation of $S_2O_5^{2-}$. Thus, it is expected that the crystallisation of solutions containing hydrogen sulphite ion will lead to the formation of solid disulphites such as $K_2S_2O_5$ and $Na_2S_2O_5$. Infrared and Raman studies, however, have indicated that, under some conditions, for example when Rb^+ and Cs^+ are the counter ions, solid hydrogen sulphites are obtained (Simon and Schmidt, 1960; Meyer et al., 1979), although the failure of these compounds to react with S_2Cl_2 in anhydrous tetrahydrofuran according to,

$$2MHSO_3 + S_2Cl_2 \rightarrow 2HCl + M_2S_4O_6$$

was regarded critically by Schmidt and Wirwoll (1960). Ammonium hydrogen sulphite (Histaune and Heicklen, 1975) and some quarternary

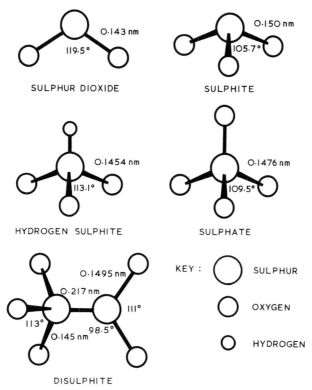

Fig. 1.1. Molecular structures and dimensions of sulphur(IV) oxoanions and of sulphate ion. Sources: SO_2 and $S_2O_5^{2-}$, Meyer and coworkers (1979); SO_3^{2-}, Larsson and Kirkegaard (1969); HSO_3^-, Johansson and coworkers (1980); SO_4^{2-}, Nord (1973). Based on Meyer and coworkers (1979) and reproduced with permission from *Spectrochim. Acta*, 1979, **35A**, page 345. © 1979 Pergamon Press Ltd.

ammonium hydrogen sulphites (Maylor *et al.*, 1972) have also been reported in the solid phase and it seems likely that one of the requirements for the crystallisation of the HSO_3^- form is the presence of a large counter ion.

Originally, two isomeric forms (**III, IV**) were suggested for hydrogen sulphite ion (Schaeffer and Köhler, 1918):

III

IV

Golding (1960) proposed the presence of **III** in dilute solutions, a result which has since been disproved (Meyer *et al.*, 1979) by the observation that the ion has the symmetry C_{3v} necessary for structure **IV**. The first reported crystallographic work on the hydrogen sulphite ion was that of Johansson and coworkers (1980), whose results are given in Fig. 1.1. It is interesting to compare the bond lengths of sulphate with the S—O bond lengths in hydrogen sulphite ion. The transition from sulphate to hydrogen sulphite involves replacement of one of the oxygen atoms in sulphate by an atom which is unable to participate in π-bonding and, therefore, the d-orbital contribution is averaged over three oxygen atoms instead of four, with a consequent increase in π-bond order and decrease in S—O bond length. The S—O bond length in hydrogen sulphite ion is similar to the value of 0.1456 nm in CH_3—SO_3^- (Charbonnier *et al.*, 1977), in which a similar bonding situation is expected (Johansson *et al.*, 1980). Meyer and coworkers (1977) considered the electron distributions around the sulphur atom of the sulphite ion with a view to establishing whether there existed any possibility of protonation on oxygen. The results, which confirm structure **IV**, are suprising on account of the normal expectation that SH bonds are weaker than OH bonds. Analysis using an extended Hückel model based on spectral atomic parameters shows that the vacant position on the sulphur atom is occupied by an S—O antibonding orbital which, when protonated, reduces the antibonding electron density and, thereby, shortens the S—O bonds in HSO_3^-.

1.5 KINETIC EFFECTS OF ISOTOPES OF SULPHUR

The presence of a significant amount of ^{34}S (4.2%) in compounds of sulphur renders the isotope potentially useful for the study of reaction mechanisms by observation of the relative rates of reaction of ^{32}S and ^{34}S species, leading to a change in the $^{32}S/^{34}S$ ratio in going from reactants to products. The measurement, with sufficient accuracy, of isotope ratios is conveniently carried out by means of mass spectrometry of sulphur dioxide prepared from the compounds in question. The theory of kinetic sulphur isotope effects has been reviewed by Saunders (1961), significant changes in isotopic composition only being observed if the bonding to sulphur changes in the rate determining step. The magnitude of the effect depends upon the differences between the vibrational frequencies of ^{32}S—X and ^{34}S—X bonds, where X denotes a species to which sulphur is bonded. If the force constants for the two bonds are identical, the vibrational energy of

S—X depends inversely on the square root of the reduced mass of the system. Hence, the bond to the heavier atom will vibrate at lower energy. Assuming that the S—X bonds are completely broken in the transition state for both isotopic forms of sulphur and, therefore, assuming that the products have identical energies, the amount of energy supplied during activation is greater for the ^{34}S-compound. Therefore, ^{34}S-compounds will react at a slower rate than those containing the lighter isotope.

An isotope effect is also expected with the use of ^{35}S as a radioactive tracer. The maximum effect expected is in the region of 5 % (Bigeleisen, 1949) and would need to be determined for the most accurate quantitative work.

1.6 PREPARATION OF SULPHUR DIOXIDE

Apart from the trivial procedure of preparing sulphur dioxide by the decomposition of sulphur(IV) oxoanions, which may be appropriate for small scale laboratory experiments,

$$Na_2SO_3 + H_2SO_4 \rightarrow Na_2SO_4 + H_2O + SO_2\uparrow$$

or

$$NaHSO_3 + H_2SO_4 \rightarrow NaHSO_4 + H_2O + SO_2\uparrow$$

or

$$Na_2S_2O_5 \xrightarrow[110-200\,°C]{air} Na_2SO_3 + SO_2\uparrow$$

the gas may be prepared by means of oxidation and reduction of commonly available sulphur compounds (Schroeter, 1966). The most straightforward oxidation reaction is the burning of sulphur in oxygen,

$$S + O_2 \rightarrow SO_2$$

or the oxidation of pyrites or metal sulphides (MS),

$$2MS + 3O_2 \rightarrow 2MO + 2SO_2$$

The most common substrate for reduction is sulphate ion and the reducing agents are typically elemental sulphur, copper or carbon. The reactions may be illustrated by the reduction of sulphuric acid,

$$S + 2H_2SO_4 \rightarrow 3SO_2\uparrow + 2H_2O$$
$$Cu + 2H_2SO_4 \rightarrow CuSO_4 + SO_2\uparrow + 2H_2O$$

or the reduction of gypsum,

$$2CaSO_4 + C \xrightarrow{900-1200\,°C} 2CaO + 2SO_2\uparrow + CO_2\uparrow$$

On the small scale, the gas is dried by bubbling through sulphuric acid and is collected by upward displacement of air, since sulphur dioxide is approximately twice as heavy as air.

1.7   SOLUTION BEHAVIOUR OF SULPHUR DIOXIDE

The solubility of sulphur dioxide in water is $3937 \, ml/100 \, g$ at $20 \, ^{\circ}C$ and $1877 \, ml/100 \, g$ at $100 \, ^{\circ}C$, the saturated solution at STP containing 53.5% by weight of the gas (Schenk and Steudel, 1968). Extensive solubility data are provided by Gmelin (1963). Solutions of gas in water obey Henry's law over a wide range of concentrations, 0.0025 molal to at least 1.5 molal, no large deviation from ideal behaviour being evident at the lower concentration suggesting that Henry's law is adhered to well below 0.0025 molal. This concentration corresponds to a partial pressure of 0.0027 for a total pressure of $1 \, atm$. The value of Henry's constant (ratio of molal concentration in solution to the pressure of the gas in the headspace) falls from 3.28 at $0 \, ^{\circ}C$ to 1.23 at $25 \, ^{\circ}C$ and 0.56 at $50 \, ^{\circ}C$ (Johnstone and Leppla, 1934). The solubility is affected by the presence of neutral salts. It is evident that both anion and cation effects are taking place, but whether they represent general medium effects or specific interactions with the gas or one of its ionic forms is not known. Weak complexes are known to occur between sulphur dioxide and halide ions in aqueous media. Formation constants of these 1:1 adducts are $0.22 \, M^{-1}$ and $0.39 \, M^{-1}$ for SO_2Br^- and SO_2I^- respectively (Salama and Wasif, 1975). The formation of complex is accompanied by a shift in λ_{max} for the absorbance of sulphur dioxide from $276 \, nm$ for the dissolved gas to $284 \, nm$ for SO_2Br^- and $344 \, nm$ for SO_2I^- in water.

The simple oxoanions of sulphur(IV) are salts of the dibasic acid, H_2SO_3, and this fact has led to the expectation that dissolution of sulphur dioxide in water is accompanied by the reaction,

$$H_2O + SO_2 \rightleftharpoons H_2SO_3$$

Although diesters of this acid are known, no evidence for the existence of the free acid has been forthcoming. Raman and infrared spectra of solutions of sulphur dioxide indicate solutions of molecular SO_2 (Falk and Giguere, 1958; Jones and McLaren, 1958; Davis and Chatterjee, 1975). Also, the similarity between spectra of dissolved SO_2 and liquid or gaseous SO_2 indicates that the interactions with water are weak. It is, therefore, concluded that dissolved sulphur dioxide exists as SO_2, often denoted as

$SO_2 . H_2O$, the limit of the amount of H_2SO_3 present being set at 3% by Falk and Giguere (1958). The results are, however, still inconsistent with the relatively strong acid behaviour of $SO_2 . H_2O$ and also the high solubility of the gas in comparison with that of carbon dioxide (Bell, 1973). Guthrie (1979) has considered the possibility of tautomeric equilibria between 'sulphurous' acid species,

$$SO_2 . H_2O \rightleftharpoons SO(OH)_2 \rightleftharpoons HSO_2(OH)$$

By means of a thermodynamic analysis based on the observation that the free energies of hydrolysis of esters of inorganic oxoacids, for which resonance effects are not important, can be calculated with useful precision from the pK_a of the corresponding oxoacid, free energy changes for each of the equilibria have been estimated. The free energy of covalent hydration of SO_2 was found to be $6.7 \pm 4.2 \, kJ \, mol^{-1}$, in satisfactory agreement with the suggested limit on the amount of the acid present at equilibrium (Falk and Giguere, 1958). The calculated pK_a value for $SO(OH)_2$ was 2·3 and that for $HSO_2(OH)$ was -2.6. The value of ΔG^0 for the reaction,

$$SO(OH)_2 \rightleftharpoons HSO_2(OH)$$

was found to be $19 \pm 5 \, kJ \, mol^{-1}$, and it is interesting that the greater stability of the species $SO(OH)_2$, than that of $HSO_2(OH)$ predicted by Guthrie (1979) is the reverse of the order of stability suggested by Meyer and coworkers (1977) from electron density studies.

The first dissociation of the dibasic acid form will be,

$$SO_2 . H_2O \rightleftharpoons HSO_3^- + H^+$$

and from ultraviolet absorption measurements the equilibrium constant, expressed as,

$$K_1 = \frac{[H^+][HSO_3^-]}{[SO_2 . H_2O]}$$

has a value of $0.0139 \pm 0.0002 \, M$ at 25 °C, corrected to zero ionic strength (Huss and Eckert, 1977). The result was in excellent agreement with conductivity data and confirmed previously reported values of $0.0139 \, M$ (Britton and Robinson, 1932; Ellis and Anderson, 1961) and $0.0142 \, M$ (Deveze and Rumpf, 1964), but was considerably lower than the often used value of $0.0172 \, M$ (Tartar and Garretson, 1941; Beilke and Lamb, 1975). In fact, the use of the higher value leads to an unacceptable anomaly in ion activities (Huss and Eckert, 1977) when activity coefficients $\gamma_\pm$ are

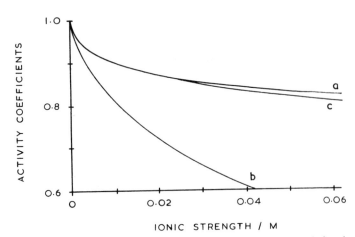

IONIC STRENGTH / M

Fig. 1.2. Relationship between activity coefficient and ionic strength for the ion pair, $H^+HSO_3^-$. Curve (a) is plotted assuming a dissociation constant of 0·0139 M for $SO_2 . H_2O$; for curve (b) this value is 0·0172 M. Curve (c) is the relationship between activity coefficient and ionic strength for sodium chloride (Huss and Eckert, 1977; Davies, 1962). Reproduced with permission from *J. Phys. Chem.*, 1977, **81**, page 2270. © 1977 American Chemical Society.

calculated using the equation proposed by Guggenheim (1935) with a specific interaction parameter, $\beta = 0\cdot3$; that is,

$$\ln \gamma_\pm = -\frac{AZ_+Z_-I^{1/2}}{1 + BaI^{1/2}} + 2\beta I$$

where A and B are the Debye–Huckel constants, Z_+ and Z_- denote the charges on positive and negative ions respectively, a is the closest approach distance and I is the ionic strength. The value of 0·3 for the specific interaction parameter compares well with that found for 1:1 electrolytes of similar size. Calculated mean ionic activity coefficients for the ion pair, $H^+HSO_3^-$, to an ionic strength of 0·06 M are shown in Fig. 1.2 to illustrate the wide differences between the predictions of Huss and Eckert (1977) and those of other workers. Superimposed on the graph is the effect of ionic strength on the activity coefficient of sodium chloride as an example of a typical 1:1 electrolyte (Davies, 1962), from which it is evident that the most recent data for the mean ionic activity coefficients of $H^+HSO_3^-$ are also the most reasonable. The pK_a for the first dissociation of 'sulphurous acid' is, therefore, 1·86.

This apparent value is close to the value of 2·3 predicted by Guthrie (1979) assuming that the major species in solution was $SO(OH)_2$. However,

TABLE 1.1

Calculated concentrations of sulphurous acid species as a function of pH for a total S(IV) concentration of 0·001 M (Guthrie, 1979)

pH	$10^4[SO_2)$ (M)	$10^4[SO(OH)_2]$ (M)	$10^4[HSO_2(OH)]$ (M)
1	8·06	0·64	3×10^{-4}
2	3·72	0·29	1×10^{-4}
3	0·58	0·05	2×10^{-5}
4	0·06	5×10^{-3}	2×10^{-6}
5	6×10^{-3}	5×10^{-4}	2×10^{-7}

Reproduced with permission from *Can. J. Chem.*, 1979, **57**, page 457. © 1979 National Research Council of Canada.

the amount of this acid actually present is too small to be detectable and it does not account for the acidity of solutions of sulphur dioxide. The potential formation of the strong acid, $HSO_2(OH)$, it is argued, resolves this difficulty even if the proportion is very small. Guthrie has also presented calculations of the expected amounts of the dibasic species present in solution, these results being shown in Table 1.1 for dilute solutions with a total sulphur(IV) concentration of 10^{-3} M. Unless differentiation between sulphurous acid and dissolved sulphur dioxide is required, the substance will, in future, be expressed in formulae as $SO_2 . H_2O$.

Although spectroscopic data for the structure of the hydrogen sulphite ion does not show any evidence for the species, $SO_2(OH)^-$, and, indeed, is supported by molecular orbital calculations, Guthrie (1979) estimated ΔG^0 for the reaction,

$$SO_2(OH)^- \rightleftharpoons HSO_3^-$$

as being $-9 \pm 8 \, kJ \, mol^{-1}$. Thus, the species, HSO_3^-, predominates and, if the estimate is accepted, the OH-form may account for some 2·5 % of the hydrogen sulphite species. Unless it is necessary to make the distinction, the hydrogen sulphite ion will be referred to as HSO_3^-.

The rate of formation of HSO_3^- was first reported by Eigen and coworkers (1961), who investigated the reaction by ultrasonic absorption measurements. Assuming that the observed relaxation process was the protonation reaction, the rate constant for

$$H^+ + HSO_3^- \rightarrow SO_2 + H_2O$$

was found to be $2 \times 10^8 \, M^{-1} \, s^{-1}$ at 20 °C and at an ionic strength of 0·1 M. However, in a subsequent paper by Wang and Himmelblau (1964), the value for the rate constant determined from the rate of transfer of radioactively labelled ^{35}S-sulphur dioxide between an aqueous solution and the gas phase was reported as nearly eight orders of magnitude smaller than that found by Eigen and coworkers. This discrepancy was pointed out by Betts and Voss (1970), who considered the rate of formation of HSO_3^- in relation to the rate of exchange of oxygen between sulphite ion and water. From measurements at high pH (8·6–11·0), the rate of exchange was found to be of second order with respect to hydrogen ion concentration and showed an order of 1·15 with respect to the total sulphite species concentration. Assuming that this non-integral kinetic order was the result of two parallel reactions:

$$HSO_3^- + H^+ \rightleftharpoons SO_2 + H_2O$$
$$2HSO_3^- \rightleftharpoons S_2O_5^{2-} + H_2O$$

the data gave a second order rate constant of $2·48 \pm 0·27 \times 10^9 \, M^{-1} s^{-1}$, at 24·7 °C and an ionic strength of 0·9 M, for the protonation of HSO_3^-. Although this result is not strictly comparable with that of Eigen and coworkers (1961) on account of the different temperature and ionic strength, it indicates that at least the order of magnitude of the rate constant is correct. The rate of protonation of HSO_3^- is, nevertheless, still slow compared with the rates of protonation of the majority of simple acids, which show second order rate constants in the region of 10^{10}–$10^{11} \, M^{-1} s^{-1}$ (Eigen et al., 1964).

The formation of HSO_3^- from gaseous sulphur dioxide is accompanied by a significant isotope effect with enrichment of the heavier isotope, ^{34}S, by approximately 1 % at 25 °C (Eriksen, 1972). The reaction offers, therefore, a convenient method for the preparation of $^{34}SO_2$ (Stachewski, 1975).

The value of the equilibrium constant for the reaction,

$$2HSO_3^- \rightleftharpoons S_2O_5^{2-} + H_2O$$

$$K = \frac{[S_2O_5^{2-}]}{[HSO_3^-]^2}$$

extrapolated to zero ionic strength at 25 °C, is found from spectro-photometric measurements to be $0·076 \, M^{-1}$ (Bourne et al., 1974), in agreement with the previously reported value of $0·07 \, M^{-1}$ (Golding, 1960; Arkhipova and Chistyakova, 1971), although the reported extinction coefficients of 4000 and $2527 \, M^{-1} cm^{-1}$ for $S_2O_5^{2-}$ at 255 nm in the two

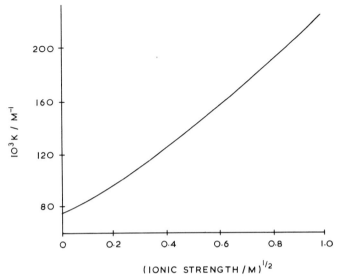

Fig. 1.3. Variation with ionic strength of the equilibrium constant for the reaction, $2HSO_3^- \rightleftharpoons S_2O_5^{2-} + H_2O$, at 25 °C. Ionic strength was varied by the addition of NaCl (Bourne *et al.*, 1974). Reproduced with permission of the copyright owner from *J. Pharm. Sci.*, 1974, **63**, page 866. © 1974 American Pharmaceutical Association.

references respectively are somewhat larger than that $(1980 \, M^{-1} cm^{-1})$ reported by Bourne and coworkers (1974).

In dilute solutions of HSO_3^- (10^{-2} M), the amount of disulphite ion is negligible, but the proportion increases rapidly with increasing HSO_3^- concentration. The presence of neutral salts, increasing the ionic strength, also favours the formation of disulphite ion. The variation of apparent equilibrium constant with ionic strength is shown in Fig. 1.3.

The rate of formation of $S_2O_5^{2-}$ is much slower than that of HSO_3^-, the second order rate constant being $7.00 \pm 0.21 \times 10^{-2} \, M^{-1} s^{-1}$ at 24.7 °C and at an ionic strength of 0.9 M (Betts and Voss, 1970). The second dissociation of sulphurous acid,

$$HSO_3^- \rightleftharpoons H^+ + SO_3^{2-}$$

$$K_2 = \frac{[H^+][SO_3^{2-}]}{[HSO_3^-]}$$

has a pK_a of 7.18 ± 0.03 at 25 °C and zero ionic strength, which represents an equilibrium constant five orders of magnitude smaller than that for the

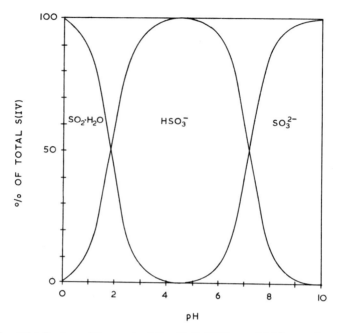

Fig. 1.4. Distribution of the species $SO_2 . H_2O$, HSO_3^- and SO_3^{2-} as a function of pH in dilute solution.

corresponding ionisation of HSO_4^- ($pK_a = 1.99 \pm 0.01$) (Smith and Martell, 1976). Using this pK_a value and also the result, $pK_a = 1.86$ for the ionisation of $SO_2 . H_2O$, the amounts of the species, $SO_2 . H_2O$, HSO_3^- and SO_3^{2-}, present in equilibrium in aqueous solution, may be calculated as a function of pH, the result being shown in Fig. 1.4. These calculations are, of course, for solutions at high dilution, under which conditions no disulphite ion is formed. Contributions from possible tautomeric species are also neglected.

Insufficient activity data are available to predict the behaviour at high concentration. The effect of increasing ionic strength on the pK_a of HSO_3^- is to reduce its value, Shapiro (1977) giving a value as low as 6.25 in concentrated salt solution. It is evident that in the normal pH range of food, that is pH 3–7, the predominant species will be hydrogen sulphite ion, with significant amounts of disulphite ion being formed as the concentration is increased, although sulphite ion is likely to become important at the higher end of this pH range and particularly at high ionic strength.

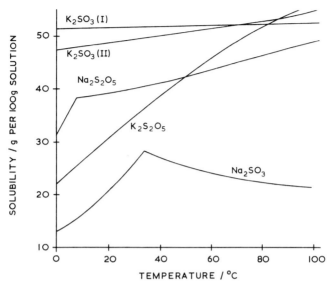

Fig. 1.5. Effect of temperature on the solubilities of K_2SO_3 (I and II refer to two different sets of data), $K_2S_2O_5$, Na_2SO_3 and $Na_2S_2O_5$ (Stephen and Stephen, 1963; Linke, 1965).

1.8 PROPERTIES OF SALTS OF SULPHUR(IV) OXOANIONS

The solubilities, as a function of temperature, of commonly available sodium and potassium salts containing sulphur(IV) are shown in Fig. 1.5, the data being representative of results reported by Stephen and Stephen (1963) and Linke (1965); where discrepancies occur, only the most consistent data have been included. The salts exhibit wide differences in their solubility characteristics. It is not clear whether the solubility of potassium sulphite is affected by temperature, two apparently conflicting sets of data being shown, but, if there is an effect, it is expected to be small. The most characteristic change in solubility behaviour is observed in the case of sodium sulphite, in which the transition,

$$Na_2SO_3 . 7H_2O \rightarrow Na_2SO_3 + 7H_2O$$

is responsible for the sharp discontinuity at 32 °C, above which temperature the solid phase is anhydrous sodium sulphite. A phase diagram for the system $Na_2SO_3–Na_2S_2O_5–H_2O$ is shown in Fig. 1.6. An example of a salt which has low solubility is calcium sulphite (0·0064 g/100 g at 30 °C) and cobalt(II) and nickel(II) sulphites are insoluble in water.

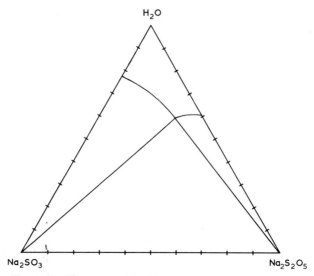

Fig. 1.6.   The system Na_2SO_3–$Na_2S_2O_5$–H_2O at 25 °C.

The activities of inorganic ions in solution are reduced through interactions with counter ions. Whereas ion pairs in 1:1 electrolytes such as $Na^+NO_3^-$ and $K^+NO_3^-$ are highly dissociated (pK, -0.6 and -0.2 respectively), those of 1:2 electrolytes such as $Na^+SO_4^{2-}$ and $K^+SO_4^{2-}$ (pK, 0.72 and 0.85 respectively), and $Na^+S_2O_3^{2-}$ and $K^+S_2O_3^{2-}$ (pK, 0.68 and 0.92 respectively) show significant stability (Sillen, 1964; Reardon, 1975) and need to be taken into account when, for example, kinetic salt effects on reactions involving these oxoanions are being studied. There are few data of this type on the formation of ion pairs with SO_3^{2-} or $S_2O_5^{2-}$ species. Since the tendency for ion-pair formation depends upon the nature of the counter ion and the magnitude and distribution of charge on the oxoanion, it is useful to compare the charge distributions on the species SO_3^{2-}, $S_2O_3^{2-}$ and SO_4^{2-}. Results for sulphite and thiosulphate have been calculated by Meyer and coworkers (1977) and are shown below:

$$(+0.009)\ S \begin{array}{l} \diagup O\ (-0.670) \\ \!\!-O\ (-0.670) \\ \diagdown O\ (-0.670) \end{array} \qquad (-0.244)\ S \overset{(+0.127)}{-\!\!-} S \begin{array}{l} \diagup O\ (-0.628) \\ \!\!-O\ (-0.628) \\ \diagdown O\ (-0.628) \end{array}$$

whilst the sulphate ion will have a symmetrical tetrahedral distribution of charge, that on each oxygen being slightly greater than -0.5 with the sulphur bearing a partial positive charge. The very similar ion-pair

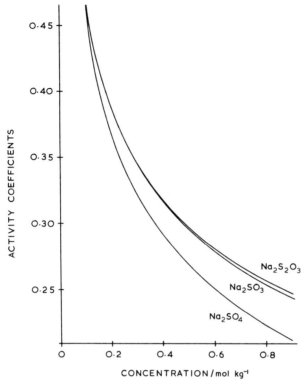

Fig. 1.7. Comparison of the relationships between activity coefficient and concentration for Na_2SO_3, Na_2SO_4 and $Na_2S_2O_3$ in water at $25\,°C$. Data from Lantzke and coworkers (1973) and Goldberg (1981).

dissociation constants for thiosulphate and sulphate suggest that the differences in charge distribution in these species are not critical for these interactions. However, mean ionic activity coefficients of solutions of sodium sulphate and of sodium thiosulphate show considerable differences (Robinson *et al.*, 1941). Since both ion-pair formation and general medium effects contribute to the observed activities of electrolytes, it is likely that, owing to the differences in charge distribution, sulphate and thiosulphate will interact differently with the solvent. Osmotic coefficients of sodium sulphite solutions in water have been measured by Lantzke and coworkers (1973) and Goldberg (1981), and their results, which show the close correspondence of experimental values with those for sodium thiosulphate, are illustrated in Fig. 1.7. It appears, therefore, that thiosulphate is a better model of sulphite behaviour from the point of view of both ion-pair interactions and general medium effects, presumably as a result of both ions having an unsymmetrical charge distribution. The mechanism of ion-pair formation has been investigated by relaxation techniques, such as those involving ultrasound absorption. In the case of formation of the ion pair, $Mg^{2+}SO_4^{2-}$, addition with stepwise displacement of water takes place (Atkinson and Petrucci, 1966; Eigen and Tamm, 1962a, b):

$$Mg^{2+} + SO_4^{2-} \rightleftharpoons Mg\,H_2O\,H_2O\,SO_4 \rightleftharpoons Mg\,H_2O\,SO_4 \rightleftharpoons MgSO_4$$

The predominant species is that in which the ions are separated by a single water molecule. The separation of acid–base pairs by a water molecule has also been demonstrated by Ralph and Grunwald (1969) and the structure of ion pairs in the case of sulphur oxoanions will, presumably, be (Young and Jencks, 1977):

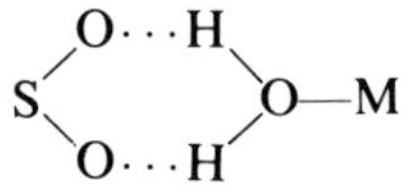

Thus, the geometry of the O—S—O bond and the net charge on the oxygens are likely to be important factors in determining the ion-pair association constants.

The only reported determination of a formation constant of an ion pair involving sulphite ion is that of Young and Jencks (1977) for KSO_3^- at an ionic strength of $1 \cdot 0\,M$, calculated from the observation of the decrease in pH of a solution containing sulphite and hydrogen sulphite ions, on the addition of potassium chloride and maintaining ionic strength with tetramethylammonium ion. The value of the formation constant was found to be $1 \cdot 42\,M^{-1}$ at 25 °C, which compares well with the value for the formation constant of KSO_4^- under these conditions.

The large size of $S_2O_5^{2-}$ is expected to result in a smaller charge density on the ion and correspondingly weaker ion pairs. Sodium dithionate, which has the symmetrical structure, $(O_3S—SO_3)^{2-}2Na^+$, shows activity coefficients which are very similar to those of $Na_2S_2O_3$ (at $0 \cdot 5$ molal salt, the activity coefficients for $Na_2S_2O_6$ and $Na_2S_2O_3$ are $0 \cdot 303$ and $0 \cdot 298$ respectively, Lantzke et al., 1973).

1.9 CHEMICAL REACTIVITY OF SULPHUR(IV) OXOANIONS

The observed chemical reactivity of oxoanions of sulphur(IV) stems from their ability to act as reducing agents or to take part in nucleophilic attack.

TABLE 1.2

Compilation of reduction potentials of sulphur oxo-anions involving S(IV) as either reactant or product

Half reaction	E^0 (V)
$SO_3^{2-} + 6H^+ + 6e^- \rightleftharpoons S^{2-} + 3H_2O$	0·231
$2HSO_3^- + 4H^+ + 4e^- \rightleftharpoons S_2O_3^{2-} + 3H_2O$	0·491
$2SO_3^{2-} + 6H^+ + 4e^- \rightleftharpoons S_2O_3^{2-} + 3H_2O$	0·705
$4H_2SO_3 + 4H^+ + 6e^- \rightleftharpoons S_4O_6^{2-} + 6H_2O$	0·509
$4HSO_3^- + 8H^+ + 6e^- \rightleftharpoons S_4O_6^{2-} + 6H_2O$	0·581
$2H_2SO_3 + H^+ + 2e^- \rightleftharpoons HS_2O_4^- + 2H_2O$	−0·056
$2HSO_3^- + 3H^+ + 2e^- \rightleftharpoons HS_2O_4^- + 2H_2O$	0·060
$2HSO_3^- + 2H^+ + 2e^- \rightleftharpoons S_2O_4^{2-} + 2H_2O$	−0·013
$2SO_3^{2-} + 4H^+ + 2e^- \rightleftharpoons S_2O_4^{2-} + 2H_2O$	0·416
$S_2O_6^{2-} + 4H^+ + 2e^- \rightleftharpoons 2H_2SO_3$	0·564
$S_2O_6^{2-} + 2H^+ + 2e^- \rightleftharpoons 2HSO_3^-$	0·455
$S_2O_6^{2-} + 2e^- \rightleftharpoons 2SO_3^{2-}$	0·026
$SO_4^{2-} + 2H^+ + 2e^- \rightleftharpoons SO_3^{2-} + H_2O$	0·031*
$SO_4^{2-} + 3H^+ + 2e^- \rightleftharpoons HSO_3^- + H_2O$	0·084*
$SO_4^{2-} + 4H^+ + 2e^- \rightleftharpoons H_2SO_3 + H_2O$	0·17†

Data compiled from Pourbaix (1966) with the exception of those marked * or † which were reported by Wedzicha and Rumbelow (1981) and by Parsons (1959) respectively.

Standard reduction potentials of a selection of oxoanions are shown in Table 1.2, the choice being restricted to reactions in which sulphur(IV) is present as either a reactant or a product. It is evident that both hydrogen sulphite and sulphite ions can be oxidised by a wide range of oxidising agents including oxygen, halate ion, hexacyanoferrate(III) ion, iodine, permanganate ion and dichromate ion. Possible oxidation products are dithionate and sulphate ion. In practical situations, the electrode potential will, in all cases except the $SO_3^{2-}/S_2O_6^{2-}$ reaction, be a function of pH as a result of two effects. First, the amounts of the relevant oxoanions and their protonated forms are a function of pH, and secondly, the position of equilibrium for the electrode reaction will also depend on pH. Thus, for example, the electrode potential, E, for the reaction,

$$S_2O_6^{2-} + 2H^+ + 2e^- \rightleftharpoons 2HSO_3^-$$

is given by,

$$E = 0·455 + \frac{RT}{2F} \ln \frac{[S_2O_6^{2-}][H^+]^2}{[HSO_3^-]^2}$$

or

$$E = 0.455 - 0.0591\,\text{pH} + 0.0295\log\frac{[S_2O_6^{2-}]}{[HSO_3^-]^2}$$

At constant $[S_2O_6^{2-}]/[HSO_3^-]$, the effect of increasing pH will be to favour the formation of dithionate. The sulphur(V) species, dithionate, is, in fact, a reasonable reducing agent, the standard reduction potential for the reaction,

$$2SO_4^{2-} + 4H^+ + 2e^- \rightleftharpoons 2S_2O_6^{2-} + 2H_2O$$

being -0.22 V (Parsons, 1959). Thus, the oxidation of sulphur(IV) species could involve a one-electron transfer,

$$S(IV) \rightarrow S(V) + e^-$$

followed by dimerisation, or an overall two-electron transfer reaction either as a simple step to give sulphur(VI),

$$S(IV) \rightarrow S(VI) + 2e^-$$

or as a stepwise oxidation through sulphur(V). An example of a reaction in which both dithionate and sulphate are produced is the oxidation of sulphur(IV) by hexaaquoiron(III), the relative amount of iron determining the principal product (Carlyle and Zeck, 1973). The related oxidation by hexacyanoferrate(III) leads only to sulphate (Lancaster and Murray, 1971).

Sulphite ion behaves as a Lewis base and is classified on the 'hard' and 'soft' convention as borderline. This method of classification lists Lewis acids which are of small size, high positive charge and without unpaired electrons as hard acids. Acids of low electronegativity and high polarisability are referred to as soft. A similar definition exists for hard and soft bases (Pearson, 1968). The proton is typical of a hard acid. A very simple statement of fact regarding the stability of acid–base complexes is that 'hard acids prefer to bind to hard bases and soft acids to soft bases'. In view of its borderline classification, sulphite ion is expected to interact widely with Lewis acids of all classifications. However, owing to its ability to accommodate $3d$-electrons, the sulphite ion may also act as an electron acceptor, and, therefore, is capable of acting as a Lewis acid when appropriate bases are available.

The large magnitude of rate constants in many oxoanion oxidation reactions, which may also be of high kinetic order, is strong evidence for the existence of a stable, but transient, complex between the reactants. If the intermediate is decomposed by collision, and the limit of the collision rate is set at $5 \times 10^{13} \, s^{-1}$, Edwards (1952) estimates that the lower limit for the value of the equilibrium constant for formation of complex is unity,

$$B^x + AO_m^- + 2H^+ \rightleftharpoons BAO_{m-1}^{x+1} + H_2O$$

where B represents a 'base' with charge x, and AO_m^- is the oxoanion containing m atoms of oxygen. Sulphite ion can act as both the base, B, and as the oxoanion.

Suggested intermediates for reactions involving sulphur(IV) species include $SO_2 . HI$ (Witekowa and Witek, 1955) and $SO_2 . I_2$ (Witekowa and Lewicki, 1963) for the oxidation by iodine; $[SO_3 . NO]^{2-}$ for the reaction between sulphite ion and nitric oxide (Drago, 1962); $[SO_2 . OOH]^-$ for the oxidation of sulphite by hydrogen peroxide in acidic solution (Hoffmann and Edwards, 1975); $[SO_2 . ClO_2]^-$ and $[SO_2 . IO_3]^-$ for the oxidation by halate ion in acid solution; and $[SO_3 . IO_2]^-$ for the oxidation by iodate in alkaline media (Edwards, 1952). On the basis of isotope exchange studies using ^{18}O, Halperin and Taube (1952) showed that the sulphate formed on oxidation of sulphite with hydrogen peroxide contained the two oxygen atoms originally present in the peroxide, even though the stoichiometry requires the net addition of only one oxygen atom. A similar result was observed in the oxidation by chlorate, in which two chlorate oxygen atoms were added to form sulphate (Halperin and Taube, 1950). Both these results are consistent with SO_2 acting as a Lewis acid accepting oxygens from the peroxide or chlorate donors. Conversely, when SO_3^{2-} is involved in reactions, it is suggested that the ion then acts as the electron donor and complexes of this type include that with nitric oxide as the acceptor (Drago, 1962), and with iodite (IO_2^-) as a result of the reaction:

$$SO_3^{2-} + IO_3^- + 2H^+ \rightleftharpoons [SO_3 . IO_2]^- + H_2O$$

When considering reaction and displacement at carbon and sulphur atoms, it is more common to refer to the nucleophilicity of the reagent than to its acid–base behaviour. Edwards and Pearson (1962) describe factors determining nucleophilic reactivities, the series of reactivity being $SO_3^{2-} > S_2O_3^{2-} > SC(NH_2)_2 > I^- > CN^- > SCN^- > NO_2^- > OH^- > N_3^- > Br^- > NH_3 > Cl^- > C_5H_5N > H_2O$. Nucleophilic order has been correlated successfully with thermodynamic properties of nucleophile and of substrate by means of four-parameter equations such as that suggested by Edwards (1954):

$$\log \frac{k}{k_o} = \alpha E_n + \beta H$$

where, for a typical displacement reaction, rate constants are defined by,

$$X^- + CH_3Y \xrightarrow{k} CH_3X + Y^-$$

$$H_2O + CH_3Y \xrightarrow{k_o} CH_3OH + Y^-$$

and,

$$E = E_{X^-} + 2.60$$

where E_{X^-} is the dimerisation potential of the nucleophile,

$$2X^- \rightleftharpoons X_2 + 2e^-$$

and,

$$H = pK_{aX^-} + 1.74$$

where pK_{aX^-} is the measure of the basicity of the nucleophile towards the proton. The constant terms, 2·60 and 1·74, are the dimerisation potential for water and log $[H_2O]$ respectively.

Whereas the E and H parameters relate to the properties of the nucleophile, α and β relate respectively to reduction and acid–base behaviour of the substrate (Davis, 1968). For nucleophilic attack at carbon, α appears to dominate the substrate parameters and, therefore, the ratio of rate constants, k/k_o, will be largely a function of αE. Thus, the so-called oxibase scale shown in Table 1.3, in which potential nucleophiles are arranged in order of increasing dimerisation potential, is a good guide to nucleophilicity. For any two nucleophilic displacement reactions on

TABLE 1.3
Some oxibase scale parameters in water at 25 °C
(Davis, 1968)

Nucleophile	E	H
H_2O	0	0
OH^-	1·65	17·48
$C_6H_5NH_2$	1·78	6·28
$S_2O_3^{2-}$	2·52	3·60
SO_3^{2-}	2·57	9·00
SH^-	2·60	8·70
CN^-	2·79	10·88
S^{2-}	3·08	14·66

Reproduced with permission from G. Nickless (Ed.), *Inorganic Sulphur Chemistry*, page 95. © 1968 Elsevier Scientific Publishing Co.

the same substrate with rate constants, k_1 and k_2, and corresponding parameters, E_1, H_1, E_2, H_2, the ratio of rate constants is given by:

$$\log \frac{k_1}{k_2} = (E_1 - E_2)\alpha + (H_1 - H_2)\beta$$

Experimental data on the rates of anion–cation combination reactions give a reasonably constant value of $\log(k_{SO_3^{2-}}/k_{OH^-})$ of 2·75–3·37 when the cations are Crystal Violet$^+$, Malachite Green$^+$, p-nitro-Malachite Green, $^+$ trianisylmethyl$^+$, p-dimethylaminophenyltropylium$^+$ and p-chlorobenzene-diazonium$^+$ and also for reactions with 2,4,6-trinitrobenzene and acrylo-nitrile (Ritchie and Gandler, 1979), suggesting similar values of α and β for these substrates. The values of $\log(k_{SO_3^{2-}}/k_{OH^-})$ for nucleophilic attack at the carbonyl group are 0·83, 0·53, 0·37 and 0·26 for p-nitrobenzaldehyde (Hine, 1971), formaldehyde, iso-butyraldehyde and dimethoxyacetone (Hine et al., 1976) respectively.

Davis (1968) has considered nucleophilicity towards sulphur atoms and, in particular, displacements at the sulphur–sulphur bond in disulphides:

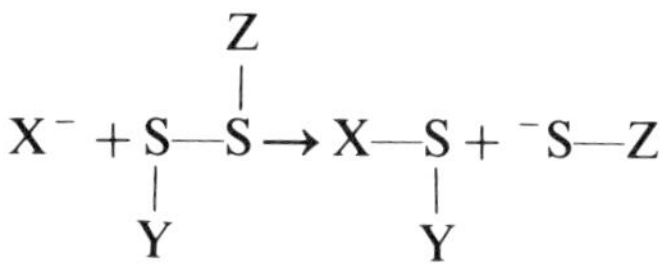

$$X^- + \overset{\overset{\textstyle Z}{\textstyle |}}{\underset{\underset{\textstyle Y}{\textstyle |}}{S}}\!\!-\!\!S \rightarrow X\!\!-\!\!\overset{}{\underset{\underset{\textstyle Y}{\textstyle |}}{S}} + {}^-S\!\!-\!\!Z$$

and notes that a nucleophile of high electrode potential is required, since the leaving group will generally have a high potential. Thus, the α term will be large and, since the leaving group is also basic, the β value of the substrate will be large and positive; therefore, good thiophiles are expected to have both large E and H values, examples being SO_3^{2-}, CN^-, HS^-, RS^- ($E_{RS^-} = 2\cdot9\,\text{V}$) and S^{2-}. Thiosulphate ion, which is a good carbon nucleophile, is a poor sulphur nucleophile. The large difference in the E value of OH^- and RS^- renders the hydroxide ion a poor species for the cleavage of disulphide bonds.

1.10 INORGANIC REACTIONS

1.10.1 Oxidation of sulphite ion in aqueous systems by oxygen

The stoichiometric equation for the oxidation of sulphite by oxygen,

$$SO_3^{2-} + \tfrac{1}{2}O_2 \rightarrow SO_4^{2-}$$

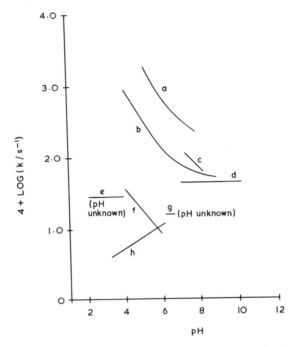

Fig. 1.8. Pseudo first order rate constant as a function of pH for the catalysed oxidation by oxygen of sulphur(IV) oxoanions in aqueous solution, reproduced from Hegg and Hobbs (1978). The curves refer to data reported as follows: (a) Fuller and Crist (1941) as modified by McKay (1971); (b) Larson and coworkers (1978); (c) Schroeter (1963); (d) Winkelmann (1955); (e) Miller and Pena (1972); (f) Brimblecombe and Spedding (1974); (g) Scott and Hobbs (1967); (h) Beilke and coworkers (1975). Reproduced with permission from *Atmospheric Environment*, 1978, **12**, page 243. © 1978 Pergamon Press Ltd.

suggests none of the complexities inherent in its mechanism. It can take place alone or through a reaction catalysed by transition metal ions. In order to illustrate the sensitivity of this reaction to the presence of trace amounts of catalyst, the data of Fuller and Crist (1941) give the rate equation for the uncatalysed reaction as,

$$\frac{d[SO_4^{2-}]}{dt} = (0 \cdot 013 + 6 \cdot 6[H^+]^{1/2})[SO_3^{2-}]$$

from experiments carried out in the range, pH 5·1–7·8. The same workers studied the copper catalysed oxidation reaction at pH 8·7, which showed a similar rate equation except that the hydrogen ion term was replaced by one containing the metal ion, that is,

$$\frac{d[SO_4^{2-}]}{dt} = (0.013 + 2.5 \times 10^6 [Cu^{2+}]) [SO_3^{2-}]$$

In both cases the concentration of sulphite is expressly that of SO_3^{2-}, the ion, HSO_3^-, being less susceptible to oxidation. Since there are no changes in rate constant of the uncatalysed reaction between pH 7 and 9, the rates may be compared at, say, pH 9 where, for the uncatalysed reaction, the rate of formation of sulphate is very nearly $0.013[SO_3^{2-}]$. The catalysed reaction will proceed at approximately three times this rate when the cupric ion concentration is 10^{-8} M, and significant effects are likely at 10^{-9} M. Thus, it is hardly surprising that the literature contains numerous and conflicting rate equations and rate constants for the oxidation of sulphite species by oxygen, although some common features are evident. The majority of data consistently show the reaction to be of first order with respect to the concentration of sulphite ion and independent of the concentration of oxygen for pH 3–7, and of first order with respect to oxygen above pH 9. The variations between the data are in the absolute magnitude of rate constants and in pH effects. These have been summarised by Hegg and Hobbs (1978), whose graph is shown in Fig. 1.8. The low rate constants reported by Beilke and coworkers (1975) and their apparently different pH behaviour was attributed by Hegg and Hobbs as due, possibly, to the use of plastic vessels and consequent presence of organic compounds which may act as inhibitors. This suggestion also indicates the extreme precautions which must be taken in order to obtain reproducible results in the system.

The literature concerning the mechanism of the oxidation reaction is divided according to whether the superoxide anion, $\cdot O_2^-$, is thought to be involved. One of the most widely accepted mechanisms for this reaction is that originally proposed by Bäckström (1934), who suggested the involvement of the species, $SO_3^{\cdot-}$ and $SO_5^{\cdot-}$, as follows:

Chain initiation

$$SO_3^{2-} + M^+ \rightarrow SO_3^{\cdot-} + M^\cdot \qquad (1)$$

Propagation

$$SO_3^{\cdot-} + O_2 \rightarrow SO_5^{\cdot-} \qquad (2)$$

$$SO_5^{\cdot-} + HSO_3^- \rightarrow HSO_5^- + SO_3^{\cdot-} \qquad (3)$$

Oxidation

$$HSO_5^- + SO_3^{2-} \rightarrow HSO_4^- + SO_4^{2-} \tag{4}$$

Termination

$$SO_3^{\cdot -} + SO_5^{\cdot -} \rightarrow S_2O_6^{2-} + O_2 \tag{5}$$

The presence of the free radicals, $SO_3^{\cdot -}$ and $SO_5^{\cdot -}$, has been confirmed by Hayon and coworkers (1972) and the mechanism allows for gaseous O_2 to be the main source of oxygen, as suggested by ^{18}O isotope exchange experiments (Winter and Briscoe, 1951; Halperin and Taube, 1952). The product, dithionate, which is formed in small quantities, is also predicted (Bassett and Parker, 1951). The mechanism, however, has a number of shortcomings. Equation 3 was included to explain the pH effect on the autoxidation but Hayon and coworkers (1972) observed that, even at pH 12, when negligible amounts of HSO_3^- were present, an efficient chain reaction was observed on flash-induced oxidation of sulphite. An additional propagating step,

$$SO_5^{\cdot -} + SO_3^{2-} \rightarrow SO_5^{2-} + SO_3^{\cdot -}$$

was, therefore, proposed. These authors, however, comment that the main difficulty encountered in the Bäckström mechanism is concerned with the effects of alcohols on the progress of the reaction. Their results showed that neither $SO_3^{\cdot -}$ nor $SO_5^{\cdot -}$ were scavenged effectively. The concentrations of alcohols required to reduce the depletion of sulphite in flash-induced oxidation to approximately 30% of its value in alcohol-free solutions were:

propan-2-ol	4×10^{-5} M
ethanol	1×10^{-4} M
methanol	2×10^{-4} M
tert-butanol	2×10^{-2} M

when the sulphite concentration was 5×10^{-4} M at pH 9.7. Therefore, these compounds should react with chain carriers at a rate sufficiently high for a significant reduction in their concentrations. A possible intermediate which could fulfil this role is $SO_4^{\cdot -}$. This suggestion is supported by the correct order of reactivity for $SO_4^{\cdot -}$ towards alcohols, second order rate constants for the reaction,

$$SO_4^{\cdot -} + ROH \rightarrow products$$

being $4\cdot6 \times 10^7$, $3\cdot5 \times 10^7$, $1\cdot1 \times 10^7$ and $9\cdot1 \times 10^5$ $M^{-1} s^{-1}$ for propan-2-ol, ethanol, methanol and tert-butanol respectively. The pH effect on the rate of oxidation can be related to the observation that $SO_4^{.-}$ reacts with SO_3^{2-} more slowly than with HSO_3^- by a factor of, at least, 2·5, and the fact that sensitivity of autoxidation to inhibition by impurities increases with pH was attributed to the reaction,

$$SO_4^{.-} + OH^- \rightarrow SO_4^{2-} + OH^{.}$$

since $OH^{.}$ is more rapidly scavenged. Direct evidence for the formation of $OH^{.}$ was obtained by observation of the spectrum of $^{.}O_3^-$ at pH 12·9. The formation of $SO_4^{.-}$ is suggested according to the following mechanism:

$$SO_5^{.-} + SO_3^{2-} \rightarrow SO_4^{.-} + SO_4^{2-}$$

Larson and coworkers (1978) gave the rate equation derived from Bäckström's mechanism, assuming a steady state approximation for intermediates as,

$$\frac{d[SO_4^{2-}]}{dt} = [M^+]^{1/2}[O_2]^{1/2}(k_1 + k_2[H^+]^{1/2})[SO_3^{2-}]$$

which shows the observed hydrogen ion and sulphite ion dependence. However, the prediction of a dependence of rate on the concentration of oxygen is inconsistent with experimental data. To resolve this difficulty, the authors suggest the involvement of $OH^{.}$ in the propagating steps, that is,

$$SO_3^{.-} + O_2 \rightarrow SO_3 + ^{.}O_2^- \tag{6}$$

$$SO_3 + OH^- \rightarrow SO_3^{.-} + OH^{.} \tag{7}$$

$$SO_3 + O_2 \rightarrow SO_5 \tag{8}$$

$$SO_5 + OH^- \rightarrow SO_5^{.-} + OH^{.} \tag{9}$$

$$SO_3^{2-} + OH^{.} \rightarrow SO_3^{.-} + OH^- \tag{10}$$

By introducing these steps, and assuming that initiation is equally possible for HSO_3^- ion by the reaction,

$$HSO_3^- + M^+ \rightarrow HSO_3 + M^{.}$$

then, providing the rate-determining step is step 6, with the rate constant for step 7 being greater than that for step 8, the steady state approximation leads to a rate equation given by,

$$\frac{d[SO_4^{2-}]}{dt} = (k_1 + k_2[H^+]^{1/2})[SO_3^{2-}][M^+]^{1/2}$$

which is consistent with observation and is unaffected by the addition of a chain propagating sequence involving $SO_4^{\cdot-}$, as suggested by Hayon and coworkers (1972).

The superoxide ion, $\cdot O_2^-$, is known to be generated in biological systems, for example by the peroxidase catalysed oxidase action on phenols (Yamazaki and Piette, 1963) and, indeed, an enzyme, superoxide dismutase, exists to destroy this species, presumably to protect tissues from undesirable oxidation. The involvement of the superoxide ion in sulphite oxidation has been suspected for some time (Fridovich and Handler, 1958, 1959, 1961) and may be demonstrated by the fact that erythrocuprein, an excellent catalyst for the disproportionation of the superoxide ion,

$$2 \cdot O_2^- + 2H^+ \rightarrow O_2 + H_2O_2$$

is a very effective inhibitor of the spontaneous oxidation in the presence of EDTA (McCord and Fridovich, 1969a, b). Similarly, tiron, a potent superoxide scavenger (Greenstock and Miller, 1975), inhibits the sulphite mediated destruction of β-carotene in the presence of oxygen (Peiser and Yang, 1979). Yang (1970, 1973) has also observed that the addition of superoxide dismutase inhibits sulphite oxidation, and these results, considered together, provide strong evidence for the intermediacy of $\cdot O_2^-$ in the chain propagating stages.

It has been suggested that the formation of superoxide ion, for the Mn^{2+} catalysed oxidation of sulphite, takes place as a result of simple electron abstraction by oxygen (Abel, 1951):

$$Mn^{2+} + O_2 \rightarrow Mn^{3+} + \cdot O_2^-$$

A propagating sequence involving both sulphite free radicals and $\cdot O_2^-$ has been postulated (Yang, 1970) as follows:

$$SO_3^{2-} + \cdot O_2^- + 3H^+ \rightarrow HSO_3^- + 2 \cdot OH$$

$$SO_3^{2-} + \cdot OH + 2H^+ \rightarrow HSO_3^- + H_2O$$

$$HSO_3^- + O_2 \rightarrow SO_3 + \cdot O_2^- + H^+$$

$$HSO_3^- + \cdot OH \rightarrow SO_3 + H_2O$$

$$2HSO_3^- \rightarrow SO_3 + SO_3^{2-} + 2H^+$$

$$SO_3 + H_2O \rightarrow SO_4^{2-} + 2H^+$$

It is significant also that the oxidation of phenols by peroxidase in the capacity of the latter to act as an oxidase enzyme can initiate sulphite oxidation (Fridovich and Handler, 1961). Although both $\cdot O_2^-$ and HSO_3^-

are able to initiate the chain propagating reactions, their maintenance depends upon the formation of $\cdot O_2^-$, and, on the basis of the proposed scheme, removal of this species interrupts the sequence.

The situation regarding the metal catalysed reaction appears to be even less well defined and the change from 'metal initiated' to 'metal catalysed' is also not clear. In summarising a large number of experimental rate laws, Hegg and Hobbs (1978) suggest that the rate equation for the catalysed oxidation of sulphite is,

$$\frac{d[SO_4^{2-}]}{dt} = k[M^+][H^+][SO_3^{2-}]$$

Although free radical mechanisms similar to those of the uncatalysed oxidation have been proposed, the formation of sulphito complexes with transition metal ions may be involved in the initiating steps (Schmidkunz, 1963; Barrie and Georgii, 1976), for example,

$$Mn(SO_3)_3^{4-} + O_2 \rightarrow Mn(SO_3)_3^{3-} + \cdot O_2^-$$

$$Mn(SO_3)_3^{3-} + SO_3^{2-} \rightarrow Mn(SO_3)_3^{4-} + SO_3^{\cdot -}$$

followed by free radical chain reactions involving these two radical products.

The chain propagation and termination steps in sulphite autoxidation are still very much 'possible' schemes rather than proven pathways; it is clear, however, that they must involve reactive free radical species. For this reason, any substance which can effectively scavenge these intermediates should act as an inhibitor to autoxidation. Schroeter (1966) has reviewed early work on the inhibition of sulphite oxidation and reports the sensitivity of the reaction to the presence of compounds such as inorganic ions, organic acids, alcohols, glycols, polysaccharides, amines, amides, aldehydes, ketones and phenols, the inhibitory action being detectable at concentrations as low as 10^{-6} M. In the case of inhibition by mannitol (Fuller and Crist, 1941) and by N,N-dimethylformamide and N,N-dimethylacetamide (Schroeter, 1963), the rate law for the inhibited reaction is given by,

$$-\frac{d[S(IV)]}{dt} = k \frac{A}{A + [I]} [S(IV)]$$

where A is a constant describing the inhibition, I represents the inhibitor and k is the rate constant for the uninhibited reaction. This rate law is in agreement with experimental data over a change in inhibitor concentration of two orders of magnitude, with a value for A of 10^{-5} M in all cases.

Since the inhibitor is required to react with free radical intermediates of oxidation, it is expected to undergo some chemical change. Thus, for example, when pyridine is used to inhibit sulphite oxidation, the mixture of products contains 3-pyridylpyridinium ions (1.7%) and N-pyridinium-sulphonic acid (22%), suggesting oxidation of pyridine by the radical intermediates, and sulphonation (Baumgarten, 1936).

As expected, solutions of sulphur(IV) oxoanions containing a large proportion of ethanol are stable to oxidation. Wedzicha and Lamikanra (1982) have studied the progress of manganese catalysed sulphite oxidation in such solutions by observing the consumption of oxygen by means of a Clark-type oxygen electrode. An interesting feature of this reaction is that the addition of glycine, which acts as an inhibitor in aqueous media, considerably enhances the rate of loss of oxygen in the presence of ethanol.

1.10.2 Oxidation with iodine

The reaction between sulphite ion and iodine is represented by the stoichiometric equation,

$$I_2 + H_2O + SO_3^{2-} \rightarrow 2I^- + SO_4^{2-} + 2H^+$$

In aqueous solution, this reaction is commonly used for the standardisation of sulphite solutions, whereas in non-aqueous media it forms the basis of the Karl Fischer titrimetric method for the determination of water (Fischer, 1935). In the latter a reagent consisting of equal volumes of anhydrous pyridine and methanol containing iodine and sulphur dioxide is used as the titrant for the sample containing an unknown quantity of water, addition being continued until an excess of iodine is present. The reaction also constitutes a pathway in the Landolt clock reaction, that is, the oxidation of sulphite by iodate (Landolt, 1887).

The reaction product, in aqueous solution, is almost exclusively sulphate, a small amount of dithionate (of the order of 0.5%) also being formed (Bassett and Henry, 1935). The oxidation proceeds rapidly and has been followed by a stop-flow kinetic technique (Bünau and Eigen, 1962) and at a rotating disc electrode (Verhoef et al., 1978). In the latter case, the iodine is produced by electrolytic decomposition of sodium iodide at the electrode surface. The pH dependence of second order rate constants from electrochemical measurements shows that sulphite ions react with iodine an order of magnitude faster than does HSO_3^-, and that, as suggested by Bünau and Eigen, the species, $SO_2 . H_2O$, is relatively unreactive. The effect of increasing the iodide concentration is a reduction in the rate of reaction: the rate constant for the reaction with I_3^- ($2.1 \times 10^5 \, M^{-1} s^{-1}$) is three orders of magnitude less than that for the reaction with molecular iodine ($1 \cdot 7 \times 10^8 \, \text{M}^{-1} \text{s}^{-1}$). The reactants are in equilibrium with a complex whose formula is suggested as, HSO_3I (Bünau and Eigen, 1962),

$$HSO_3^- + I_3^- \rightleftharpoons HSO_3I + 2I^-$$

$$HSO_3^- + I_2 \rightleftharpoons HSO_3I + I^-$$

Hydrolysis of the intermediate leads to the expected products,

$$HSO_3I + H_2O \rightarrow HSO_4^- + 2H^+ + I^-$$

The effective reagent in these oxidations is, presumably, I^+, the formation of which from molecular iodine is expected to be easier than from I_3^-, according to the following equation in which X represents the sulphur(IV) oxoanion involved in the reaction:

$$X^- \curvearrowright I \cdot \overset{\curvearrowleft}{\cdot} I \rightarrow XI + I^-$$

This explanation is in agreement with the observed greater effectiveness, in all cases, of iodine as an oxidising agent than I_3^-.

1.10.3 Oxidation with hydrogen peroxide

The stoichiometric reaction for the oxidation of sulphur(IV) oxoanions by hydrogen peroxide is,

$$SO_2 + H_2O_2 \rightarrow H_2SO_4$$

or,

$$HSO_3^- + H_2O_2 \rightarrow SO_4^{2-} + H^+ + H_2O$$

and the almost exclusive formation of sulphate as the reaction product (Albu and Schweinitz, 1932; Higginson and Marshall, 1957; Mader, 1958) indicates that this is a heterolytic process, homolysis leading to the expectation of some dithionate. A detailed study of this reaction has been carried out by Hoffmann and Edwards (1975) over the pH range 4–8. In the pH range 4–6, where HSO_3^- is the predominant species, the reaction requires the addition of one hydrogen ion in the rate-determining step, whereas when SO_3^{2-} is the more abundant species, the requirement is for two hydrogen ions. The existence of general acid catalysis also renders any Brönsted acid (HA) capable of fulfilling the role of hydrogen donor. Incorporating the fact that both oxygens of the hydrogen peroxide appear on sulphate (Halperin and Taube, 1952), a nucleophilic displacement by

H_2O_2 on sulphur(IV) was suggested as the reaction by which the intermediate was formed. The rate-determining step was either the slow formation of this intermediate,

$$A^- + H_2O_2 + H_2O \cdot SO_2 \xrightarrow{\text{slow}} \begin{array}{c} ^-O \\ \diagdown \\ \diagup \\ O \end{array} S{-}OOH + HA$$

or the slow reaction with the Brönsted acid,

$$HSO_3^- + H_2O_2 \rightleftharpoons \begin{array}{c} ^-O \\ \diagdown \\ \diagup \\ O \end{array} S{-}OOH + H_2O$$

$$\begin{array}{c} ^-O \\ \diagdown \\ \diagup \\ O \end{array} S{-}OOH + HA \xrightarrow{\text{slow}} H_2SO_4 + A^-$$

Although a heterolytic mechanism has been proposed for this reaction, some involvement of free radicals cannot be ruled out. Flockhart and coworkers (1971) have demonstrated, by ESR, the presence, in mixtures of hydrogen peroxide and $SO_2 \cdot H_2O$, of the HSO_3^- radical, in a flow system in which the time between mixing and observation was of the order of 70 ms. For equal concentrations of the reactants (0·1 M) in water, the observed radical concentration was increased to some extent on replacement of SO_2 by $NaHSO_3$. Free radicals are presumably formed by the following primary reaction:

$$HOOH + SO_2 \rightarrow {}^{\cdot}OH + HSO_3^-$$

and also by,

$${}^{\cdot}OH + SO_2 \rightarrow HSO_3^-$$

1.10.4 Reaction with sodium nitroprusside
The reaction between sulphur(IV) oxoanions and nitroprusside ion leads to a product which is red in colour, and the latter is greatly enhanced by the addition of a concentrated solution of zinc sulphate (Boedeker, 1861). Ultimately zinc sulphitonitroprusside is precipitated (Pavolini, 1930). Continuous variation curves obtained from experiments using mixtures of nitroprusside ion and sulphite ion (pH 9) show that a relatively unstable 1:1 complex is formed with an extinction coefficient of $3100 \pm 400 \, \text{M}^{-1} \text{cm}^{-1}$ at 475 nm (Moser et al., 1965).

In concentrated solution the reaction involves ligand substitution to give sulphitopentacyanoferrate(II) ion (Hoffmann, 1900),

$$Fe(CN)_5NO^{2-} + SO_3^{2-} \rightarrow Fe(CN)_5SO_3^{5-} + NO^+$$

whilst at low concentrations sulphitonitroprusside ion is formed. However, it is reasonable to expect that the sulphitonitroprusside ion is an intermediate in this reaction, therefore the preferred alternative mechanism, at low concentrations, is,

$$Fe(CN)_5NO^{2-} + SO_3^{2-} \overset{K}{\rightleftharpoons} Fe(CN)_5NOSO_3^{4-}$$

Highly charged ions are capable of forming relatively stable ion pairs even with alkali metal ions and, therefore, it is not surprising that both sodium and potassium ions stabilise complex formation, presumably through formation of $Fe(CN)_5NOSO_3M^{3-}$ (M = Na, K). This would also be true for zinc, giving $Fe(CN)_5NOSO_3Zn^{2-}$. The equilibrium constant, K, for the formation of complex without the agency of metal ion is found to be $0·001 \text{ M}^{-1}$, indicating very weak association, whereas for the following metal catalysed reaction:

$$Fe(CN)_5NO^{2-} + SO_3^{2-} + M^+ \overset{K}{\rightleftharpoons} Fe(CN)_5NOSO_3M^{3-}$$

it is $0·13$ and $0·39 \text{ M}^{-2}$ for sodium and potassium ions respectively (Moser et al., 1965). The corresponding value of the equilibrium constant for the reaction involving zinc does not appear to have been published. The equilibrium constants for ion-pair formation between the tetravalent complex and the alkali metal ions are $8·6$ and 36 M^{-1} for sodium and potassium ions respectively.

1.10.5 Oxidation by transition metal ions

Evidence for the formation of acid–base complexes as the reactive entities in the oxidation of sulphur(IV) oxoanions is presented in Section 1.9. The formation of such complexes necessitates close approach of sulphur(IV) to its partner and, therefore, requires either the penetration of the inner coordination sphere or attachment to bridging ligands. The ability of transition metals to form coordination complexes is very well known for ligands with lone pairs of electrons and it is, therefore, not surprising that many of the mechanisms cited for metal ion oxidations of sulphur(IV) oxoanions involve electron transfer in inner sphere complexes. The sulphite ion is expected to form two types of coordination complex: with sulphur

TABLE 1.4

Stoichiometry of the reaction between Cr(VI) and S(IV) at pH 4·7 (Haight *et al.*, 1965)

10^2[Cr(VI)] (M)	10^2[S(IV)] (M)	$\dfrac{[\text{Cr(VI)}]}{[\text{S(IV)}]}$	$\dfrac{[\text{Cr(III)}]_{\text{produced}}}{[\text{S(IV)}]_{\text{consumed}}}$
0·4	2·43	0·16	0·512
0·5	2·02	0·25	0·559
1·41	5·91	0·24	0·604
1·56	1·97	0·79	0·640
0·89	0·99	0·90	0·636
1·71	1·35	1·27	0·642
0·92	0·68	1·35	0·628

Reproduced with permission from *J. Amer. Chem. Soc.*, 1965, **87**, page 3838. © 1965 American Chemical Society.

bonded closest to the metal, that is as a monodentate ligand, or through oxygens as a bidentate ligand.

One common feature of metal ion oxidations of sulphur(IV) oxoanions is their variable stoichiometry, which is a direct consequence of the formation of dithionate (involving a one-electron oxidation of two sulphur(IV) atoms) and sulphate (a two-electron oxidation of sulphur(IV)). This may be demonstrated readily by the use of permanganate and dichromate for the standardisation of sulphur(IV) oxoanion solutions. For an iodimetrically standardised solution with a concentration of 0·394 M, titration with permanganate and dichromate gives 0·383 M and 0·357 M respectively (Haight *et al.*, 1965). A more detailed result for oxidation by chromium(VI), giving the stoichiometric ratios as a function of concentration, is shown in Table 1.4, and may be explained by considering the competition between two processes which constitute the overall reaction:

$$2\text{HCrO}_4^- + 4\text{HSO}_3^- + 6\text{H}^+ = 2\text{Cr}^{3+} + 2\text{SO}_4^{2-} + \text{S}_2\text{O}_6^{2-} + 6\text{H}_2\text{O}$$

$$2\text{HCrO}_4^- + 3\text{HSO}_3^- + 5\text{H}^+ = 2\text{Cr}^{3+} + 3\text{SO}_4^{2-} + 5\text{H}_2\text{O}$$

The limiting stoichiometries are, of course, 0·50 and 0·67 respectively for the two reactions. In the first step, a complex, $\text{O}_2\text{SOCrO}_3^{2-}$, is formed by the reaction of HCrO_4^- and HSO_3^-, and the equilibrium constant for the formation of this product is estimated to be of the order of 30–40 M^{-1}. Subsequent reaction of intermediate with SO_2 in solution leads to an

TABLE 1.5

The formation of dithionate in the reaction between Fe(III) and S(IV) for an initial $[Fe(III)] = 1 \cdot 09 \times 10^{-2}$ M and $[H^+] = 0 \cdot 1$ M (Carlyle and Zeck, 1973)

$10^2[S(IV)]$ (M)	$[S(V)]^a/[Fe(III)]_{initial}$
1·09	0·46
2·46	0·67
4·92	0·81
7·38	0·78
12·3	0·83

[a] Expressed as gram atoms S(V) per litre.

activated complex from which $SO_3^{\cdot -}$ may be expelled leading either to dithionate by dimerisation, or to sulphate by one-electron oxidation with $HCrO_4^-$. Alternatively, $CrSO_6^{2-}$ may react with more $HCrO_4^-$ leading to an activated complex which is capable of expelling sulphate ion directly. In all cases, the chromium-containing product is Cr(V), which may be converted to Cr(III) by,

$$Cr(V) + S(IV) \rightarrow Cr(III) + S(VI)$$

A large quantity of dithionate is also produced in the reaction between sulphur(IV) oxoanions and iron(III), as also illustrated in Table 1.5 (Carlyle and Zeck, 1973), from which it is evident that the formation of dithionate product is favoured at the higher molar ratios of S(IV):Fe(III). The theoretical maximum ratio $[S(V)]:[Fe(III)]_0$ is unity for the following reaction:

$$S(IV) + Fe(III) \rightarrow S(V) + Fe(II)$$

Clearly, of the two overall contributing processes, that leading to dithionate is favoured in all the cases shown:

$$2Fe^{3+} + HSO_3^- + H_2O = 2Fe^{2+} + SO_4^{2-} + 3H^+$$

$$2Fe^{3+} + 2HSO_3^- = 2Fe^{2+} + S_2O_6^{2-} + 2H^+$$

The mechanism probably involves the formation of an inner sphere complex,

$$Fe^{3+} + HSO_3^- = FeSO_3^+ + H^+$$

This may subsequently decompose to yield free radical intermediates. The suggestion that iron(III) forms an inner sphere complex with sulphite is in accord with other oxidations by the former species, which take place by means of an analogous mechanism (Sutin, 1966) to that proposed for reactions involving sulphur(IV). However, the concentration of such an adduct, in the latter case, appears to be too low for its detection spectrophotometrically.

In the case of hexaaquoiron(III) considered above, the water ligands are labile and there would appear to be no hindrance to substitution. If, however, the oxidising agent is a complex of iron(III) with ligands other than water, then two possibilities must be considered. First, there is a mechanism involving ligand substitution, and secondly, a mechanism which involves the transfer of electrons by way of the ligands which serve as a bridge between oxidising agent and reducing agent. In the case of oxidations by hexacyanoferrate(III), the complex is non-labile and the lack of any substitution by sulphur(IV) oxoanions is confirmed by experiments using radiolabelled cyanide ion in the medium, when it is found that the label is not incorporated into the ferrocyanide product (Wiberg et al., 1968). The differences between the oxidation of sulphur(IV) by hexa-aquoiron(III) and hexacyanoiron(III) are also borne out by the fact that in the latter case only sulphate ion is produced. Murray and his group (Lancaster and Murray, 1971; Murray, 1974) have shown that this reaction proceeds by way of complex formation between the cyanide ligands and sulphite ion, thus:

$$Fe(CN)_6^{3-} + SO_3^{2-} \rightleftharpoons [Fe(CN)_5(CNSO_3)]^{5-}$$

$$[Fe(CN)_5(CNSO_3)]^{5-} + Fe(CN)_6^{3-} \rightleftharpoons [Fe(CN)_5(CNSO_3)]^{4-} + Fe(CN)_6^{4-}$$

$$[Fe(CN)_5(CNSO_3)]^{4-} + H_2O \rightarrow Fe(CN)_6^{4-} + SO_4^{2-} + 2H^+$$

The formation of an adduct between the reducing agent and the ligand may be compared with the formation of the sulphite adduct of nitroprusside ion (Moser et al., 1965), since NO^+ is isoelectronic with CN^- (Swinehart, 1967).

The examples considered so far have not included any cases in which stereospecific effects of ligands may be identified. A case is that of the internal oxidation–reduction of trans-$Co(en)_2SO_3OH_2^+$ (Murray and Stranks, 1970). If the medium is strongly acidic, at elevated temperature disproportionation leads to dithionic acid as the major product:

$$Co(en)_2SO_3OH_2^+ + 3H^+ \rightarrow Co_{aq}^{2+} + HSO_3^- + 2enH$$

The HSO_3^- species dimerises to dithionate. There is a tendency for increased redox behaviour in complexes in which the reducing ligand is *trans* to a water molecule. Thus, *trans*-$Co(en)_2SO_3OH_2^+$ is the only sulphito-complex which is reported to have undergone such disproportionation; *trans*-$Co(NH_3)_4ASO_3^+$, $Co(CN)_5SO_3^{4-}$ and *trans*-$Co(en)_2(SO_3)_2^{4-}$ all undergo conventional aquation and substitution reactions but not disproportionation. Another complex, $Fe(SO_3)_3^{3-}$, which should contain no aquo-ligands if it is assumed that sulphito-ligands are bidentate, is quite stable in the presence of excess free sulphite (Danilczuk and Swinarski, 1961). However, $FeSO_3^+$ and $Fe(SO_3)_2^-$, both of which must contain aquo-ligands which are probably *trans* to the sulphito-ligands, undergo disproportionation to give dithionate.

The presence of a sulphito-ligand in a complex influences markedly the stability of the other ligands. The S-bonded sulphito-ligand exhibits a pronounced *trans* labilising effect in a variety of cobalt(III) complexes and this behaviour is well illustrated by the exchange of ammonia in the solvent with $Co(NH_3)_5SO_3^+$, according to the following equation:

$$Co(NH_3)_5SO_3^+ + {}^{15}NH_3 \rightarrow Co(NH_3)_4{}^{15}NH_3SO_3^+ + NH_3$$

Here it is found that only one ligand exchanges with the medium, whereas in the case of *cis*-$Co(NH_3)_4(SO_3)_2^-$ two ammonia molecules are labile. The result confirms the specificity of the effect for the *trans* ligand and also demonstrates the independent behaviour of the sulphito-ligands (Richards and Halperin, 1976).

Although the redox decomposition of tetramminecobalt(III) complexes with the remaining coordination sites filled *trans* with either aquo- and sulphito-, or with two sulphito-groups, appears to involve labilisation of ligands (Thacker *et al.*, 1974; Scott, 1974), more recently Eldick and Harris (1980) have found that sulphitopentamminecobalt(III) shows only a small degree of conversion to the *trans*-aquo complex during redox decomposition. The formation of an oxygen-bonded sulphito-complex was inferred in this case and the main reaction product was sulphate ion, consistent with a change in mechanism. It is argued that the oxygen-bonded sulphito-ligand is probably a far more effective bridging ligand than S-bonded sulphite and is capable of effectively oxidising sulphur(IV) oxoanions.

1.10.6 The sulphite/ammonia and sulphite/hydroxylamine systems

Sulphur(IV) oxoanions react readily with nitrite ion to form products which are sulphonates of either hydroxylamine or of ammonia. The reason for the formation of the two series of compounds may be appreciated by an examination of the structures of nitrous acid and nitrite ion:

nitrous acid   nitrite ion   hydroxylamine

It is likely that reaction with sulphur(IV) oxoanions will take place as a result of addition of HSO_3^- to the nitrogen–oxygen double bond, or by replacement of OH^-. Compounds with an O-sulphonate (or sulphate) grouping have also been prepared. When the products contain a nitrogen–oxygen bond, and are, therefore, derived from hydroxylamine, there exists the possibility of isomerism. There are two possible monosulphonates (N-bonded and O-bonded), two disulphonates (N,N-bonded and N,O-bonded) and one trisulphonate (N,N,O-bonded). Since the products apparently derived from ammonia do not contain a nitrogen–oxygen bond, the only such product, which may be formed directly from nitrite ion and sulphur(IV) oxoanions, is the trisulphonate (N,N,N-bonded). The interrelationship between this series of compounds, hydroxylamine and ammonia, is shown in Fig. 1.9 after Audrieth and coauthors (1940), not all of the intermediates shown being necessarily possible. The chemistry of these compounds has also been extensively reviewed by Burton and Nickless (1968).

The reaction of alkali metal nitrites, in the cold, with either an excess of sulphur dioxide or alkali metal hydrogen sulphite leads to the formation of

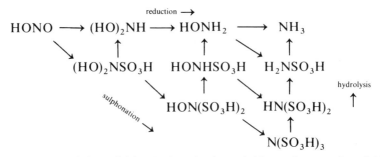

Fig. 1.9. The nitrite–sulphite, hydroxylamine–sulphite and ammonia–sulphite systems and their interrelationships (Audrieth *et al.*, 1940). Reproduced with permission from *Chem. Rev.*, 1940, **26**, page 53. © 1940 American Chemical Society.

hydroxylamine N,N-disulphonate. At elevated temperatures, complete substitution occurs at the nitrogen atom with the formation of the all-N-bonded trisulphonate. Thus, the primary reaction of nitrous acid (or nitrite), denoted as sulphonation in Fig. 1.9, involves only formation of N-bonded sulphonates. It seems that the primary product is unknown. All these compounds undergo acid hydrolysis; thus, hydroxylaminemono-sulphonate is formed by acid hydrolysis of the N,N-disulphonate:

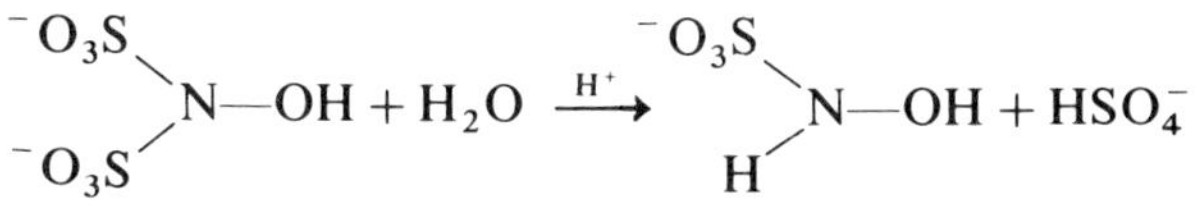

$$\begin{array}{c} ^-O_3S \\ \diagdown \\ N{-}OH + H_2O \xrightarrow{\;H^+\;} \\ ^-O_3S \diagup \end{array} \quad \begin{array}{c} ^-O_3S \\ \diagdown \\ N{-}OH + HSO_4^- \\ H \diagup \end{array}$$

The subsequent conversion of the monosulphonate to hydroxylamine is slow under these conditions. Perhaps the most familiar product shown in the scheme is sulphamic acid, NH_2SO_3H. This may be prepared by acid hydrolysis of the trisulphonate; careful hydrolysis in neutral solution also allows preparation of the intermediate disulphonate. Further treatment of sulphamic acid with mineral acid leads to complete hydrolysis. On the other hand, its reaction with nitrous acid leads to the production of nitrogen:

$$HONO + NH_2SO_3H \rightarrow H_2SO_4 + N_2 + H_2O$$

The preparation of compounds containing O-bonded sulphonate groups requires oxidising conditions. Hydroxylamine trisulphonate may be prepared by oxidation of hydroxylamine N,N-disulphonate using lead peroxide. Since the N-bonded sulphonate group appears to be more labile than the corresponding O-bonded compound, the former offers a means for the production of N,O-bonded disulphonates and the O-monosulphonate. Thus, hydroxylaminetrisulphonate is rapidly hydrolysed in aqueous acid to the disulphonate:

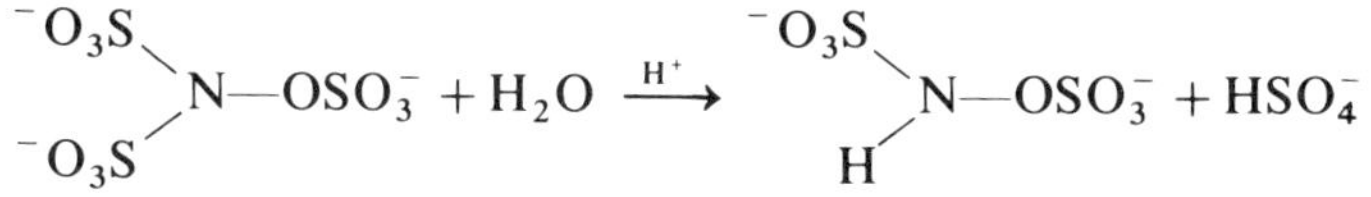

$$\begin{array}{c} ^-O_3S \\ \diagdown \\ N{-}OSO_3^- + H_2O \xrightarrow{\;H^+\;} \\ ^-O_3S \diagup \end{array} \quad \begin{array}{c} ^-O_3S \\ \diagdown \\ N{-}OSO_3^- + HSO_4^- \\ H \diagup \end{array}$$

A further slow hydrolysis yields monosulphonate. It is also of some interest to compare further the reactivity of the two isomeric hydroxylamine disulphonates. The N,O-disulphonate is hydrolysed very slowly in boiling alkali to sulphamic acid, whereas the N,N-disulphonate is hydrolysed in the cold to give sulphite and nitrite ion:

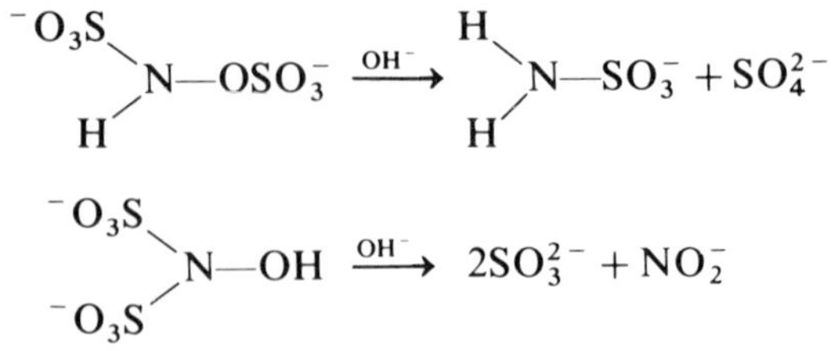

$$\ce{^{-}O_3S\diagdown} \quad \ce{N-OSO_3^-} \xrightarrow{OH^-} \quad \ce{H\diagdown} \quad \ce{N-SO_3^- + SO_4^{2-}}$$

$$\ce{^{-}O_3S\diagdown} \ce{N-OH} \xrightarrow{OH^-} 2SO_3^{2-} + NO_2^-$$

1.10.7 Reduction of sulphur(IV) oxoanions

The most common reducing agent for sulphur(IV) oxoanions is zinc, the reaction being one of the normal methods of preparation of dithionites:

$$2HSO_3^- + 2H^+ + 2e^- \rightarrow S_2O_4^{2-} + 2H_2O$$

Alternatively, this reaction may be successfully carried out using sodium amalgam, metallic iron, electrolysis (Oloman, 1970) or sodium borohydride (Mochalov et al., 1973). Dithionite ion is an extremely good reducing agent. It is used biochemically to prepare the reduced forms of several enzymes, co-enzymes and electron-transfer proteins and will reduce azo-, diazo-, diazonium-, C-nitroso-, N-nitroso- and nitro-compounds, imines, pyridinium salts, oximes, quinones, aldehydes and ketones (Vries and Kellogg, 1980). In the case of aldehydes an adduct, reminiscent of aldehyde–hydrogen sulphite adducts, is formed reversibly, for example:

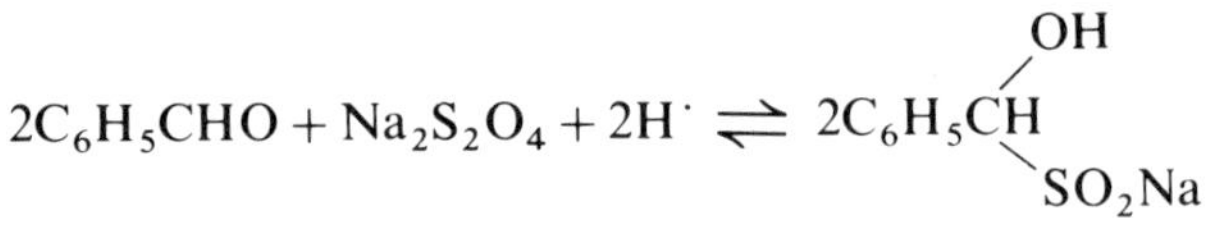

$$2C_6H_5CHO + Na_2S_2O_4 + 2H^. \rightleftharpoons 2C_6H_5\overset{\displaystyle OH}{\underset{\displaystyle SO_2Na}{CH}}$$

Solid dithionites are stable but, in aqueous solution, the ion undergoes spontaneous decomposition, the rate of which is increased with decreasing pH. The concentration–time curve for this reaction is typical of an autocatalytic process, that is, one with a slow initial step or induction period followed by a phase of increasing rate. A solution of sodium dithionite at pH 4 and 27 °C, at a concentration of $2 \cdot 5 \times 10^{-3}$ M, is completely decomposed in 20 min, and the reaction is catalysed by the presence of sulphur(IV) oxoanions, sulphide and thiosulphate ions. The decomposition reaction has been described by the following equations:

$$2H_2S_2O_4 \rightarrow S + 3SO_2 + 2H_2O$$

$$3H_2S_2O_4 \rightarrow H_2S + 5SO_2 + 2H_2O$$

The final composition of the products is the result of parallel and consecutive reactions involving intermediates, hydrogen sulphide and atomic sulphur, with sulphur dioxide, which eventually yield thiosulphate and/or polysulphides and polythionates (Cermak and Smutek, 1975).

Both zinc and cadmium are thought to stabilise dithionite, while transition metal ions are generally believed to increase the rate of decomposition (Oloman, 1970). Decomposition of stabilised dithionite solutions differs from that of sodium dithionite, the zinc salt showing an almost linear concentration–time relationship (Lem and Wayman, 1970).

1.11 ORGANIC REACTIONS

1.11.1 Reaction with carbonyl groups

The reaction between sulphur(IV) oxoanions and carbonyl compounds is a widely studied process and has been of some synthetic importance, especially for the isolation and purification of carbonyl compounds as crystalline derivatives. The particular success of its use in this respect is the facile decomposition of the adducts above pH 7, with regeneration of the original carbonyl compound.

Nucleophilic attack at the carbonyl group takes place as a result of the partial positive charge on the carbon, aldehydes being more reactive in this respect than ketones.

When solutions containing carbonyl compounds and sulphur(IV) oxoanions are mixed, the adducts form spontaneously according to the equation,

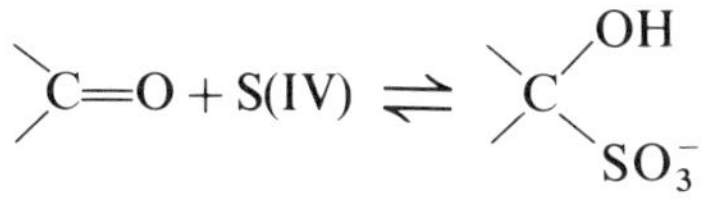

$$\mathrm{\Large{\diagdown}C}\!\!=\!\!O + S(IV) \rightleftharpoons \mathrm{\Large{\diagdown}C}\!\!\diagup\!\!\begin{array}{l}OH\\[4pt]SO_3^-\end{array}$$

The sulphonate structure of this product has been confirmed by spectroscopic studies and also on the basis of the sulphur isotope effect (Sheppard and Bourns, 1954). The equilibrium constant for this reaction is practically unaffected by pH in the range pH 2–6, as illustrated, in Fig. 1.10, by the formation of 1-hydroxy-2-methylpropanesulphonate from iso-butyraldehyde (Green and Hine, 1974), in which the apparent equilibrium constant is defined as,

$$K_{obs} = \frac{[\text{adduct}]}{[iso\text{-PrCHO}][\text{free sulphite}]}$$

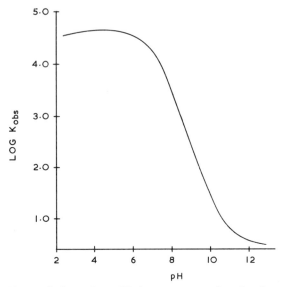

Fig. 1.10. The variation of equilibrium constant for the formation of the hydroxysulphonate adduct of isobutyraldehyde with pH, in water at 25 °C (Green and Hine, 1974). Reproduced with permission from *J. Org. Chem.*, 1974, **39**, page 3898. © 1974 American Chemical Society.

and where the given concentrations are analytical and include all states of protonation or ionisation of the reactants and products in question. This result is typical, except in the absolute value of the equilibrium constant, for adducts involving a wide range of carbonyl compounds such as benzaldehyde (Kokesh and Hall, 1975), 1,3-dimethoxyacetone (Hine *et al.*, 1976) and glucose (Vas, 1949), for which extensive pH dependent equilibrium constant data are available. As the pH is reduced below pH 2, the adducts become less stable (Stewart and Donnally, 1932*a*). The rate of formation or dissociation of the adduct is of equal importance to the equilibrium data. First order rate constants for the decomposition of *iso*-butyraldehyde (Green and Hine, 1974) and 1,3-dimethoxyacetone adducts (Hine *et al.*, 1976) show inverse dependence upon hydrogen ion concentration in the range pH 4–8. From earlier kinetic work of Stewart and Donnally (1932*b*) on the rate of decomposition of the benzaldehyde adduct, this decrease in apparent first order rate constant continues to approximately pH 2, below which it begins to rise. More detailed data, showing this effect for the dissociation of the adduct with *p*-methoxyaceto-phenone (Young and Jencks, 1977), are shown in Fig. 1.11. Thus, it is evident that the reaction with sulphur(IV) oxoanions shows three types of behaviour according to pH. At pH < 2 the adducts become increasingly labile with decreasing pH, both from the point of view of equilibrium constant and kinetics. In the approximate range pH 3–6 the complexes have pH independent stability constants but become increasingly labile with rising pH. At pH > 6 dissociation is favoured on the grounds of both equilibrium constant and rate of dissociation.

The first step in addition must involve attack by the nucleophile at the carbon atom:

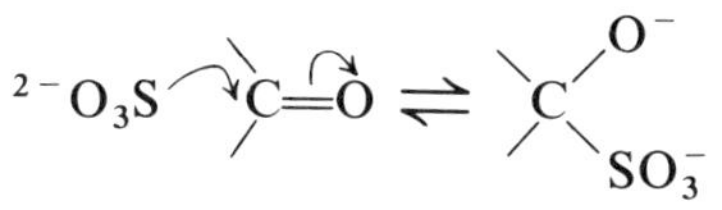

$$^{2-}O_3S \, \diagdown C{=}O \rightleftharpoons \diagdown C \diagup \begin{smallmatrix}O^-\\ SO_3^-\end{smallmatrix}$$

Subsequent reactions involve protonation at either or both negatively charged positions (Young and Jencks, 1977). For acetophenone, pK_a values are as follows:

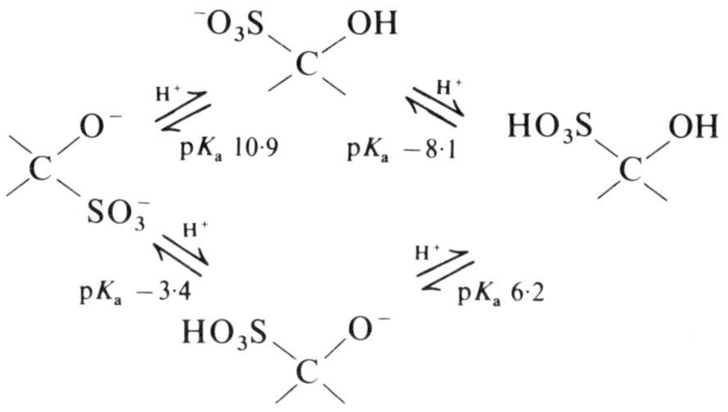

$$^-O_3S \diagdown C \diagup OH \quad \xrightarrow[\ pK_a\ 10{\cdot}9\ ]{H^+} \quad \xleftarrow[\ pK_a\ -8{\cdot}1\ ]{H^+} \quad HO_3S \diagdown C \diagup OH$$

$$\diagdown C \diagup \begin{smallmatrix}O^-\\ SO_3^-\end{smallmatrix} \qquad \qquad$$

$$\xleftarrow[\ pK_a\ -3{\cdot}4\ ]{H^+} \qquad \qquad \xrightarrow[\ pK_a\ 6{\cdot}2\ ]{H^+}$$

$$HO_3S \diagdown C \diagup O^-$$

Thus, in the pH range 2–7, protonation leads exclusively to the product, $R_2C(OH) \cdot SO_3^-$. In order to gain some insight into the pH dependence of the overall equilibrium constant, Kokesh and Hall (1975) have calculated, as follows, the constants for the reactions of HSO_3^- and SO_3^{2-} with benzaldehyde:

$$HSO_3^- + PhCHO \underset{}{\overset{K_1}{\rightleftharpoons}} PhCH(OH)SO_3^-$$

$$H^+ \Big\updownarrow K_{a1} \qquad\qquad\qquad H^+ \Big\updownarrow K_{a2}$$

$$SO_3^{2-} + PhCHO \underset{}{\overset{K_2}{\rightleftharpoons}} PhCH(O^-)SO_3^-$$

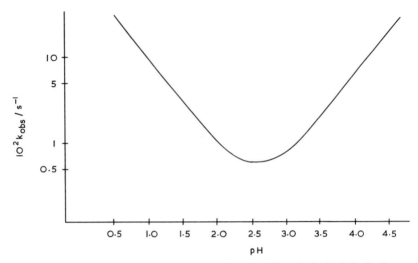

Fig. 1.11. The variation of rate constant for the dissociation of the hydroxy-sulphonate adduct of p-methoxyacetophenone with pH at 25 °C (Young and Jencks, 1977). Reproduced with permission from *J. Amer. Chem. Soc.*, 1977, **99**, page 1208. © 1977 American Chemical Society.

The observed equilibrium constant then becomes,

$$K_{obs} = K_1 \frac{(1 + K_{a2}/[H^+])}{(1 + K_{a1}/[H^+])}$$

which adequately describes the data between pH 2–12, when $K_1 = 2.4 \times 10^{-7}$ M, $K_2 = 0.912$ M and $K_{a2} = 2.11 \times 10^{-11}$ M, for an ionic strength of 1·0 M at 21 °C. Thus, when $[H^+] > K_{a1}$ (i.e. pH < 6),

$$K_{obs} = K_1 \left(1 + \frac{K_{a2}}{[H^+]}\right) \approx K_1$$

When pH < 10 and the denominator of the expression can no longer be ignored, the variation in observed equilibrium constant becomes,

$$K_{obs} = \frac{K_1}{(1 + K_{a1}/[H^+])}$$

The complete equation is applicable above pH 10. At pH < 2 the weakening of the adduct is, presumably, due to the formation of $SO_2 . H_2O$, which does not behave as an efficient nucleophile. The only other reaction which may compete with addition of sulphur(IV) oxoanion is the addition of water or hydroxide ion to the carbonyl group (Green and Hine, 1974; Hine *et al.*, 1976):

$$\begin{array}{c}R \\ \diagdown \\ R \diagup\end{array} C=O + H_2O \rightleftharpoons \begin{array}{c}R \\ \diagdown \\ R \diagup\end{array} C \begin{array}{c}OH \\ \diagdown \\ OH\end{array}$$

$$\Big\updownarrow H^+$$

$$\begin{array}{c}R \\ \diagdown \\ R \diagup\end{array} C=O + OH^- \rightleftharpoons \begin{array}{c}R \\ \diagdown \\ R \diagup\end{array} C \begin{array}{c}O^- \\ \diagdown \\ OH\end{array}$$

The equilibrium constant of hydration of 1,3-dimethoxyacetone is 0·4 at 25 °C and the pK_a for the adduct is 13·2.

In order to understand the kinetic behaviour of the decomposition of hydroxysulphonates, it is necessary to identify the rate-determining step as the expulsion of the sulphite ion, that is:

$$OH^- + \begin{array}{c}^-O_3S \\ \diagdown \\ \diagup \end{array} C \begin{array}{c}OH \\ \diagdown\end{array} \overset{fast}{\rightleftharpoons} \begin{array}{c}^-O_3S \\ \diagdown \\ \diagup \end{array} C \begin{array}{c}O^- \\ \diagdown\end{array} \overset{slow}{\longrightarrow} \begin{array}{c}\diagdown \\ \diagup \end{array} C=O + SO_3^{2-}$$

In the case of *p*-methoxyacetophenone, Young and Jencks (1977) calculate the rate constant for the slow step to be at least an order of magnitude less than that for loss of H^+: the ratio of the two constants is of the order of 100 for *p*-methoxy-, *p*-chloro- and *p*-nitrobenzaldehydes. Since the overall rate of reaction depends upon the concentration of intermediate, the observed inverse dependence on $[H^+]$ may readily be explained. One important feature of this reaction is that of general base catalysis by both nitrogen- and oxygen-containing bases, which give a Brönsted β value of 0·94 ± 0·05 for both *p*-methoxyacetophenone and *p*-methoxybenzaldehyde. The implication of such a high β value is that proton transfer to the base is almost complete in the transition state and, therefore, it is possible to write the following equation:

$$\begin{array}{c}^-O_3S \\ \diagdown \\ \diagup \\ R_1 \end{array} C \begin{array}{c}OH \\ \diagdown \\ H\end{array} + B \rightleftharpoons BH^+ . [R_1HC(O)SO_3]^{2-} \rightleftharpoons \text{products}$$

Having considered the factors affecting the addition of sulphur(IV) oxoanions to a given carbonyl compound, the effect of structure on the reactivity of different substances will now be discussed. The addition reaction will be subject to steric factors since the attacking group approaches in a suitable way to form a bond with the orbital associated with the π-bond to oxygen. Since this orbital is at right angles to the plane of the sp^2-hybridised carbon atom, approach of sulphite must also be perpendicular to this plane (Jencks, 1964). Furthermore, the sulphonate group must be accommodated satisfactorily in the final product in which the carbon is sp^3-hybridised:

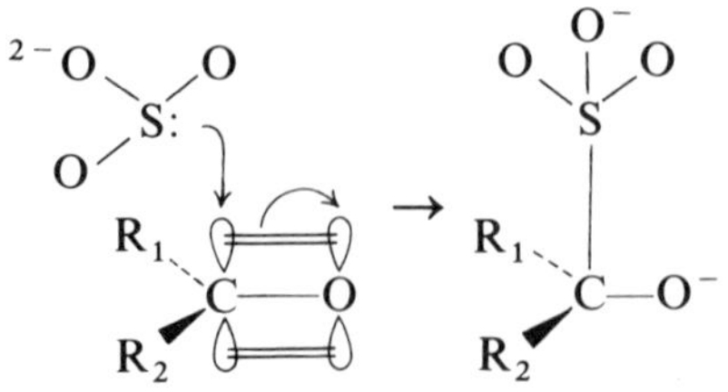

Other important contributions arising from R_1 and R_2 are their effects on charge density at the carbonyl group. With the simplest ketones, the predominant effects are likely to be those of hyperconjugation:

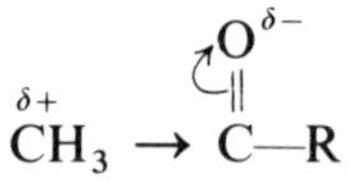

for which the order of electron-donating ability is $CH_3— > C_2H_5— > (CH_3)_2CH— > (CH_3)_3C—$, decreasing with increasing length of aliphatic carbon chain. The importance of hyperconjugation is clearly illustrated by the order of reactivity of the following series of compounds:

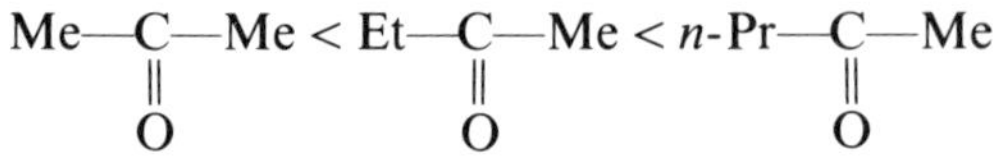

with apparent equilibrium constants of, respectively, $4·6 \times 10^5$, $19·7 \times 10^5$ and $33·3 \times 10^5 \, \text{M}^{-1}$ at pH 4 for the formation of adduct. The absence of a methyl group in the α-position leads to a considerable increase in reactivity, the equilibrium constant for the formation of adduct with diethylketone being $1·96 \times 10^7 \, \text{M}^{-1}$, under the same conditions (Geneste *et al.*, 1971).

It is also expected that aliphatic aldehydes and ketones will be more reactive than their aromatic analogues towards sulphite ion since, in the case of aromatic compounds, the benzene ring acts as an electron source,

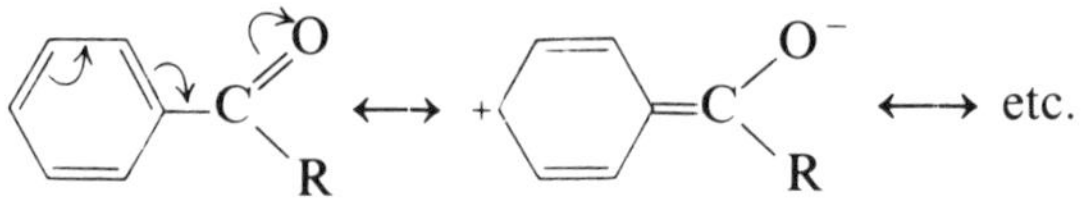

The equilibrium constant for the formation of the adduct with benzaldehyde is in the region of $10^4 \, \text{M}^{-1}$ at pH 4 (Kokesh and Hall, 1975).

The presence of electron-withdrawing groups in the aromatic nucleus increases the reactivity of the carbonyl group, as demonstrated by a positive Hammett ρ value for the equilibrium constant for the reaction of sulphur(IV) oxoanions with p-substituted benzaldehydes (Geneste $et\ al.$, 1972). Resonance effects which stabilise the carbonyl group will tend to de-stabilise the transition state and, conversely, where a substituent de-stabilises the carbonyl group, the addition product will be favoured. Extended linear free energy relationships incorporating separate polar and resonance contributions have been applied to the reactions of substituted acetophenones with sulphur(IV) oxoanions. It appears that the hydroxide ion catalysed decomposition reaction is favoured by both electron withdrawal through a polar effect and by electron donation through resonance. This observation may be explained if the individual steps for the reaction are taken separately. First, the loss of H^+ from the adduct is considered:

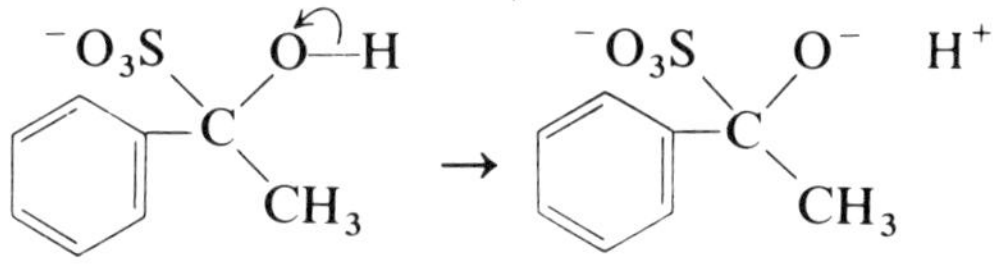

This step is favoured by electron withdrawal into the ring with consequent stabilisation of the intermediate. The developing carbonyl compound is able to accept electrons through resonance, and electron donation, therefore, facilitates the expulsion of sulphite (Young and Jencks, 1979). The overall effect on the rate of decomposition of both p-substituted benzaldehydes (Geneste $et\ al.$, 1972) and substituted acetophenones is that the rate constant is rather insensitive to the substituent compared with the variation in rate constant for the forward reaction. The latter closely follows the equilibrium constant for the formation of adduct, values increasing by approximately an order of magnitude as the substituent is changed from p-OCH_3 to p-Br.

In the case of simple aliphatic ketones, in which only the electron-donating effects of the methyl group are considered to be significant (Geneste $et\ al.$, 1971), the overall variation in equilibrium constant at pH 4 results from changes in both forward and reverse rate constants, the contribution from the reverse process (the decomposition of adduct) being some 20–40% greater than that from the forward process.

A steric effect may be evident in aliphatic aldehydes and ketones as a result of hindrance either to an approach of sulphite ion or to development of the transition state, or owing to the effects of stereochemistry on the

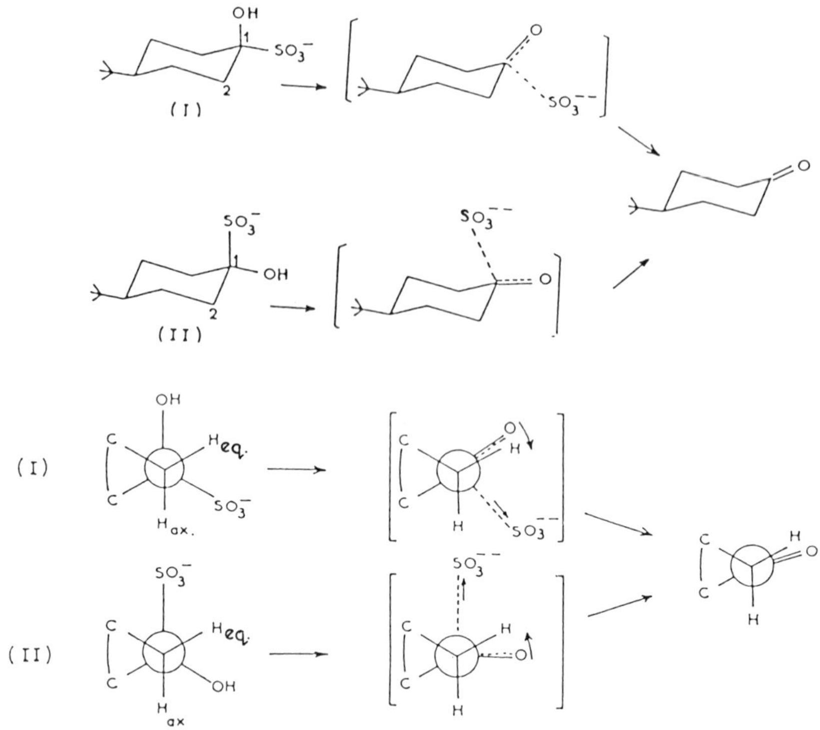

Fig. 1.12. Decomposition of the hydroxysulphonate of 4-*tert*-butyl-cyclo-hexanone. (a) Stereochemistry of alternative mechanisms. (b) Newman projections showing positions of groups at carbons 1 and 2 for both mechanisms (Lamaty and Roque, 1969). Reproduced with permission from *Rec. Trav. Chim.*, 1969, **88**, page 135. © 1969 Royal Netherlands Chemical Society.

extent of hyperconjugation. An interesting demonstration of a steric effect is given by Lamaty and Roque (1969) for the formation of isomeric adducts with 4-*tert*-butylcyclohexanone (Fig. 1.12). The bulky tertiary butyl group is normally equatorial and this, therefore, fixes the conformation of the ketone as well as that of the adducts. When reaction is allowed to proceed (pH 4, 25 °C), the isomer with —SO_3^- in the axial position (II) is formed in only 10% yield, and this product decomposes ten times faster than the corresponding adduct with —SO_3^- in the equatorial position (I). Thus, sulphite ion attacks preferentially in the equatorial position, probably as a consequence of the larger size of the sulphonate group compared with the hydroxyl group. The higher rate of decomposition of compound (II) may be accounted for by considering the change in stereochemistry at the carbonyl group as the reaction proceeds and the presence of hyperconjugation, which aids the expulsion of sulphite ion. Referring to the Newman projections shown in Fig. 1.12, it is evident that loss of sulphite at the transition state is aided by donation of electrons from the neighbouring axial hydrogens through hyperconjugation, since the latter effect is greatest when the hydrogen in question is at right angles to the plane of the developing carbonyl group. However, in the case of the decomposition of compound (I), the angle between the developing carbonyl group in the transition state and the axial hydrogens is always greater than 90°, whereas that made with the equatorial hydrogens is smaller. Thus, hyper-conjugation does not facilitate the expulsion of sulphite in this case. Moreover, in the case of compound (I), the developing carbonyl group passes through an eclipsed position with respect to the equatorial hydrogens, depending upon the extent to which the transition state resembles the products of decomposition. No corresponding steric effect exists for compound (II). Thus, a reasonable explanation is afforded for the lability of compound (II). A study of the hydrogen isotope effect (in positions 2 and 5) on the rate of reaction suggests that the main contribution to the difference in reactivity of the two isomers is the effect of hyperconjugation.

1.11.2 Reaction with carbon–carbon double bonds

The reaction of sulphur(IV) oxoanions with carbon–carbon double bonds may proceed by both heterolytic and homolytic routes, depending upon the nature of the unsaturated compound. In the case of addition to α,β-unsaturated compounds in which the double bond is attached to an electron-withdrawing group, a simple heterolytic mechanism is observed. The electron-withdrawing groups most commonly found in conjugation with carbon–carbon double bonds are aldehyde, keto, carboxylic acid and ester. In reactions with a nucleophile, the following sequence of events is required (Sykes, 1965):

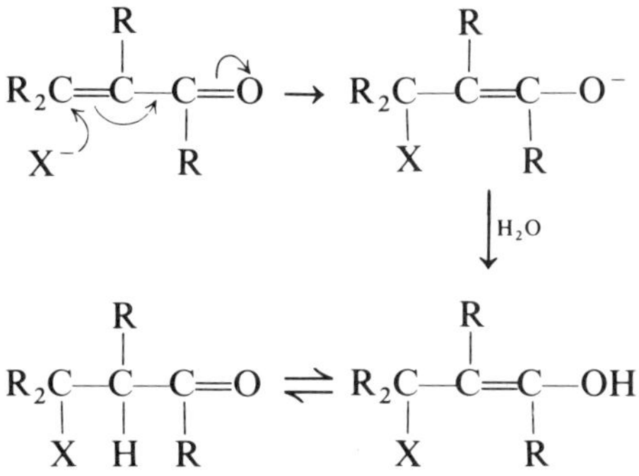

In the case of a poor nucleophile, prior activation by protonation of the oxygen is required to give a carbonium ion of the type,

$$R_2C=C-\overset{+}{C}-OH \longleftrightarrow R_2\overset{+}{C}-C=C-OH$$
$$\quad\quad\ \ |\ \ |\quad\quad\quad\quad\quad\quad |\ \ |$$
$$\quad\quad\ \ R\ \ R\quad\quad\quad\quad\quad\quad R\ \ R$$

In each case, the product is the result of overall 1,4-addition. The reaction occurs readily with ketones, acids and esters, but the carbonyl carbon atom of aldehydes is sufficiently positive to allow direct attack leading to 1,2-addition.

When the nucleophile is a sulphur(IV) oxoanion, the reaction is found to proceed over a wide range of pH: the sulphonation of the sodium salt of crotonic acid may be taken as an example. In this case, the conversion takes place in weakly acidic aqueous media and is unaffected by the presence of oxygen or antioxidants,

$$RCH=CH-C=O \xrightarrow{S(IV)} RCH-CH_2-C=O$$
$$\quad\quad\quad\quad |\quad\quad\quad\quad\quad\ \ |\quad\quad\quad\quad |$$
$$\quad\quad\quad\quad OH\quad\quad\quad\quad\ SO_3^-\quad\quad OH$$

In the case of addition to carboxylic acids, there are, presumably, two effects which contribute to the observed pH dependence of the rate of reaction. First, for addition of sulphur(IV) oxoanions, there is the expected change in oxoanion composition. Secondly, the development of negative charge on the carboxyl group resulting from ionisation of the acid tends to disfavour the reaction. The most powerful nucleophile is expected to be the sulphite ion on account of its high charge and the exposed sulphur atom. From studies on the rate of addition of sulphur(IV) oxoanions to acrylonitrile, which is expected to proceed by a similar mechanism, it is evident that the reaction is favoured as pH is increased above 7, a maximum rate occurring around pH 9. Since there are no changes to acrylonitrile in this pH range, the results suggest that the effective sulphonating agent is, indeed, SO_3^{2-} (Morozov et al., 1972). A reduction of rate at pH > 9 was attributed to the alkaline hydrolysis of the nitrile.

In the case of crotonic acid, there is little reaction below pH 2 or above pH 9, relative to the rate of reaction at intermediate pH. The maximum rate occurs at around pH 5·6 (Schenk and Danishefsky, 1951; Critchfield and Johnson, 1956). Assuming that the rate of this reaction is proportional to the concentrations of sulphite ion and free acid only, and that the other species are relatively unreactive, the rate equation, in terms of the concentration of the total sulphur(IV) oxoanion and crotonic species, which are invariant, becomes, as a function of $[H^+]$,

$$\text{Rate} = k \left(\frac{[H^+]}{K_a + [H^+]} \right) \left(\frac{K_a'}{K_a' + [H^+]} \right) [A][S(IV)]$$

K_a and K_a' are the dissociation constants of crotonic acid and HSO_3^- respectively, k is a rate constant and A represents the sum of concentrations of organic acid and anion. The equation neglects any contribution from $SO_2 . H_2O$ or sulphurous acid species. Differentiation with respect to $[H^+]$ and equating to zero yields the following expression for hydrogen ion concentration for maximum rate:

$$[H^+] = (K_a K_a')^{1/2}$$

or

$$pH = \frac{pK_a + pK_a'}{2}$$

If the pK_a for HSO_3^- is 7·2 and that for crotonic acid is 4·7 (Heilbron et al., 1965), the maximum rate is expected at pH 5·9, in excellent agreement with experimental data.

An interesting application of this reaction is its use in the determination of α,β-unsaturated compounds, the principle of the method being the measurement of the hydroxide ion liberated (Critchfield and Johnson, 1956), for example,

$$CH_2=CHCN + SO_3^{2-} + H_2O \rightarrow {}^-O_3SCH_2CH_2CN + OH^-$$

Another example of a reaction in which attack takes place in position 2 is the addition of sulphur(IV) oxoanions to acrolein (Finch, 1962). Initial attack is rapid and leads to the formation of the expected adduct,

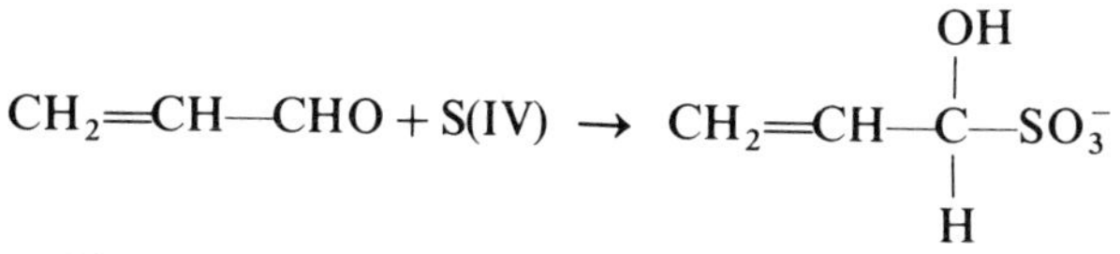

$$CH_2=CH-CHO + S(IV) \rightarrow CH_2=CH-\overset{\displaystyle OH}{\underset{\displaystyle H}{\overset{|}{\underset{|}{C}}}}-SO_3^-$$

At pH < 5, this adduct slowly adds a second sulphur(IV) oxoanion to yield the fully sulphonated product,

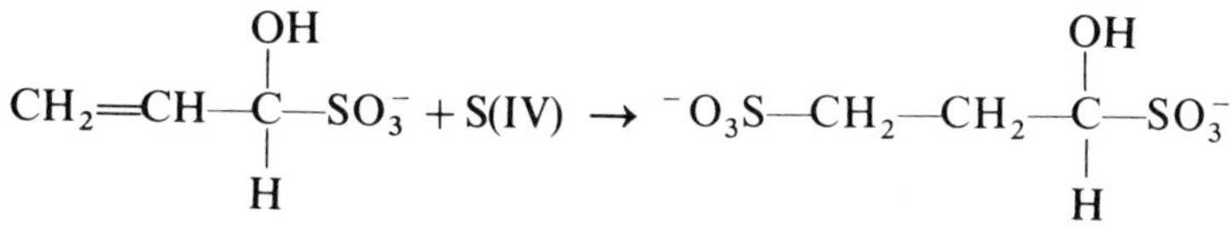

$$CH_2=CH-\overset{\displaystyle OH}{\underset{\displaystyle H}{\overset{|}{\underset{|}{C}}}}-SO_3^- + S(IV) \rightarrow {}^-O_3S-CH_2-CH_2-\overset{\displaystyle OH}{\underset{\displaystyle H}{\overset{|}{\underset{|}{C}}}}-SO_3^-$$

When the reaction is carried out at pH > 5·2, there appear to be two mechanisms, one of which involves apparent disproportionation of the carbonyl adduct:

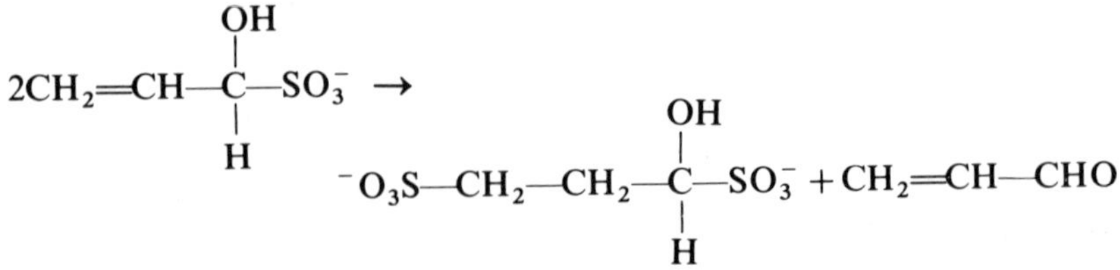

$$2CH_2{=}CH{-}\underset{\underset{H}{|}}{\overset{\overset{OH}{|}}{C}}{-}SO_3^- \;\rightarrow\; {}^-O_3S{-}CH_2{-}CH_2{-}\underset{\underset{H}{|}}{\overset{\overset{OH}{|}}{C}}{-}SO_3^- + CH_2{=}CH{-}CHO$$

The other mechanism involves further reaction of the acrolein produced in this way. The disproportionation reaction implies either that the carbonyl–sulphur(IV) adduct can act as a sulphonating agent, or that, at pH > 5·2, the complexes become labile and dissociate liberating sulphur(IV) oxoanions. As the pH is raised, these become more capable of sulphonation at position 4.

When addition of sulphur(IV) oxoanions to non-conjugated double bonds takes place, the reaction is known to proceed by a free radical mechanism and the main products are a result of anti-Markownikoff addition. Oxidising conditions are required for this reaction, which may be initiated by any method leading to production of the sulphite free radical. Kharasch and coworkers (1939a, b) suggested that the reaction proceeded by a mechanism involving this radical intermediate:

$$SO_3^{\cdot-} + CH_2{=}CHR \;\rightarrow\; {}^-O_3SCH_2\dot{C}HR$$

$$ {}^-O_3SCH_2\dot{C}HR + HSO_3^- \;\rightarrow\; {}^-O_3SCH_2CH_2R + SO_3^{\cdot-}$$

A convenient method of producing $SO_3^{\cdot-}$ is the oxidation of hydrogen sulphite ion by Ce^{4+} (Ozawa et al., 1971). When solutions in which these radicals are being produced are mixed with allyl alcohol, the ESR signal due to $SO_3^{\cdot-}$ is completely quenched, indicating reaction of this species with the organic compound. In the sulphonation of methacrylic, crotonic and fumaric acids by this method, secondary free radicals, produced from attack by $SO_3^{\cdot-}$, have been identified from ESR spectra (Ozawa et al., 1972); for example, the addition of crotonic and fumaric acid leads, respectively, to the following intermediates:

$$\underset{\underset{SO_3^-}{/}}{\overset{\overset{CH_3}{\backslash}}{H{-}C}}{-}\underset{\underset{COOH}{\backslash}}{\overset{\overset{H}{/}}{\dot{C}}} \qquad\qquad \underset{\underset{SO_3^-}{/}}{\overset{\overset{COOH}{\backslash}}{H{-}C}}{-}\underset{\underset{COOH}{\backslash}}{\overset{\overset{H}{/}}{\dot{C}}}$$

The sulphite radical, in which the charge is localised on the sulphur atom (Chantry *et al.*, 1962), behaves as a sulphur radical leading to the formation of a C—S bond (Ozawa *et al.*, 1972). This behaviour is in contrast to that of the sulphate ion radical ($SO_4^{\cdot-}$) which behaves as an oxygen radical and which tends to form carbon–oxygen bonds on reaction with carbon–carbon double bonds (Norman *et al.*, 1970; Ozawa *et al.*, 1972).

Hydrogen abstraction from HSO_3^- in the sulphonation of allyl alcohol, according to the Kharasch mechanism above, is supported by a large kinetic solvent isotope effect, $k_{H_2O}/k_{D_2O} = 4.11$ (Miyata *et al.*, 1975*a*), but it appears that the substrate may also interact with the secondary radicals in a termination reaction (Miyata *et al.*, 1975*b*).

$$HOCH_2\dot{C}HCH_2SO_3^- + CH_2{=}CHCH_2OH \rightarrow$$
$$HOCH_2CH_2CH_2SO_3^- + CH_2{=}CH{-}\dot{C}HOH$$

A variation of this reaction takes place when the substrate for sulphonation is the hydrophobic species, 1-dodecene (Sakumoto *et al.*, 1975). If the sulphonation is attempted in a solution containing *tert*-butanol:water (1:1) using γ-irradiation as the initiator, it seems that no chain sulphonation takes place until some of the reaction product, 1-dodecanesulphonate, has accumulated; the addition of this product allows the reaction to proceed correctly. Although the state of the hydrocarbon in this solution has not been investigated, it is likely to be present in the form of micelles, access to which by the polar free radical is promoted by the addition of a surfactant, such as the sulphonated product. Sulphonation takes place at the surface of these micelles and the hydrophilic region so formed, with its associated secondary radical, will interact more closely with the aqueous phase, and it is not surprising, therefore, that a termination reaction involves formation of disulphonate, thus:

$$CH_3(CH_2)_9\dot{C}HCH_2SO_3^- + SO_3^{\cdot-} \rightarrow CH_3(CH_2)_9\overset{\displaystyle \overset{SO_3^-}{|}}{C}HCH_2SO_3^-$$

The sulphite free radical is expected to interact with hydrogen ions to form $HSO_3^{\cdot}$,

$$HSO_3^{\cdot} \rightleftharpoons H^+ + SO_3^{\cdot-}$$

and, therefore, the sulphonation reaction may be affected by pH if either charge preferences exist, or if the reactivity of $HSO_3^{\cdot}$ differs from that of $SO_3^{\cdot-}$. Sakumoto and coworkers (1976) calculate the pK_a for the dissociation of $HSO_3^{\cdot}$ as 4.5; that is, it is a stronger acid than HSO_3^-, and considerable changes in the relative amounts of the protonated and unprotonated species are thus expected over the pH range 3–6. The relative reactivities of the two species towards allyl alcohol and methallyl alcohol are similar, the ratio of HSO_3^- reactivity to that of SO_3^{2-} being 1·32 and 0·93 in the two cases respectively. A small pH effect is, therefore, expected although an optimum pH for addition to cyclohexene and propene, using oxygen, is close to 6 (Mayo and Walling, 1940), and for the sulphonation of 1-dodecene (Norton et al., 1968), the pH is 5–7. It is necessary, however, to dissociate the pH dependence of the initiation reaction, for example autoxidation of sulphur(IV) oxoanions, with any pH effect associated with the free radical additions.

1.11.3 Reaction with hydroperoxides

Sulphur(IV) oxoanions are effective reducing agents for hydroperoxides, the reaction proceeding smoothly when, for example, tert-butyl hydroperoxide and sodium sulphite are mixed at 25 °C (Anbar et al., 1962). The literature to 1960 has been reviewed by Davies (1961). It is evident that when SO_3^{2-} is the predominant oxoanion, the reduction of hydroperoxide always leads to the formation, in good yield, of the corresponding alcohol according to the stoichiometric equation:

$$SO_3^{2-} + ROOH \rightarrow SO_4^{2-} + ROH$$

However, when the pH is lowered, the reaction becomes more complex and ketones, olefins and ethers are formed in place of alcohols. Whilst the mechanism at high pH appears to be heterolytic, there is evidence to suggest the involvement of free radicals in acid solutions.

In order to ascertain the participation of alkoxy species rather than alkyl intermediates in the conversion of hydroperoxide to alcohol, Davies and Feld (1956) followed the conversion of optically active hydroperoxides at pH 4·8, when it was found that the asymmetric centre remained unaffected during the course of the reaction. Thus, the process involves cleavage of the oxygen–oxygen bond of the hydroperoxide, a conclusion which is further supported by the observation that when tert-butyl, 1-phenylethyl and cumene hydroperoxides were reduced in ^{18}O-water at high pH, the alcohols which were formed contained oxygen of normal isotopic composition (Davies, 1961). Hence, the most likely mechanism is that of nucleophilic displacement, by SO_3^{2-}, of an alkoxy group,

$$^{2-}O_3S \frown O \frown OR \rightarrow {}^-O_3S . OH + {}^-OR$$
$$\quad\quad\quad\quad | \\ \quad\quad\quad\quad H$$

The same reaction scheme was also suggested, independently, with apparently no reference to earlier work, by Anbar and coworkers (1962), for the reduction of *tert*-butyl hydroperoxide.

At pH 5, the use of ^{18}O-water leads to alcohol which has approximately half of its oxygens derived from the solvent, and it is suggested by Davies (1961) that half of the reaction may proceed via an intramolecular rearrangement of a peroxosulphurous acid ester, formed from the initial reaction of hydroperoxide with HSO_3^-:

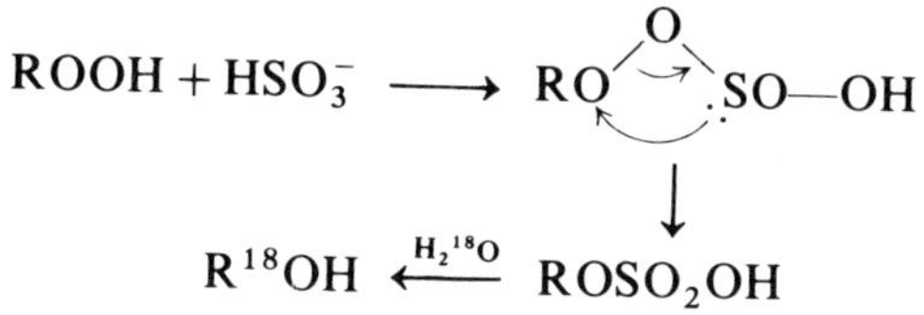

$$ROOH + HSO_3^- \longrightarrow RO\overset{O}{\underset{}{\diagup}}SO-OH$$

$$R^{18}OH \xleftarrow{H_2^{18}O} ROSO_2OH$$

Anbar and coworkers (1962) also report small amounts (5%) of organic sulphate at pH 4–6. The reduction of *tert*-butyl hydroperoxide is accompanied by the formation of acetone (the yield increasing from approximately 40% at pH 4–6 to around 70% at pH 0–1) and methanol. It is possible that this reaction is effected by an acid catalysed decomposition, which does not involve the sulphur(IV) species, as suggested by Kharasch and coworkers (1951) for cumene hydroperoxide (Husbands and Scott, 1979):

$$(CH_3)_3COOH \xrightarrow{H^+} (CH_3)_3CO^+ + H_2O$$

$$(CH_3)_2C{=}O + CH_3OH \leftarrow (CH_3)_2\overset{OH}{\underset{}{C}}OCH_3 \xleftarrow{H_2O} (CH_3)_2\overset{+}{C}OCH_3$$

However, a free radical mechanism involving the sulphur(IV) species is also likely. Both *tert*-butyl hydroperoxide and cumene hydroperoxide lead, on reduction with sulphur(IV) oxoanions, to the production of the free radical, $Me\dot{S}O_2$, identified by ESR, over the pH range 2–5 (Flockhart *et al.*, 1971). A possible reaction sequence for the formation of $Me\dot{S}O_2$, which would also allow for acetone production, is the initial formation of the oxy-radical, followed by reaction with sulphur(IV) oxoanion, illustrated below for the *tert*-butyloxy radical:

$$Me_3CO^\cdot \rightarrow Me_2CO + Me^\cdot$$

$$Me^\cdot + SO_2 \rightarrow Me\dot{S}O_2$$

$$Me_3CO^\cdot + SO_2 \rightarrow Me_3CO\dot{S}O_2 \rightarrow Me_2CO + Me\dot{S}O_2$$

When alkenes are present, it is generally observed that new free radicals are formed in reactions between the alkene and the sulphur-containing radical.

The initial formation of the alkoxy radical, by reaction of hydroperoxide with two-electron reducing agents (such as HSO_3^-), is thought to take place by a mechanism similar to that observed for one-electron reductions (Kharasch *et al.*, 1951). Davies (1961) suggests a mechanism for the reaction of hydroperoxide with HSO_3^-, analogous to that of the Fe^{2+}-mediated decomposition of hydroperoxide (Kharasch *et al.*, 1951), that is,

$$ROOH + HSO_3^- \rightarrow RO^{\cdot} + OH^- + HSO_3^{\cdot}$$

The formation of $HSO_3^{\cdot}$ (or $SO_3^{\cdot-}$) in the *tert*-butyl hydroperoxide–SO_2 system has been confirmed by ESR (Flockhart *et al.*, 1971).

1.11.4 Addition of sulphur(IV) oxoanions to Schiff's bases

The ability of sulphur(IV) oxoanions to add to Schiff's bases has been known for over a century (Schiff, 1866), and the simplest example is the reaction between ammonia and the hydroxysulphonate of formaldehyde (Backer and Mulder, 1933, 1934):

$$H_2C \underset{SO_3^-}{\overset{OH}{\diagup}} + NH_3 \rightarrow H_2NCH_2SO_3^- + H_2O$$

It is only recently, however, that the structures of the products have been completely elucidated (Clark and Miles, 1978). In the case of aromatic aldehydes, a detailed study is not possible owing to insufficient stability of the Schiff's bases, but the products derived from substituted benzaldehydes may be isolated. The most likely reaction between a Schiff's base and sulphur(IV) oxoanions is addition to the $C{=}N$ bond,

$$RCH{=}NR' + HSO_3^- \rightarrow \underset{SO_3^-}{\overset{\displaystyle RCHNHR'}{|}}$$

This is the case for an aliphatic amine and, in an excess of water, the product is hydrolysed to the alcohol, that is, the hydroxysulphonate:

$$\underset{SO_3^-}{\overset{\displaystyle RCH{-}NHR'}{|}} + H_2O \rightleftharpoons \underset{SO_3^-}{\overset{\displaystyle RCHOH}{|}} + R'NH_2$$

The reaction product, in the case of an aromatic amine, appears to be stable to hydrolysis and exists as an amine salt:

$$RCH{=}NR' + R'\overset{+}{N}H_3HSO_3^- \rightleftharpoons \underset{\substack{| \\ SO_3^-\overset{+}{N}H_3R' \\ V}}{RCHNHR'}$$

Strongly electron-withdrawing groups in the aldehyde, and a sterically hindered amine assist hydrolysis during the reaction and lead to the hydroxysulphonate. The only aliphatic aldehyde investigated was *iso*-butyraldehyde with an aromatic amine, and the product typical of an aromatic amine was observed.

An application, suggested by Adams and Garber (1949), of this reaction is its use for the resolution of asymmetric aldehydes and ketones, when an optically active base is used in the reaction,

$$\underset{(+\text{ or }-)}{\overset{(\pm)}{RCHO}} + R'NH_3SO_3^- \rightarrow \underset{\substack{| \\ SO_3^-\overset{+}{N}H_3R' \\ (+\text{ or }-)}}{\overset{(\pm,(\pm)}{RCHOH}}$$

Ingles (1959*a*) successfully prepared the addition compound of glucose, aniline and sulphur(IV) oxoanion by reaction of D-glucose with aniline hydrogen sulphite in methanol. The crystalline product, which contained two moles of amine per mole of glucose, was assigned the structure, **V** above, expected for reactions of aromatic amines, with $R = CH_2OH(CHOH)_4-$ and $R' = C_6H_5-$; mannose and galactose behaved similarly. The configuration of C_1 is β for compounds involving glucose and galactose with a wide range of aromatic amines. The products undergo hydrolysis in water and in acidic and alkaline solutions. In acid media decomposition proceeds to give glucose as the final product, whereas in alkaline solution the glycosylamine is stable. The configuration at C_1 of the product of hydrolysis is α for both glucose and galactose, indicating that a Walden inversion has taken place at this position during the hydrolysis (Ingles, 1959*b*). When the amine was cyclohexylamine, the product contained only one cyclohexylamine group and was considered to be the corresponding salt of the hydroxysulphonate adduct.

1.11.5 Reaction with isothiocyanate

Many isothiocyanates are important food flavouring agents and their reaction with sulphur(IV) oxoanions is, therefore, important. The reaction may be summarised by the following stoichiometric equation:

$$RNCS + KHSO_3 \rightarrow RNHCS . SO_3K$$

Effectively, the mechanism involves addition across the $C{=}N$ bond, and is believed to be similar to that of addition to the carbonyl group (Backer *et al.*, 1935). An alternative structure, $RNHCSOSO_2K$, suggested earlier, has since been shown to be incorrect (Sankaran and Narasimha Rao, 1977). From work carried out on the product formed in reactions between a number of substituted benzylisothiocyanates and HSO_3^-, it has been found that the compounds are stable in the dry state and in neutral or acidic aqueous solution in the absence of air. Exposure to air leads to the formation of carbylamines and thiosulphate ion,

$$RNHCS . SO_3^- \rightarrow RN{=}C: + S_2O_3^{2-} + H^+$$

Alkaline conditions promote dissociation of the product to give the sulphite ion and organic products.

1.11.6 Formation of anionic σ-adducts with aromatic compounds

In an aromatic system the π-electrons confer susceptibility towards electrophilic attack and, hence, this is the normal type of reactivity associated with the benzene ring. However, if sufficiently electron-withdrawing groups are present, nucleophilic attack may occur and one of the best known nucleophilic substitution reactions is the hydroxylation of nitrobenzene to give *o*-nitrophenol (Sykes, 1965):

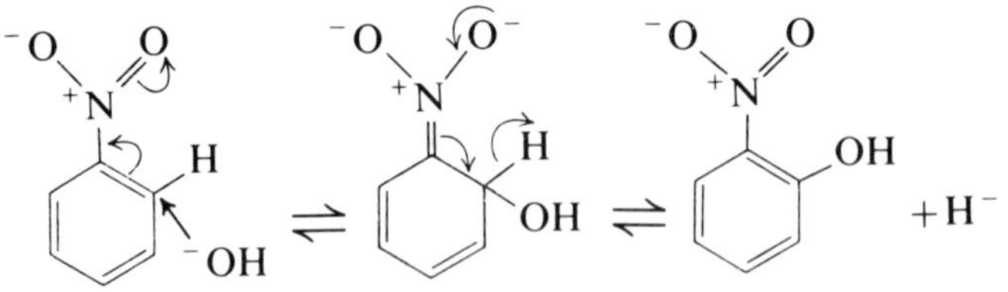

As with normal electrophilic substitution, this reaction proceeds via a σ-complex intermediate but, in this case, the complex is anionic. The positions *ortho* and *para* to the electron-withdrawing group are activated towards this type of reaction. The presence of a formal positive charge on the nucleus will also facilitate the addition of a nucleophile, illustrated by the addition of methoxide ion to substituted pyrylium cations to form the σ-adduct (Doddi *et al.*, 1979):

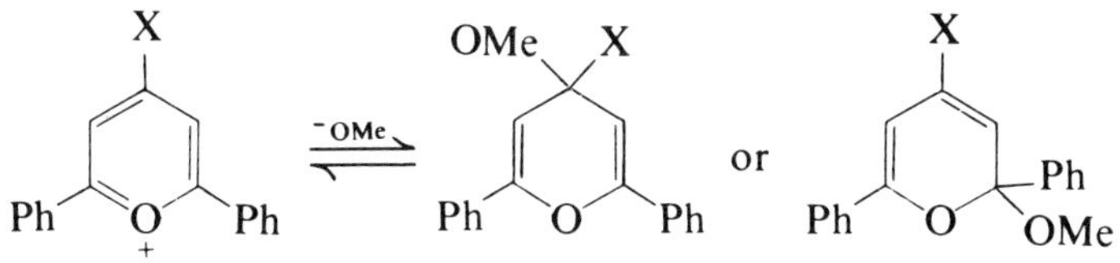

(where X = H or OMe)

In the case in which the reaction stops at the σ-complex stage, the product is often referred to as a Meisenheimer complex. Typical of aromatic systems activated towards the formation of Meisenheimer complexes is 1,3,5-trinitrobenzene, for which extensive data are available (Crampton, 1969). From results obtained in water as solvent, the sulphite ion is approximately one hundred times more effective than hydroxide ion in forming the adduct. Addition to trinitrobenzene ultimately leads to *cis/trans* isomeric di-adducts (Bernasconi and Bergstrom, 1973), the equilibrium constant for the second addition being lower than that for the first by a factor of 500 (Crampton, 1967):

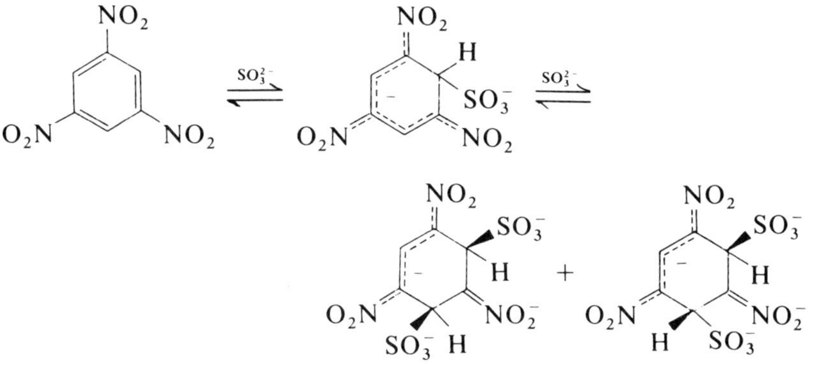

Suitably activated molecules present in food systems include the 2-phenylbenzopyrylium salts, which form the basis of anthocyanin pigments and which will be discussed in Chapter 6. The powerful electron-withdrawing effect of the nitro-group in the Meisenheimer complex illustrated above renders the adduct a strong acid. Thus, addition of $H^+SO_3^{2-}$ to the ring does not take place. However, in cases in which such extensive electron delocalisation does not take place, the anionic σ-complex may be converted to the addition product.

1.11.7 Addition to pyridine compounds

The structure of pyridine has much in common with that of benzene, both molecules having considerable aromatic stability. The effect of nitrogen, however, is to withdraw electrons from the ring, leading to overall deactivation of the molecule towards electrophilic attack. The reactivity of pyridine may be understood by a consideration of the following canonical structures:

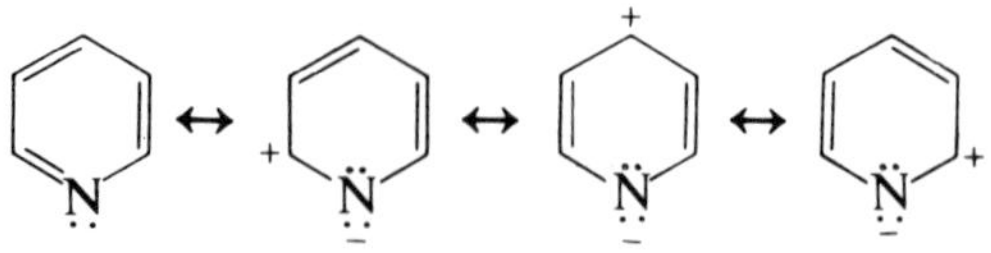

As a result of this electron withdrawal and the development of the partial positive charge in positions 2, 4 and 6, the normal reactions of pyridine are those of nucleophilic addition and substitution (Roberts and Caserio, 1965). In rare cases pyridine will undergo electrophilic substitution in position 3. The course of the reaction with nucleophiles involves the following primary step:

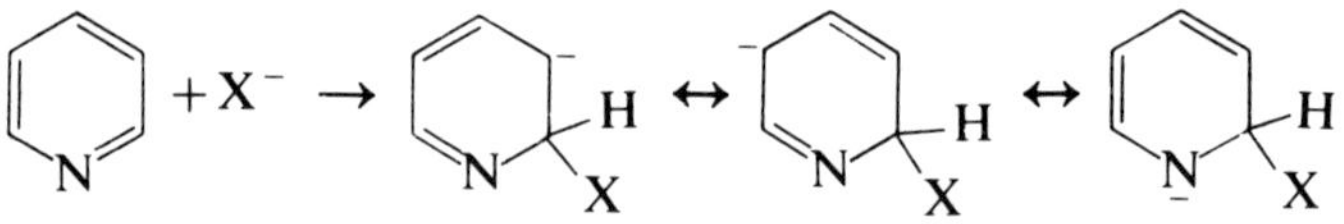

The reaction may be completed by the loss of hydride ion to give a substitution product, or it may stop at the stage of the addition compound, formed by 1,2- or 1,4-addition of the acid, HX, for example:

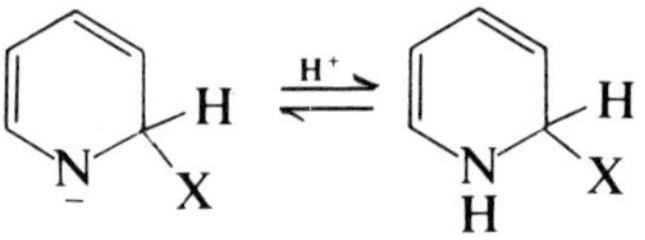

The simplest naturally occurring pyridine compound is nicotinic acid (pyridine-3-carboxylic acid). N-substituted nicotinamide is the reducible part of the coenzyme, nicotinamide adenine dinucleotide (NAD^+):

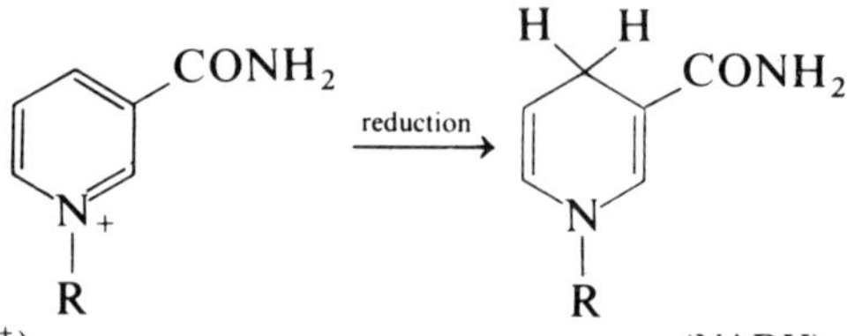

The chemistry of dihydropyridines has been extensively reviewed by Eisner and Kuthan (1972). NAD^+ also readily adds sulphur(IV) oxoanion in position 4 (Pfleiderer *et al.*, 1960):

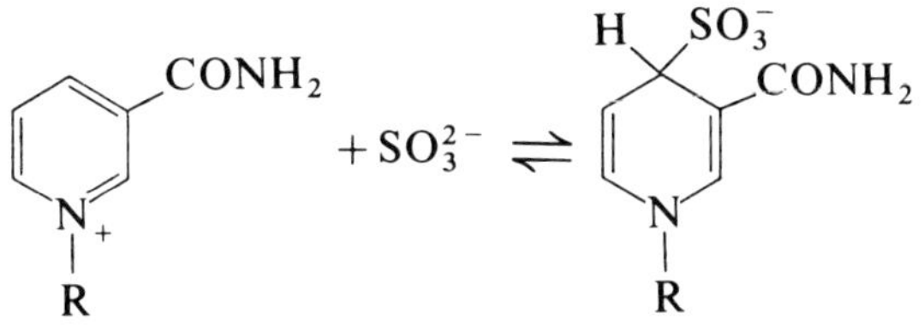

The reaction is accompanied by an increase in absorbance at 320 nm, the addition compound having an extinction coefficient, which is apparently independent of buffer composition, of $4750\,M^{-1}\,cm^{-1}$ at this wavelength (Shih and Petering, 1973), from measurements made at pH 7·4, whereas NAD^+ absorbs at 320 nm with extinction coefficients of $190\,M^{-1}\,cm^{-1}$ and $88\,M^{-1}\,cm^{-1}$ in tris buffer and in phosphate buffer respectively. The dependence of the equilibrium constant of this reaction upon pH is consistent with sulphite ion as the nucleophile, as shown in the equation, and suggestions that the reaction is the result of the addition of hydrogen sulphite ion (Meyerhof *et al.*, 1938; Colowick *et al.*, 1951; Ciaccio, 1966) are incorrect. The pH independent equilibrium constant is given by Shih and Petering (1973) as:

$$K = \frac{[NAD.SO_3^-]}{[NAD^+][SO_3^{2-}]} = \frac{[NAD.SO_3^-][1 + ([H^+]/K_a)]}{[NAD^+][S(IV)]}$$

where [S(IV)] is the total sulphur(IV) oxoanion concentration in the reaction mixture and K_a is the dissociation constant of hydrogen sulphite ion. The value of the equilibrium constant for the formation of product is quoted as $36\,M^{-1}$ at 25 °C, which may be compared favourably with a value of $50\,M^{-1}$ given by Pfleiderer and coworkers (1960) from measurements at pH 8. The addition reaction is rapid, requiring stop-flow kinetic measurements, with a second order rate constant of $2·4 \times 10^3\,M^{-1}\,s^{-1}$ at pH 8. The first order rate constant of $35\,s^{-1}$ for the decomposition of the adduct gives a kinetically determined equilibrium constant of $69\,M^{-1}$ for the reaction (Parker *et al.*, 1978).

Sulphite ion will bind to NAD^+ when the latter is bound to an enzyme such as lactate dehydrogenase and the adduct is spectroscopically similar to that formed when the NAD^+ is free (Parker *et al.*, 1978). The reduced form of nicotinamide adenine dinucleotide also reacts with sulphur(IV) oxoanions. Two distinct reactions, one involving a free radical oxidation and the other proceeding by way of an ionic mechanism, may be resolved. The ionic reaction results in hydration of the 5,6-double bond of NADH by way of a reactive complex involving both sulphur(IV) oxoanion and the substrate, which subsequently undergoes general acid catalysed hydration (Tuazon and Johnson, 1977).

1.11.8 Addition to pyrimidine compounds

The presence of a single nitrogen atom in an aromatic ring (pyridine) leads to deactivation of the ring towards electrophilic attack and this situation is enhanced when two nitrogen atoms are present in a 1,3-arrangement, as in the case of pyrimidine. The latter may undergo nucleophilic attack in the 2-, 4- and 6-positions, demonstrated by the following canonical structures (Roberts and Caserio, 1965):

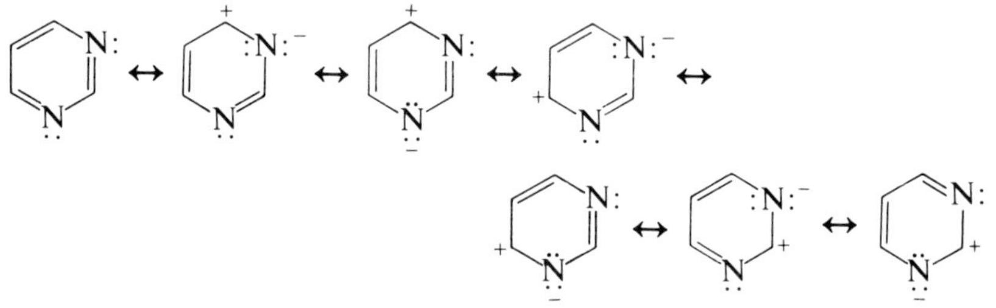

The importance of substituted pyrimidines to biological systems has recently prompted a great deal of interest in their reactions with sulphur(IV) oxoanions. The simplest substituted pyrimidine studied is 2-aminopyrimidine (Pitman and Ziser, 1970). The pK_a of the 2-aminopyrimidinium ion is 3·8 at 25 °C and, over the range pH 3–5, it is the protonated pyrimidine ring which is involved in addition. The pH dependence of this reaction, in this pH range, and the observation that the reaction of 2-amino-1-methylpyrimidine is independent of pH are consistent with this conclusion. In the case of 2-amino-1-methylpyrimidine, the charge on the pyrimidinium ion is fully developed and is independent of pH. Kinetic studies show that the sulphite ion, as predicted, is the most effective nucleophile, the second order rate constant involving this species being some three orders of magnitude greater than that observed for hydrogen sulphite ion.

In the range pH 3–5, the addition proceeds as follows:

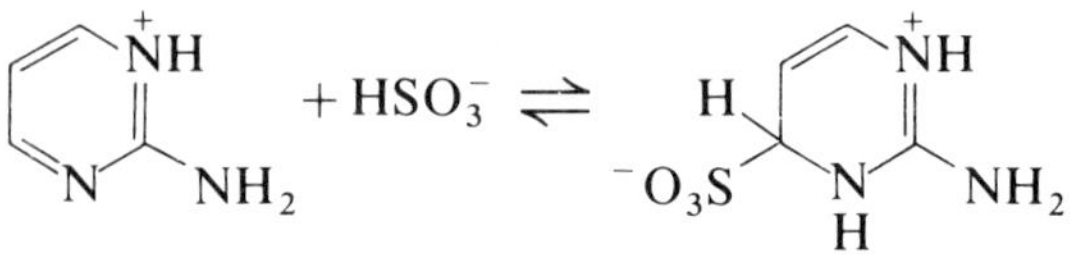

Whilst only one product is formed from 2-aminopyrimidine, the reaction of 2-amino-1-methylpyrimidine with HSO_3^- leads to the formation of two possible isomeric products, arising from 3,4- and 3,6-addition, in the ratio $0.8 : 1.0$ respectively.

Two of the most widely studied substrates for the addition of sulphur(IV) oxoanions are uracil and cytosine:

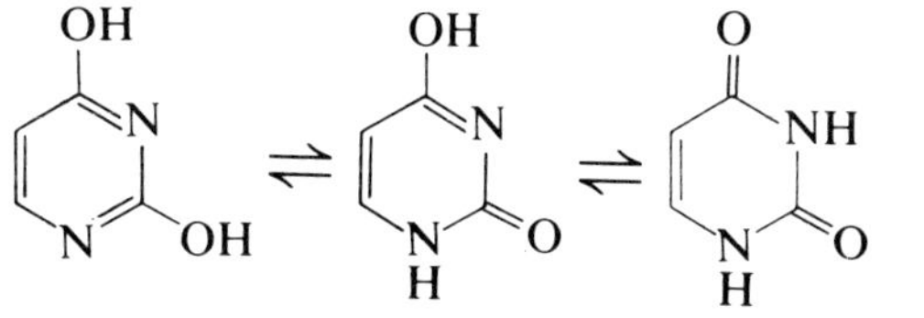

uracil   predominant form

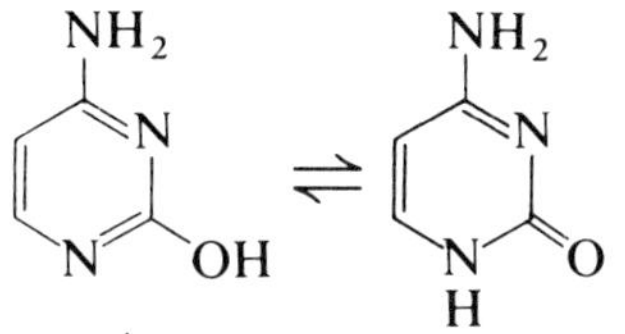

cytosine

In both cases the addition reaction is limited by the availability of only position 6 for the sulphonate group, and the evidence suggests that the reaction is, in fact, a 5,6-addition of $H^+SO_3^{2-}$ (Shapiro *et al.*, 1970; Hayatsu *et al.*, 1970):

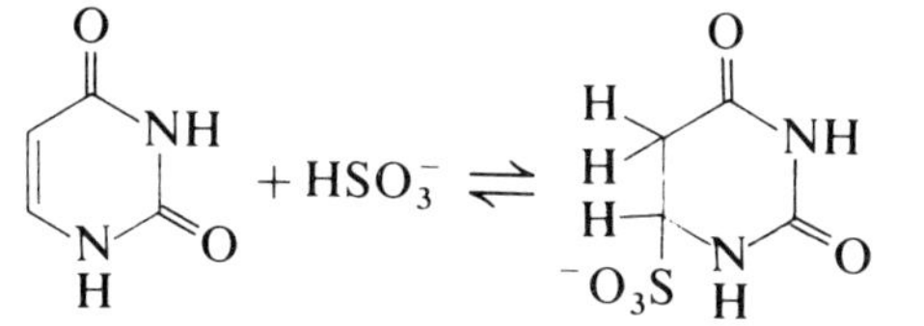

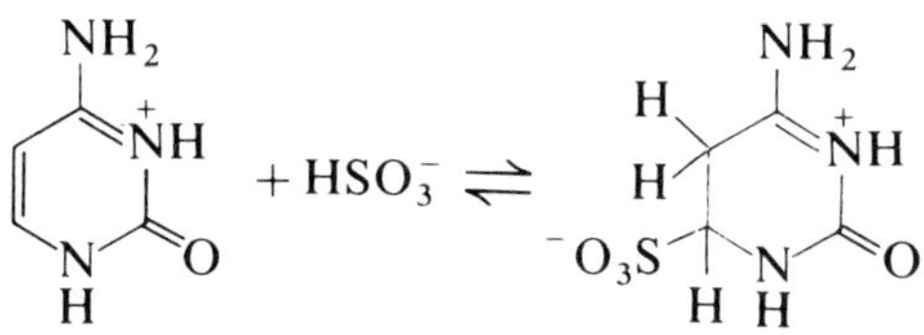

In the case of uracil the nucleophile has been demonstrated to be the sulphite ion (Shapiro *et al.*, 1970).

In natural products these bases are found as 1-N-substituted compounds in which the substituent is either D-2-deoxyribofuranose or D-ribofuranose, according to whether the base is part of the DNA or RNA molecule. The effect of the substituent upon the extent and rate of addition is small for uracil, the equilibrium and rate constants being identical for methyl or ribose substituents (Pitman and Jain, 1979). An interesting reaction is that between the substituted cytosine and the sulphur(IV) oxoanion. In accordance with the equation shown above for cytosine, addition to cytidine (ribose substituted cytosine) shows the expected pH dependence, the extent of addition falling off with increasing pH (Shapiro et al., 1974), although there is also a contribution from the reaction with unprotonated pyrimidine species:

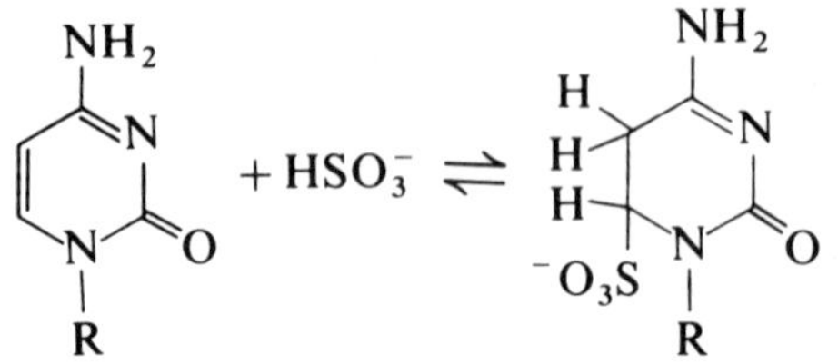

$$ \text{(cytosine)} + HSO_3^- \rightleftharpoons \text{(sulphonate adduct)} $$

The equilibrium constant for the reaction with the protonated species is some three times greater than that with the unprotonated pyrimidine. The interesting feature is that both substituted and unsubstituted cytosine subsequently undergo a deamination reaction, the cytosine derivative being converted to the corresponding uracil derivative:

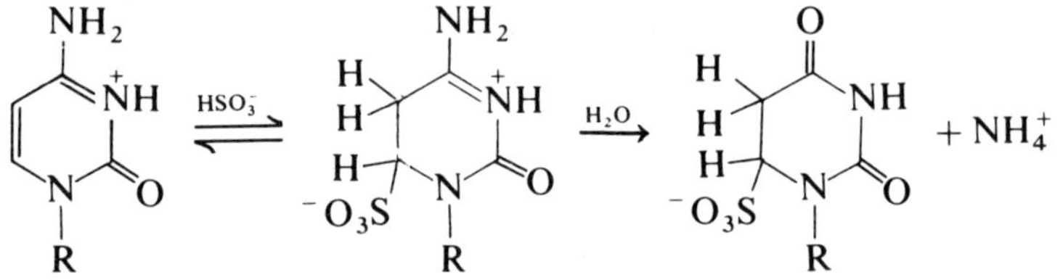

$$ \text{(protonated cytosine)} \underset{HSO_3^-}{\rightleftharpoons} \text{(sulphonate adduct)} \overset{H_2O}{\longrightarrow} \text{(uracil derivative)} + NH_4^+ $$

This reaction has a very marked pH dependence, showing a maximum rate at approximately pH 5.

The presence of a hydroxymethyl side-chain in position 5 of cytosine and uracil leads to substitution in the side-chain rather than the ring (Hayatsu and Shiragami, 1979). This is illustrated below for the case of sulphonation of 5-hydroxymethylcytosine:

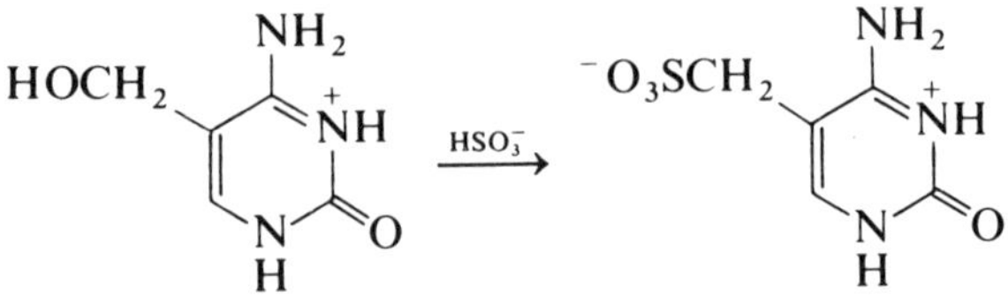

$$ HOCH_2\text{-(cytosine)} \overset{HSO_3^-}{\longrightarrow} {}^-O_3SCH_2\text{-(cytosine)} $$

The optimum pH for this reaction is pH 4·5, and that for the reaction of hydroxymethyluracil is pH 6–7. The reaction product with 5-hydroxy-methylcytosine deaminates only very slowly.

An important heterocyclic system based on pyrimidine is that of purine:

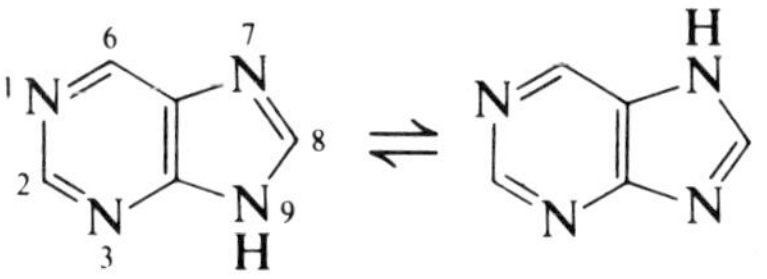

When 6-unsubstituted purines are dissolved in aqueous solutions of hydrogen sulphite ion, their ultraviolet absorbance spectra exhibit a hypso-chromic shift, which may be attributed to the formation of adducts by 1,6-addition (Pendergast, 1975):

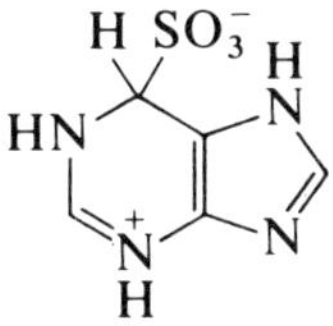

The reaction is favoured by low pH, the apparent equilibrium constant being $11 \, \text{M}^{-1}$ at pH 2 and $4·2 \, \text{M}^{-1}$ at pH 3. This pH dependence is completely explained in terms of the protonated base and hydrogen sulphite ion as reactants (Pitman and Sternson, 1976).

Pitman and Sternson (1976) investigated correlations between the equilibrium constants for the addition of hydrogen sulphite ion to heterocyclic nitrogen compounds and their polarographic half wave potentials. In addition to pyrimidine, purine and substituted pyrimidines, the investigation included quinazoline and adenine, both of which are pyrimidine compounds, and acridine:

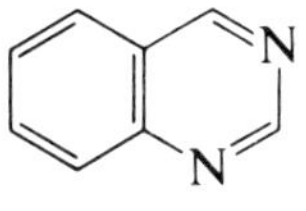

Quinazoline 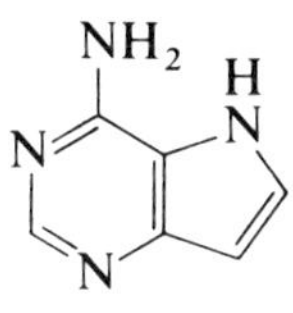   Adenine 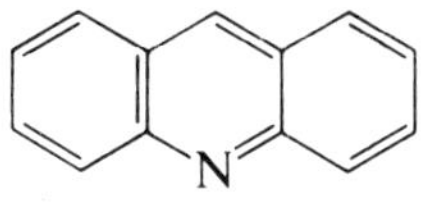   Acridine

The data for these compounds are summarised in Fig. 1.13 and the regression line drawn through the points has the equation:

$$\log K = 16·2 E_{1/2} + 14·8$$

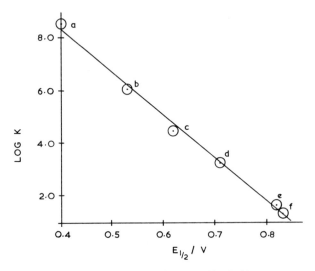

Fig. 1.13. The relationship between polarographic half wave potential and equilibrium constant for formation of adduct with sulphite ion for: (a) quinazolinium ion; (b) acridinium ion; (c) pyrimidinium ion; (d) 2-aminopyridinium ion; (e) 2-amino-4-methylpyrimidinium ion; (f) purine ion (Pitman and Sternson, 1976). Reproduced with permission from *J. Amer. Chem. Soc.*, 1976, **98**, page 5237. © 1976 American Chemical Society.

$E_{1/2}$ is the polarographic half-wave potential in 1 M hydrochloric acid. Assuming that the free energy changes for the addition reaction (ΔG_C^0) and for one-electron reduction (ΔG_R^0) are linearly related such that,

$$\Delta G_C^0 = \alpha \Delta G_R^0 + \beta$$

where α and β are constants, then,

$$\log K = \frac{\alpha n F E^0}{2 \cdot 303 RT} - \frac{\beta}{2 \cdot 303 RT}$$

where n is the number of electrons transferred. The values of $E_{1/2}$ at pH 0 are expected to be close to the standard electrode potential, E^0, because the reactions are essentially reversible at this pH. The slope of the regression line (16·2) is very close to the value of 16·9 for $nF/2 \cdot 303RT$, when $n = 1$, at 25 °C. Thus, the theoretical approach is consistent with the experimental result if $\alpha = 1$. The reactivity of these heteroaromatic systems towards sulphur(IV) oxoanions may, therefore, be predicted and, in addition, the relationship suggests similarities in the mechanisms of addition of sulphite ion and of reduction. Thus, both processes may involve localisation of electrons in the same double bond in which aromatic character is lost, for example:

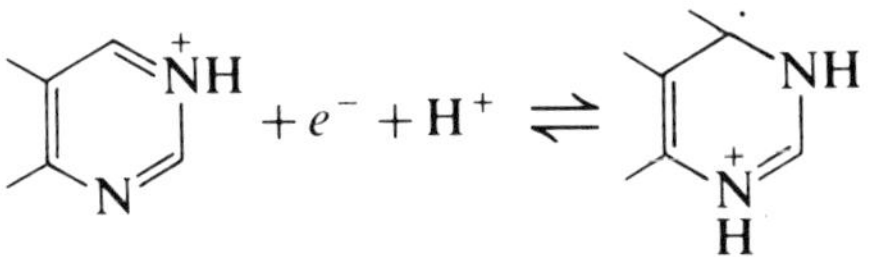

and,

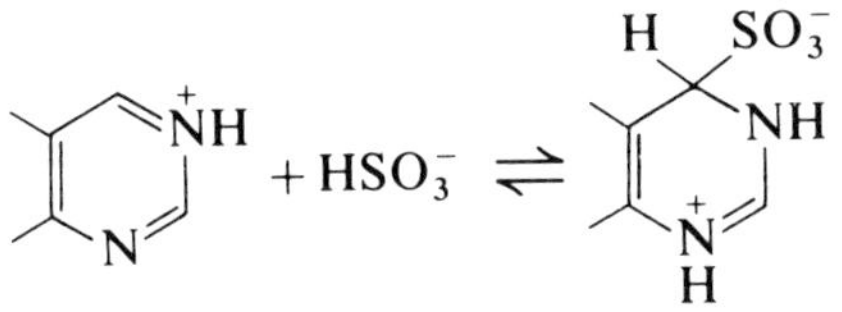

1.11.9 Reaction with disulphide bonds

The ability of sulphite ion to carry out nucleophilic attack at the S—S bond of disulphides has already been explained. The reaction is of the S_N2 type and, by analogy with S_N2 reactions at carbon atoms, it is expected to proceed by way of a trigonal bipyramidal transition state, with the entering and leaving groups at the vertices:

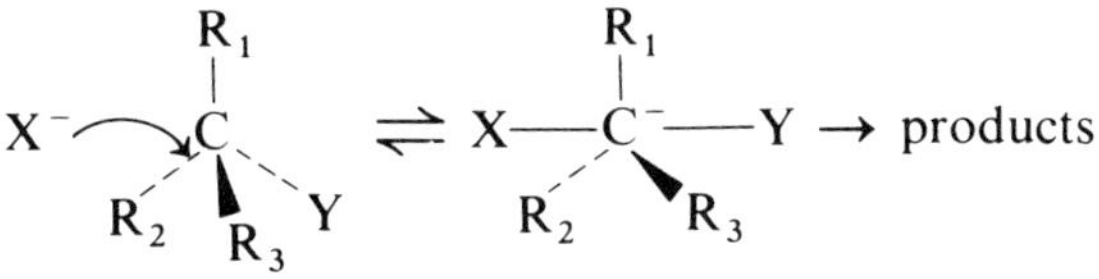

In the case of attack at the S—S bond, two lone pairs of electrons at the sulphur atom under attack will provide this stereochemistry for the transition state (Parker and Kharasch, 1959):

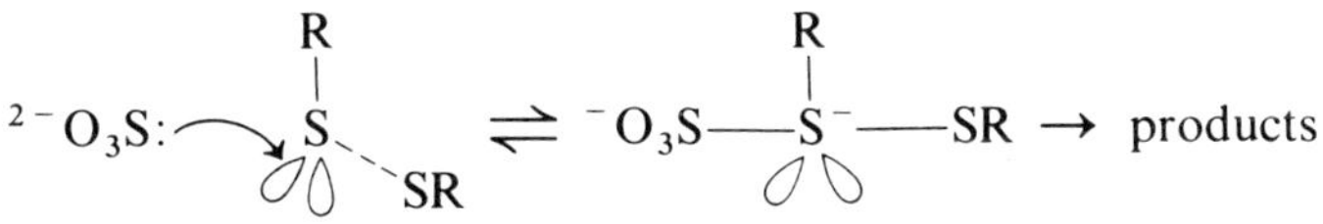

The products are S-sulphonates or thiosulphonates and are often known as Bunte salts. From these considerations it is possible to identify three factors which will contribute to the progress of disulphide cleavage.

1. Attack by sulphite ion will take place from the rear with respect to the leaving group. At this stage the nucleophile will approach close to the group, R, and will, therefore, be affected by the size and charge of the latter.

2. Since attack at the sulphur atom is nucleophilic, it will be favoured by electron withdrawal from the site of attack and disfavoured by a high electron density. Furthermore, if the products are stabilised, with respect to reactants, by inductive or resonance effects, the reaction will proceed, and, conversely, the reverse of this effect will favour the reactants.

3. In the case of unsymmetrical disulphides, the sulphonate group may appear preferentially on one half of the disulphide owing to effects arising from 1 and 2 above.

The overall reaction is reversible, since the resulting thiol may act as a nucleophile, displacing sulphite ion:

$$RSSR' + SO_3^{2-} \rightleftharpoons RSSO_3^- + R'S^-$$

or,

$$RSSR' + SO_3^{2-} \rightleftharpoons R'SSO_3^- + RS^-$$

Biochemically, the most important disulphide is the amino acid, cystine, which is readily attacked by sulphur(IV) oxoanions to give cysteine-S-sulphonate. Cystine will also undergo this reaction when linked as part of a peptide chain, one of the simplest cysteine-containing peptides being oxidised glutathione, which is frequently employed as a model compound. Stricks and coworkers (1955) have considered the possible reactions taking

TABLE 1.6

Equilibrium constants for the reaction of disulphide compounds with sulphur(IV) oxoanions determined at 25 °C (Stricks *et al.*, 1955)

Equilibrium constant	Disulphide		
	Cystine*	Glutathione	Dithioglycolic acid*
K_1	0·010	0·039	0·000 47
K_2	0·000 52	0·006 2	
K_3	0·016	0·025	
K_4	0·15	0·037	
K_5	4·1	2·6	1·4

Data marked * taken from Stricks and Kolthoff (1951). Dithioglycolic acid has no amino group and K_1 and K_5 refer to the reactions of this compound in alkaline and in weakly acidic media respectively. Reproduced with permission from *J. Amer. Chem. Soc.*, 1955, **77**, page 2061. © 1955 American Chemical Society.

place when such disulphide compounds react with sulphur(IV) oxoanions at various pH values:

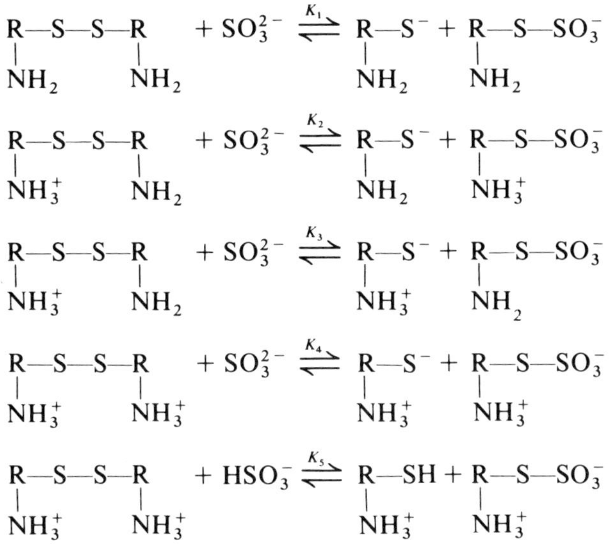

$$\underset{\underset{NH_2}{|}}{R-S-S-}\underset{\underset{NH_2}{|}}{R} \quad + SO_3^{2-} \overset{K_1}{\rightleftharpoons} \underset{\underset{NH_2}{|}}{R-S^-} + \underset{\underset{NH_2}{|}}{R-S-SO_3^-}$$

$$\underset{\underset{NH_3^+}{|}}{R-S-S-}\underset{\underset{NH_2}{|}}{R} \quad + SO_3^{2-} \overset{K_2}{\rightleftharpoons} \underset{\underset{NH_2}{|}}{R-S^-} + \underset{\underset{NH_3^+}{|}}{R-S-SO_3^-}$$

$$\underset{\underset{NH_3^+}{|}}{R-S-S-}\underset{\underset{NH_2}{|}}{R} \quad + SO_3^{2-} \overset{K_3}{\rightleftharpoons} \underset{\underset{NH_3^+}{|}}{R-S^-} + \underset{\underset{NH_2}{|}}{R-S-SO_3^-}$$

$$\underset{\underset{NH_3^+}{|}}{R-S-S-}\underset{\underset{NH_3^+}{|}}{R} \quad + SO_3^{2-} \overset{K_4}{\rightleftharpoons} \underset{\underset{NH_3^+}{|}}{R-S^-} + \underset{\underset{NH_3^+}{|}}{R-S-SO_3^-}$$

$$\underset{\underset{NH_3^+}{|}}{R-S-S-}\underset{\underset{NH_3^+}{|}}{R} \quad + HSO_3^- \overset{K_5}{\rightleftharpoons} \underset{\underset{NH_3^+}{|}}{R-SH} + \underset{\underset{NH_3^+}{|}}{R-S-SO_3^-}$$

Equilibrium constants, K_1-K_5, for the formation of products are shown in Table 1.6, from which it is apparent that, even in the case of formation of the most stable product, the reaction is far from quantitative. The results, with the exception of the anomalously greater value of K_3 compared with K_2 ($K_3/K_2 = 3\cdot4$), reflect the trend in apparent equilibrium constant, K, as pH is varied, where K is defined by:

$$K = \frac{\sum [RSH][RSSO_3^-]}{\sum [RSSR][S(IV)]}$$

The summation signs imply the total concentration of all ionic forms of each species. Thus, in alkaline solutions, when K_1 is most important, the apparent equilibrium constant for the formation of cysteine-S-sulphonate is of the order of $0\cdot01$ increasing to a value of the order of 4 (the value of K_5) at pH 4–5, when hydrogen sulphite ion predominates. In the case of dithioglycolic acid, a carboxyl group, which carries a negative charge at all values of pH studied, is attached to the α-carbon atom. The low value of equilibrium constant (K_1) in alkaline media probably reflects the effect of the negative charge on the approaching nucleophile, and the increase in stability at lower pH is observed as in the case of the sulphur amino acids.

TABLE 1.7
The effect of net charge on the rate of disulphide
cleavage by sulphite ion; k_2 is the second order rate
constant for the cleavage reaction at 25 °C

Disulphide	Net charge	k_2 ($\text{M}^{-1}\text{s}^{-1}$)
Cystine	0	18·3
	−1	1·18
	−2	0·05
Oxidised glutathione	−2	2·66
	−3	0·75
	−4	0·17
Homocystine	0	2·6
	−2	0·33

Data taken from Cecil and McPhee (1955) and
McPhee (1956).

A demonstration of what appears to be an effect of overall charge on the
rate of cleavage of symmetrical disulphides has been given by Cecil and
McPhee (1955) and McPhee (1956). Their data are summarised in
Table 1.7. It is evident that, in all cases, the presence of a net negative
charge on the substrate for nucleophilic attack reduces the rate of reaction,
this effect depending upon the magnitude of the charge.

When the S-sulphonate product is dissolved in solutions containing
labelled sulphite ion, nucleophilic attack by sulphite leads to exchange
(Fava and Pajaro, 1956):

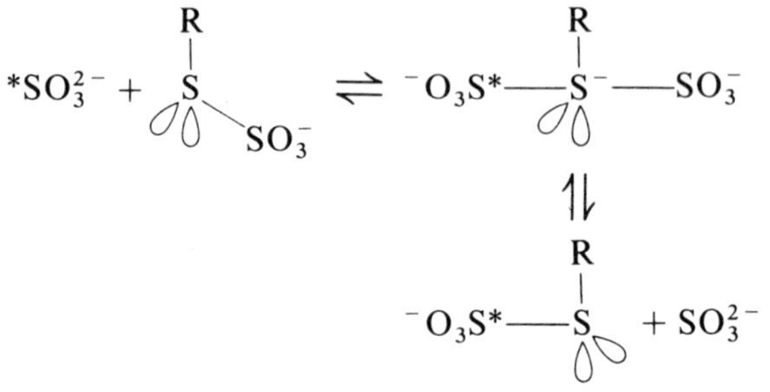

$$*SO_3^{2-} + \underset{\overset{\displaystyle |}{\underset{\displaystyle SO_3^-}{S}}}{\overset{\displaystyle R}{}} \rightleftharpoons {}^-O_3S^* \!\!—\!\!\overset{\overset{\displaystyle R}{\displaystyle |}}{\underset{\displaystyle}{S^-}}\!\!—SO_3^-$$

$$\updownarrow$$

$$ {}^-O_3S^*\!\!—\!\!\overset{\overset{\displaystyle R}{\displaystyle |}}{\underset{\displaystyle}{S}} + SO_3^{2-}$$

The effect of steric hindrance by the group, R, on this nucleophilic
displacement has been demonstrated very clearly by Fava and his
coworkers (Fava and Pajaro, 1956; Fava and Iliceto, 1958), who measured
the rate of exchange of sulphite ion with a series of alkyl thiosulphonates, in

which the alkyl group was methyl, ethyl, *iso*-propyl and *tert*-butyl. The relative rates of exchange in water at pH 7·9 (25 °C) were as follows:

R =	Methyl	Ethyl	*iso*-Propyl	*tert*-Butyl
Relative rate constant	100	50	0·7	0·0006

These results were compared with the exchange of bromide ion with the corresponding alkyl bromides for which the relative rate constants were: 100; 65; 3·3; 0·0015 respectively. Evidence is, thereby, provided for a similar transition state for substitution at carbon and at sulphur.

The ease of disulphide cleavage also depends upon the sulphur–sulphur bond length (Davis *et al.*, 1963). An inverse correlation exists between activation energies for the reaction of disulphides with cyanide and sulphite ion and the S—S bond length, interpreted with the aid of molecular orbital calculations as being a consequence of the destabilisation of the acceptor anti-bonding orbital by the stronger and shorter bond. As the nucleophile approaches, the anti-bonding orbital accepts the electron pair whilst the sulphur–sulphur bond breaks. An unstable antibonding orbital renders this change less likely to take place.

Rensburg and Swanepoel (1967*a*) studied the cleavage of unsymmetrical disulphides by sulphite ion in order to observe whether there was preferential formation of thiosulphonate from one part of the disulphide molecule. Straight chain dialkyldisulphides with acidic and basic substituents two carbon atoms removed from the disulphide group showed pH independent random attack to form both products in equal amounts. The test compounds were:

$$NH_2CH_2CH_2—S—S—CH_2CH_2COOH$$
$$NH_2CH_2CH_2—S—S—CH_2CH_2OH$$
$$NH_2CH_2CH_2—S—S—CH_2CH_2CH(NH_2)COOH$$

For the case of a secondary *β*-carbon atom the disulphides studied were:

$$CH_3CH_2—S—S—CH_2CH(NH_2)COOH$$
$$NH_2CH_2CH_2—S—S—CH_2CH(NH_2)COOH$$
$$HOOCCH_2CH_2—S—S—CH_2CH(NH_2)COOH$$
$$NH_2CH_2CH_2—S—S—CH_2CH(COOH)NHCOCH_3$$
$$NH_2CH_2CH_2—S—S—CH_2CHCONHCH_2COOH$$
$$| $$
$$NHCOCH_2CH(NH_2)COOH$$
$$NH_2CH_2CH_2—S—S—CH_2CH(CH_3)_2$$

In these experiments the ratio of S-sulphonate produced from the straight chain part to that from the remainder of the molecule was in the range, 1·8–2·35:1, and was independent of pH in the range pH 5–9·5. Thus, the observed effect is mainly one of steric hindrance controlled by the presence of the secondary carbon atom in the β-position. When this carbon atom is tertiary, as in the case of the disulphide,

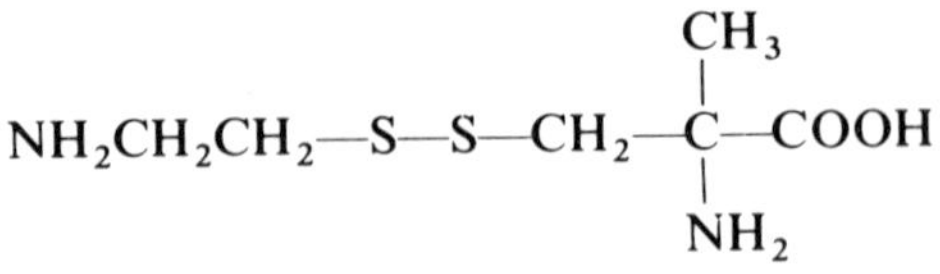

$$NH_2CH_2CH_2-S-S-CH_2-\underset{\underset{NH_2}{|}}{\overset{\overset{CH_3}{|}}{C}}-COOH$$

the proportion of straight-chain to branched product is 30:1. As expected, the steric effect is further exaggerated when the substituent occurs on the α-carbon atom, illustrated by corresponding ratios of 20:1, and greater than 100:1, for the following compounds respectively:

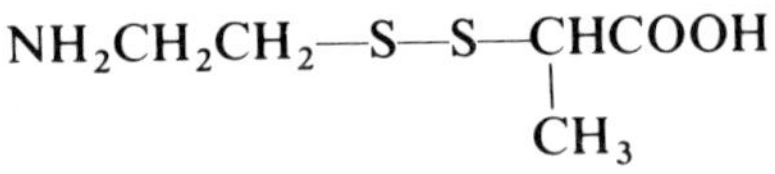

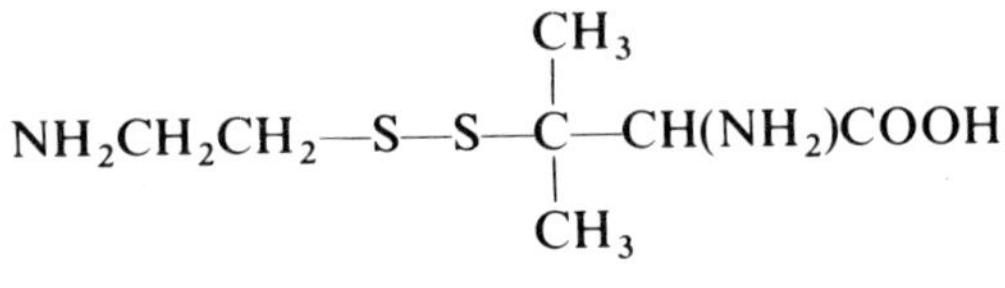

$$NH_2CH_2CH_2-S-S-\underset{\underset{CH_3}{|}}{CH}COOH$$

$$NH_2CH_2CH_2-S-S-\underset{\underset{CH_3}{|}}{\overset{\overset{CH_3}{|}}{C}}-CH(NH_2)COOH$$

Complete conversion of disulphide to the equivalent quantity of S-sulphonate is achieved by removal of the thiol product by, for example, treatment with mercuric chloride or organomercuric halide; reactions which form a convenient method for the analysis of disulphide groups:

$$2RSSR + Hg^{2+} + 2SO_3^{2-} \rightarrow RSHgSR + 2RSSO_3^-$$

$$RSSR + R'HgX + SO_3^{2-} \rightarrow RSHgR' + RSSO_3^- + X^-$$

The reactions also proceed in the direction of product formation when the thiol which is produced is stabilised with respect to the disulphide; for example, 5,5'-dithiobis(2-nitrobenzoic acid) and 4,4'-dithiopyridine react with sulphite ion to form the following products respectively, at pH 7 (Humphrey et al., 1970):

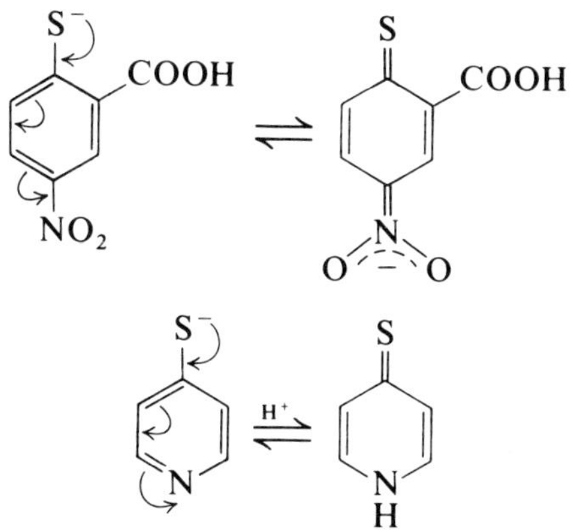

A continuous variation curve for 4,4'-dithiopyridine (total concentration of disulphide + sulphite $= 10^{-4}$ M, pH 7) is triangular, the experimental point at a concentration ratio of 1:1 lying at the apex, suggesting no significant dissociation of the product. A similar situation exists in the case of 5,5'-dithiobis(2-nitrobenzoic acid).

A situation where it is necessary to displace the equilibrium between disulphide and S-sulphonate to the side of S-sulphonate is during the preparation of the latter. Such preparative reactions have been reviewed by Klayman and Shine (1968). Conversion of disulphide to S-sulphonate as the sole product is possible if the cleavage is carried out in the presence of an oxidising agent:

$$RSSR + 2SO_3^{2-} \rightarrow 2RSSO_3^- + 2e^-$$

If the oxidising agent used is phosphotungstic acid, the reaction forms the basis of the Folin colorimetric method for the quantitative determination of cysteine and cystine. A catalytic amount of cupric ion in the presence of oxygen is also suitable (Segel and Johnson, 1963) for purposes of oxidation. However, the most satisfactory synthetic process is probably the classical method first used by Bunte (1874), in which an alkyl halide is allowed to react with sodium thiosulphate:

$$RCl + Na_2S_2O_3 \rightarrow RS_2O_3Na + NaCl$$

There are no simple or general explanations for the reported reactivity of sulphur(IV) oxoanions with disulphide groups in proteins. Effects of pH and medium are combinations of often poorly defined changes in protein structure and, therefore, each case is likely to be unique. The type of disulphide involved, that is, intermolecular or intramolecular, is also important; for example, when bovine serum albumin is treated with sulphite ion, there is no apparent reaction, whereas the addition of mercuric chloride and guanidine leads to the cleavage of seventeen disulphide bonds. Alternatively, almost all of the available disulphide groups of trypsin are available to sulphite ion without the need of denaturation of the protein (Cecil and Wake, 1962). An intermediate situation exists in the case of insulin, in which two out of three disulphide groups are attacked by the nucleophile, the third requiring guanidine. The reaction, however, is pH dependent in the absence of denaturing agent, being slow at pH 2·8–5·5 with splitting of only one disulphide group, and faster at pH 6·2–7·2 with two disulphide groups being split; cleavage of the third disulphide group is possible at pH 7·2–9·0. The addition of such a reagent allows total cleavage to take place at pH 6·5–7·0 (Cecil and Loening, 1957). The presence of such specific effects renders difficult the making of generalisations regarding the reactivity of sulphur(IV) oxoanions towards proteins, but the use of this reagent in identifying two types of disulphide bond is clearly demonstrated. Examples of specific

TABLE 1.8

Formation of cysteine-S-sulphonate and cysteine by the cleavage of unsymmetrical disulphides using sulphite ion

Disulphide	[CySSO$_3^-$]/[CySH]	
	Without urea	In 6·7 M urea
G—S—S—Cy	1	—
BPA—S—S—Cy	9	1
Ur—S—S—Cy	4	1
Hm—S—S—Cy	2	1
La—S—S—Cy	9	1
Pa—S—S—Cy	2	1

The reaction was carried out in the presence of HgCH$_3$I and cysteine was, therefore, measured as CySHgCH$_3$ (Rensburg and Swanepoel, 1967b). G, reduced glutathione; BPA, bovine plasma albumin; Ur, urease; Hm, haemoglobin; La, lactalbumin; Pa, papain. Reproduced with permission from *Arch. Biochem. Biophys.*, 1967, **121**, page 730. © 1967 Archives of Biochemistry and Biophysics.

reactions with proteins will be considered in later chapters, as appropriate, but a simple case, that of a protein existing as a disulphide with cysteine, is worthy of further consideration at this stage. Rensburg and Swanepoel (1967b) allowed various proteins to react with cysteine, then measured the amounts of cysteine-S-sulphonate and cysteine formed when the unsymmetrical disulphides underwent reaction with sulphite ion. The proportions of the two products are given in Table 1.8, from which it is evident that, for all proteins, without denaturation by urea, the sulphite ion preferentially attacked at cysteine, probably for steric reasons. Treatment with urea completely removed this specificity. Also, there appears to be no preferential attack at any particular sulphur atom in glutathione-S-S-cysteine, an observation consistent with a previous conclusion reached by these authors (Rensburg and Swanepoel, 1967a) that the effect of substituents in the alkyl groups attached to the disulphide bond depends upon the type of substitution (that is, whether substitution occurs on the α- or β-carbon atom, and whether the latter is primary, secondary or tertiary) rather than upon the nature of the substituent. Thus, for a half cystine residue, the effect on the cleavage of the S—S bond is the same whether the amino and carboxyl groups are free or incorporated as part of a peptide chain.

Acid hydrolysis of alkyl thiosulphonates leads to the formation of the corresponding thiol and hydrogen sulphate ion:

$$RSSO_3^- + H_2O \xrightarrow{H^+} RSH + HSO_4^-$$

This reaction is, to some extent, similar to the acid catalysed hydrolysis of sodium arylsulphates, in which the decomposition is promoted by protonation on the oxygen attached to carbon (Kice and Anderson, 1966):

$$ArOSO_3^- + H^+ \underset{}{\overset{fast}{\rightleftharpoons}} Ar\overset{+}{\underset{|\ H}{O}}{-}SO_3^- \xrightarrow[H_2O]{slow} ArOH + H_2SO_4$$

An analogous mechanism for the acid hydrolysis of thiosulphonates involves protonation at the thiol sulphur rather than at the sulphonate group (Kice $et\ al.$, 1966):

$$RSSO_3^- + H^+ \underset{}{\overset{fast}{\rightleftharpoons}} R{-}\overset{+}{\underset{|\ H}{S}}{-}SO_3^- \underset{}{\overset{slow}{\rightleftharpoons}} \text{transition state}$$

Thus, when cysteine-S-sulphonate is subjected to hydrolysis with 2·5 M hydrochloric acid for 24h at 96 °C, the compound is converted quantitatively to sulphate, and no sulphur dioxide is liberated on acidification of the sulphonate (Segel and Johnson, 1963).

1.11.10 Reaction with quinones

Quinones are conjugated cyclic diketones rather than aromatic compounds, although they show greater stability than is expected on the basis of bond energies, the stabilisation energy of p-benzoquinone being approximately 21 kJ mol^{-1}. The most common structures are those of 1,2- and 1,4-quinones, the former being the more reactive. Because they are α,β-unsaturated ketones, it is predicted that quinones will give 1,4-addition products in the same way as open-chain compounds, and this behaviour may be illustrated by the addition of sulphite ion to p-benzoquinone (Roberts and Caserio, 1965):

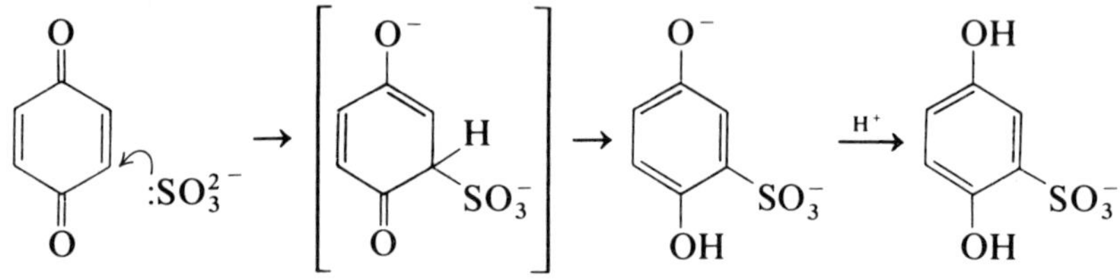

During this reaction the quinone is reduced to hydroquinone, a process involving a two-electron transfer from the reducing agent. The product may, in fact, be prepared in two ways. The first is the direct reaction implied by the scheme shown above, whilst the second is through reaction with hydroquinone in the presence of oxidising agent. The fact that the latter gives a high product yield (alkaline pH) whereas in the former case the yield is low and is accompanied by a coloured by-product, suggests that the reaction may be that of semiquinone or its dimer, rather than quinone (LuValle et al., 1958). The formation of semiquinone takes place as a result of the one-electron oxidation of hydroquinone, or one-electron reduction of quinone (Pryor, 1966):

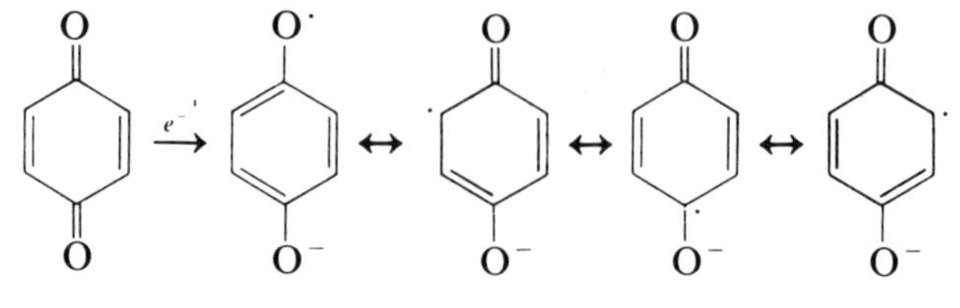

It is possible that a reduction of this type could be accomplished by sulphite ion. The ability of the system to delocalise the unpaired electron renders the semiquinone moderately stable.

The overall reaction is of considerable interest in photographic science, in which hydroquinone is used in developer as a reducing agent for silver bromide (Mason, 1975):

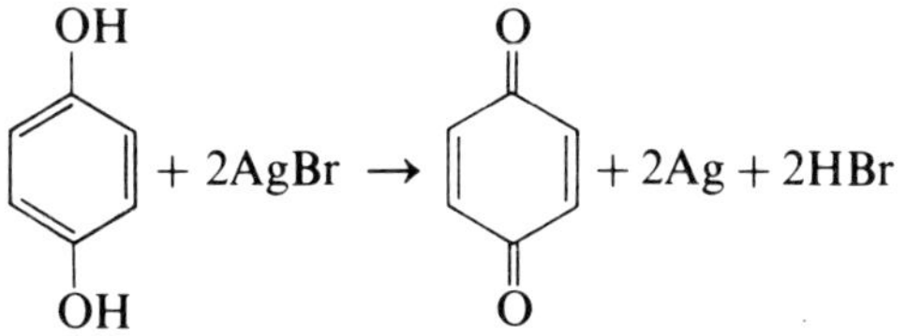

In the presence of sulphite ion the reaction becomes,

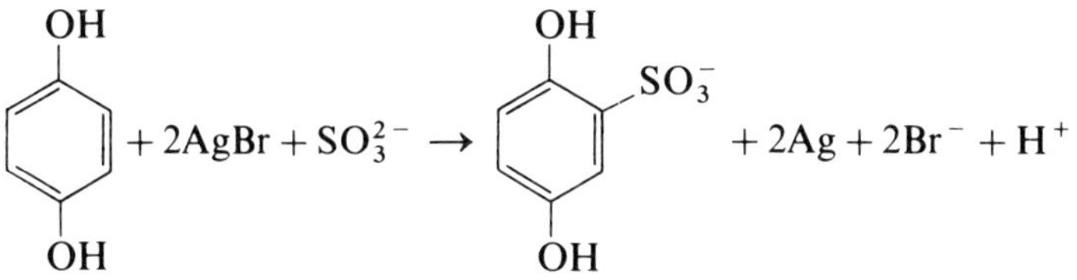

Thus, sulphite ion serves to prevent the formation of the highly coloured benzoquinone product or compounds derived from this by subsequent polymerisation.

A naturally occurring 1,4-quinone is vitamin K_1 which is, in fact, a substituted 1,4-naphthoquinone:

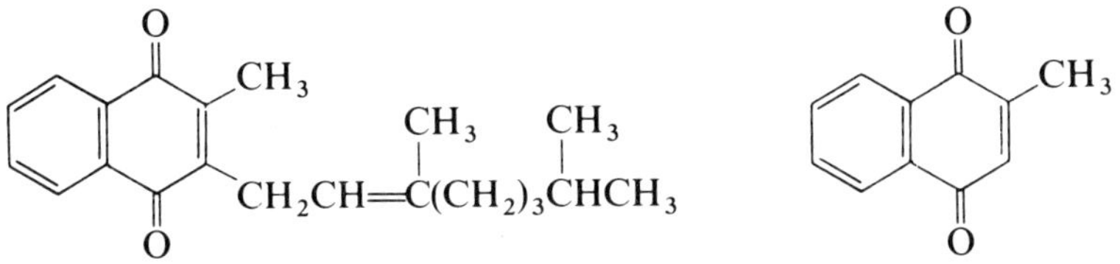

Vitamin K_1

Vitamin K_3

A synthetic substitute for this vitamin is 2-methyl-1,4-naphthoquinone (vitamin K_3), which may be administered in a water-soluble form as an addition compound with hydrogen sulphite ion, and whose effectiveness, presumably, results from the decomposition of the adduct in the alkaline environment of the gut. The reaction of vitamin K_3 with sulphur(IV) oxoanions is one of the few reactions which can be used to demonstrate kinetic and thermodynamic control of a reaction (Greenberg et al., 1971).

Two reaction products may be formed:

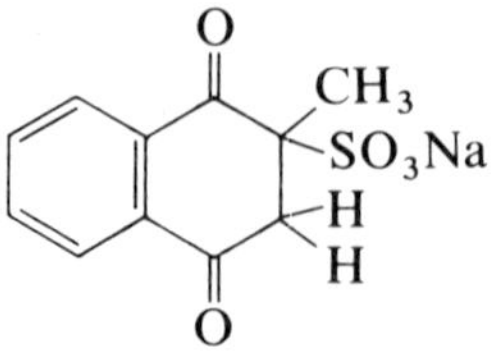

VI 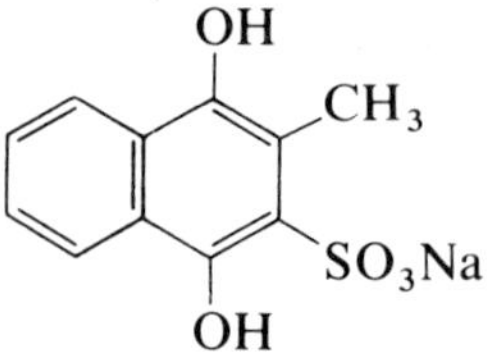   VII

If the reactants are mixed and warmed for five minutes, the main product is
VI resulting from addition at carbon 2, whereas heating the reactants under
reflux for an hour leads to the formation of the thermodynamically more
stable product VII. Kalus and Filby (1976) have found that treatment of
vitamin K_3 with sulphite ion leads to a high yield (88–92 %) of the radical
anion, presumably as a result of the following reaction:

$$K_3 + SO_3^{2-} \rightarrow K_3^{\cdot -} + SO_3^{\cdot -}$$

Two radical anions are possible:

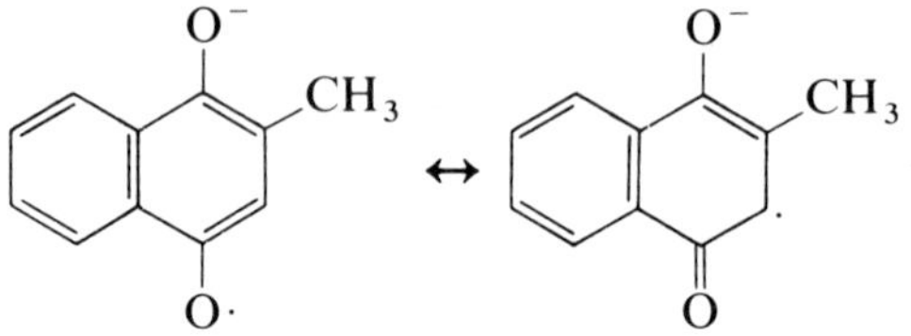

VIII

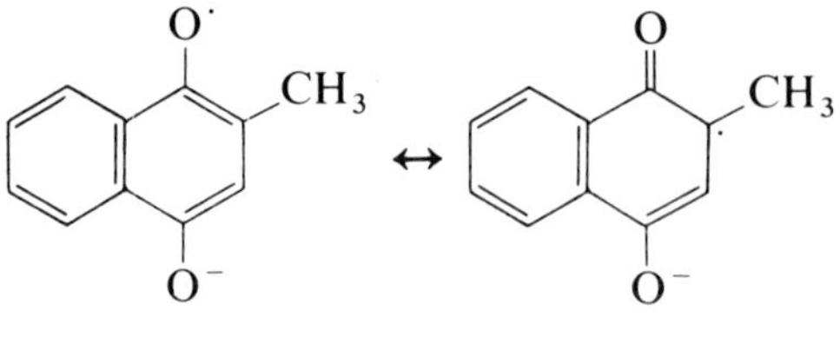

IX

The stability of free radicals increases as the degree of substitution is
increased at the carbon atom bearing the unpaired electron, that is, *tert*-
butyl > *iso*-propyl > ethyl > methyl. Therefore, the preferred intermediate
will be compound IX which is stabilised by hyperconjugation from the
methyl group to the *p*-orbital which carries the unpaired electron and which
is perpendicular to the plane of the ring.

The low temperature product may then be obtained as follows:

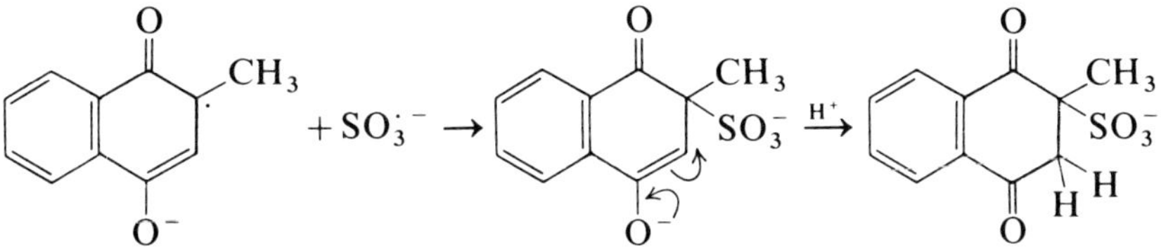

The accumulation of the intermediate radical species implies that the rate-determining step is attack by the sulphite radical on this intermediate. This is, in effect, a radical combination reaction, for which the activation energy is expected to be low.

Aromatisation will only result from addition at carbon 3 and, since this involves the bond making and breaking processes of a heterolytic reaction, such as that of addition to benzoquinone above, it will require a significant activation energy. Thus, at low temperatures, the radical mechanism predominates. Temperature has only a small effect on the rate of this reaction, whereas the high temperature dependence of the heterolytic mechanism renders the latter dominant as the temperature is increased.

The low temperature addition product, I, may be decomposed by the action of alkali. The reaction is possibly heterolytic (Greenberg et al., 1971), although unidentified free radicals have also been observed during the process. A simple heterolytic mechanism for alkaline decomposition is,

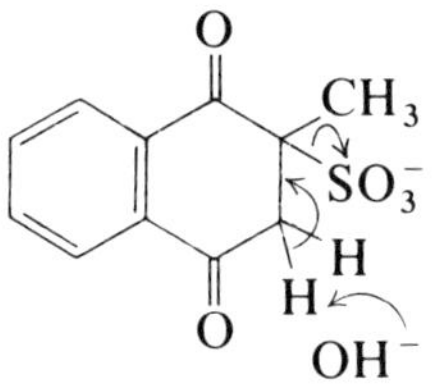

If the pH independent equilibrium constant for the formation of product I is given as,

$$K = \frac{[\text{Adduct}]}{[\text{Vit } K_3][\text{SO}_3^{2-}][\text{H}^+]}$$

the experimental data yield a value of $3 \cdot 1 \times 10^{14} \, \text{M}^{-2}$ at 25 °C, which, though relatively large, is still sufficiently small to allow significant dissociation of the adduct at pH 7 in dilute solution (Shih and Petering, 1973).

In food products the most important quinones are those which are formed by oxidation of *ortho*-diphenols:

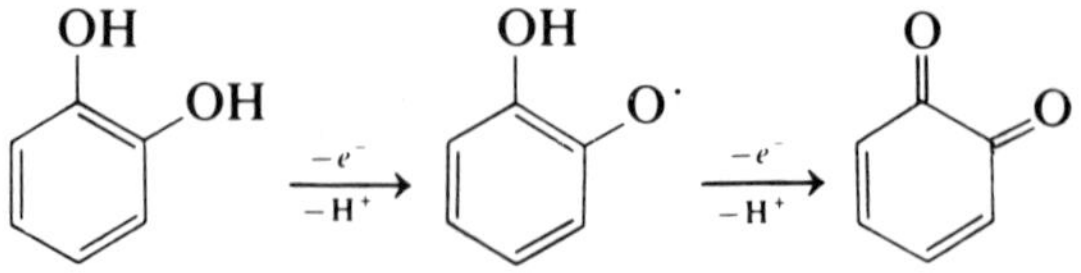

If the oxidation is carried out in the presence of benzenesulphinic acid, nucleophilic addition takes place to give the 4-substituted product and, when it is carried out in the presence of aniline, the amino group attacks at positions 4 and 5 to give the corresponding disubstituted product (Pugh and Raper, 1927):

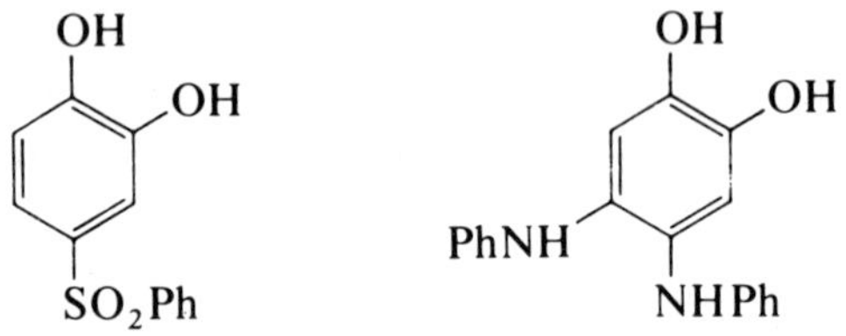

Therefore, the quinones or semiquinones, which are equally capable of leading to the observed reaction products, are activated towards nucleophilic attack in, at least, the 4- and 5-positions and, therefore, the expected products of their reaction with sulphur(IV) oxoanions are:

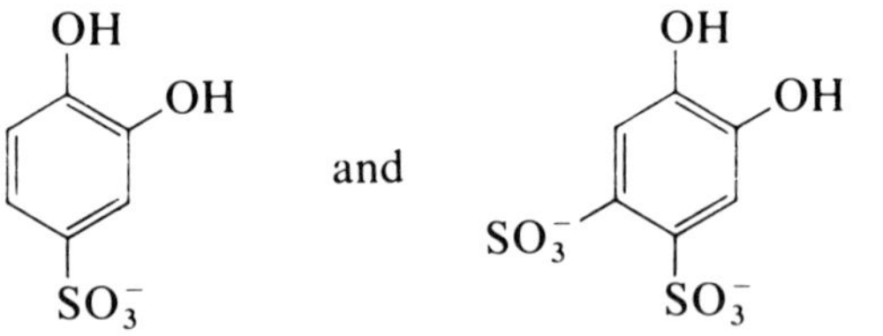

In their study of the determination of *ortho*-diphenols by means of the Folin–Ciocalteu colorimetric assay, Somers and Ziemelis (1980) found very serious interference in the presence of sulphur(IV) oxoanions. The determination depends upon the specific reduction by the phenolic compounds, of a mixture of phosphotungstic ($H_3PW_{12}O_{40}$) and phosphomolybdic ($H_3PMo_{12}O_{40}$) acids to complex oxides of tungsten and molybdenum, in which the metals exhibit lower valency states. The process is accompanied by the development of a blue colour ($\lambda_{max} = 765\,nm$). The colour yield when sulphur(IV) oxoanion is mixed with the reagent under appropriate conditions (pH 9) is low but, when also mixed with diphenol, excessive colour production occurs. If the reaction with Folin–Ciocalteu reagent is,

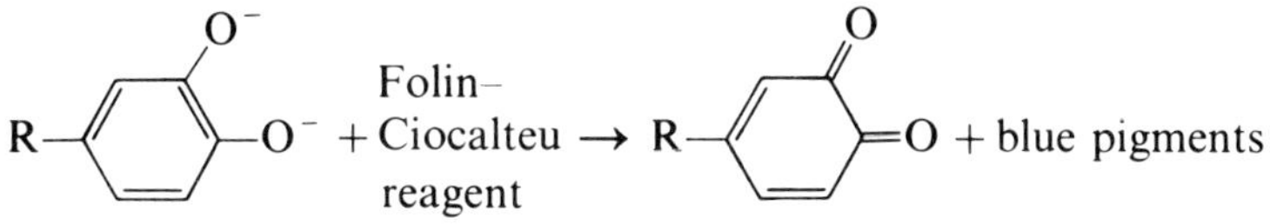

the effect of sulphite ion may be interpreted as that of reductive sulphonation of the quinone product, thus:

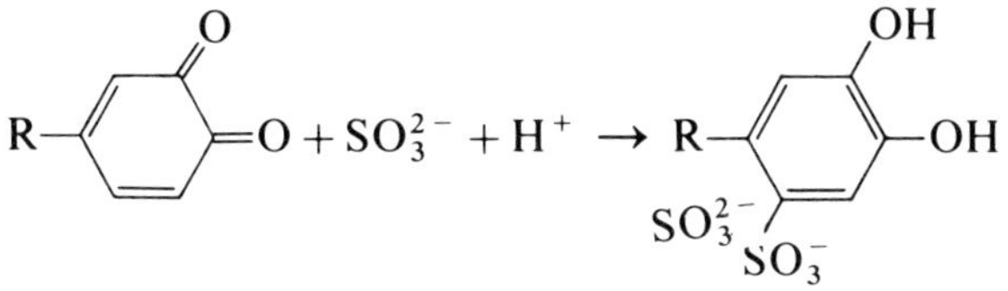

and, after successive oxidations, the disulphonate product may be formed as follows:

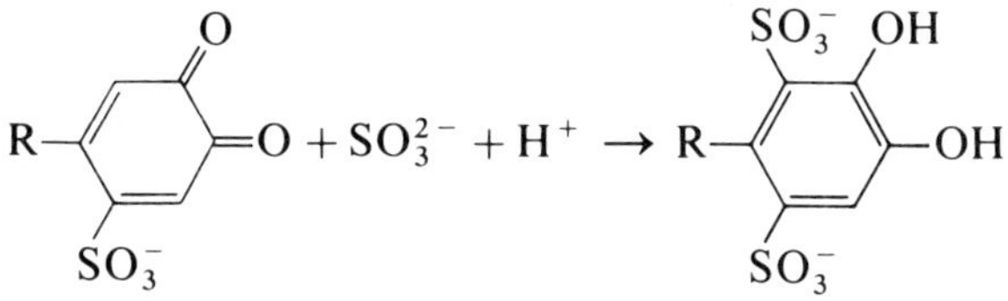

1.11.11 Reduction of azo-compounds

It is generally accepted that the reaction between an aromatic azo-compound and sulphur(IV) oxoanions is one of reduction, the azo-compound being converted to either the hydrazo- or the amine product. According to Biilman and Blom (1924), and also Connant and Pratt (1926), the reduction may be reversible, although the latter authors claim that, if one of the aromatic rings has a hydroxyl group in position 2, rearrangement of the reduced form is possible, leading to irreversible formation of amine:

$$ArNHNHAr'OH \rightarrow ArNH_2 + Ar' \overset{O}{\underset{NH}{\diagdown}}$$

Compounds in which the hydroxyl group is in position 4, may be reduced to hydrazo-compounds, which are relatively stable. The azo food dye, disodium 4-hydroxy-3-(4-sulpho-1-naphthylazo)naphthalene-1-sulphonate (Carmoisine), is reduced by sulphur(IV) oxoanions with the transfer of two electrons, the sulphur(IV) being completely oxidised to sulphate ion (Wedzicha and Rumbelow, 1981). The rate of loss of azo-compound, in the range of pH 4–5·3, is given by:

$$-\frac{d[\text{azo}]}{dt} = k \frac{[\text{HSO}_3^-][\text{azo}^{2-}]}{[\text{H}^+]}$$

This rate equation is consistent with a reaction in which one molecule each of sulphur(IV) oxospecies and azo-compound make up the activated complex, probably with the elimination of H^+ when hydrogen sulphite ion is the reactant:

$$\text{HSO}_3^- + \text{azo}^{2-} \overset{\text{slow}}{\rightleftharpoons} [\text{azo} . \text{SO}_3]^{4-} \xrightarrow[\text{H}_2\text{O}]{\text{fast}} \text{hydrazo}^{2-} + \text{SO}_4^{2-}$$

1.12 PRODUCTION OF FREE RADICALS BY MECHANISMS OTHER THAN OXIDATION

The dithionite ion, $S_2O_4^{2-}$, which may be regarded as a dimer of the radical, SO_2^-, undergoes thermal decomposition on heating to yield its component free radicals. The SO_2^- radical may also be detected when sodium disulphite is heated above 150 °C. The thermal decomposition of $Na_2S_2O_5$ leads, over a wide temperature range (100–400 °C), to SO_2 and SO_3^{2-}, presumably as a result of a simple electron transfer disproportionation reaction (Foerster and Hamprecht, 1926):

$$^-O_2S\text{---}SO_3^- \rightarrow SO_2 + SO_3^{2-}$$

The suggestion that homolytic cleavage of the S—S bond takes place does not wholly account for the formation of SO_2^-, since SO_3^- cannot be detected, but it is possible that the SO_2^- free radical is formed in all thermal decomposition reactions which lead to the evolution of sulphur dioxide (Janzen, 1972). Although the conditions required for the thermal decomposition of $Na_2S_2O_5$ are rather severe when compared with the thermal treatment of foods, Janzen and DuBase (1965) have found that the adducts of aldehydes and ketones with hydrogen sulphite ion are less stable to heat, and also produce SO_2^-. The least stable compound studied was the adduct of acetaldehyde, for which a cage decomposition mechanism may

TABLE 1.9

Aldehyde- and ketone-hydrogen sulphite adducts which decompose when heated to form SO_2^- (Janzen and DuBase, 1965)

Compound	Temperature (°C)	Amount SO_2^- formed
Acetaldehyde	Room temperature–140	+ + +
Acetone	80–160	+
Cyclohexanone	120–140	+ +
Diacetyl	110–130 (2 radicals)	+ +
Di-isopropylidene acetone (phorone)	120–140	+
Benzaldehyde	140	+ +
Anisaldehyde	140	+
2,4-Dichlorobenzaldehyde	160	+ +
Benzalacetone	80–130	+

Reproduced with permission from *Tetrahedron Letters*, 1965, page 2524. © 1965 Pergamon Press Ltd.

be operative, since some rearrangement is necessary for the formation of SO_2^- rather than SO_3^- :

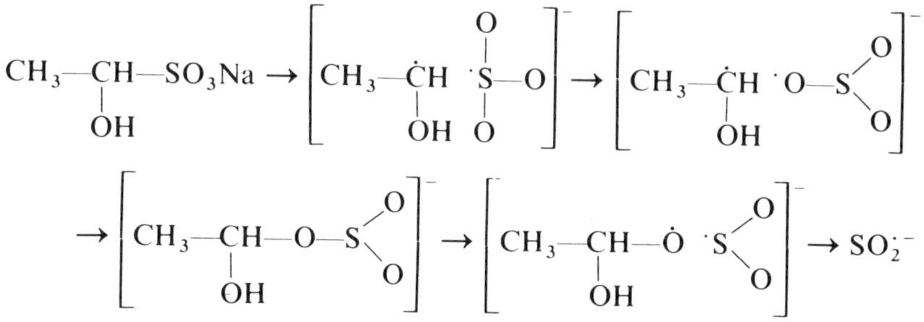

$$CH_3-\underset{\underset{OH}{|}}{CH}-SO_3Na \rightarrow \left[CH_3-\overset{\cdot}{C}H \underset{\underset{O}{|}}{\overset{O}{\overset{||}{S}}}-O \atop \underset{}{OH} \right]^- \rightarrow \left[CH_3-\overset{\cdot}{C}H \cdot O-S\overset{O}{\underset{O}{\diagdown}} \atop \underset{}{OH} \right]^-$$

$$\rightarrow \left[CH_3-\underset{\underset{OH}{|}}{CH}-O-S\overset{O}{\underset{O}{\diagdown}} \right]^- \rightarrow \left[CH_3-\underset{\underset{OH}{|}}{CH}-\overset{\cdot}{O} \cdot S\overset{O}{\underset{O}{\diagdown}} \right]^- \rightarrow SO_2^-$$

Other hydrogen sulphite adducts which have been shown to undergo this type of reaction are listed in Table 1.9.

2

Analytical

2.1 INTRODUCTION

The main requirement for the analysis of sulphur(IV) oxoanions by the food manufacturer is the reliable and rapid determination of the additive in foods, in order to verify compliance with regulations concerning residual levels which may be present at the time of sale. The use of instrumental methods involving the minimum of chemical work-up offers obvious possibilities for automated in-line analysis.

In the research field there are at least two requirements: first, the reliable determination of the additive, and secondly, the separation and identification of products formed in reactions between the additive and food components. Methods of identification of products may be broadly divded into those which assess the overall amounts of organic and inorganic products formed, or which categorise the organic products according to their labile and non-labile sulphite content, and those methods which require the use of chromatographic techniques for the isolation of products prior to structural identification.

The analysis of the unchanged additive in aqueous solutions and food is complicated by the fact that sulphur(IV) oxoanions may exist in a variety of forms depending upon pH and concentration. Thus, there are methods which may be used to determine total sulphur(IV) oxoanion in the sample, regardless of its ionic form, and those which lead to information regarding the amounts of the individual species. Since equilibrium between ionic forms of sulphur(IV) oxospecies is rapidly established, the latter methods are generally physical in nature and typically involve some form of spectrophotometric analysis without affecting the composition of the mixture. When the food is a dehydrated product, further complications arise. The addition of water to dissolve sulphur(IV) oxoanions will, inevitably, alter the distribution of the ionic species originally present in the sample, and may also lead to changes in composition with respect to labile covalent compounds. Analytical data obtained in such cases may not reliably be extrapolated to the original food material since, first, neither the value nor the significance of pH in this environment is known, and secondly, the additive is often present at concentrations well in excess of those at which activity coefficient data are available. The situation may be further complicated by the inhomogeneity of the system.

2.2 QUALITATIVE ANALYSIS OF SULPHUR(IV) OXOANIONS

A number of qualitative chemical tests for the presence of sulphur(IV) oxoanions are shown in Tables 2.1 and 2.2 after Vogel (1954) and Feigl and Anger (1972). There are, of course, many other possible tests, based upon quantitative analytical reactions, which will be described later, but those listed here provide adequate positive identification.

The tests may be divided into two types: those which are designed to detect the presence of gaseous sulphur dioxide formed on acidification of a solution of sulphur(IV) oxoanions, and those which involve reactions with the ions in solution. When mixtures containing sulphur(IV) oxoanions are acidified to pH < 0.5, the major part of the oxospecies is in the form of $SO_2 . H_2O$. There are large discrepancies between reported data on the threshold of sulphur dioxide detection by smell. McCord and Witheridge (1949) state the threshold to be 9 μg/litre, corresponding to approximately 3 parts per million (ppm) by volume. However, Leonardos and coworkers (1969) quote a result nearly an order of magnitude lower at 0·47 ppm. Assuming the higher threshold value, this concentration of sulphur dioxide in the atmosphere above an aqueous solution will be in equilibrium with a concentration of the gas in solution of 3×10^{-6} molal, assuming a value of Henry's constant in the region of unity (Section 1.7). Thus, the limit of detection of sulphur dioxide in solution by smell is, theoretically, 0·2 ppm if the volume of solution is much greater than that of the headspace, or 0·4 ppm if these two volumes are equal.

Reactions which are most interesting chemically are those involving nickel(II) hydroxide and cobalt(II) azide, in which the change is one of

TABLE 2.1

Tests for the identification of gaseous sulphur dioxide liberated from solutions of sulphur(IV) oxoanions by treatment with acid
(Vogel, 1954; Feigl and Anger, 1972)

Reagent	Test	Colour change	Sensitivity
Acidified $K_2Cr_2O_7$ paper	Hold over mouth of test tube	Orange → green	
KIO_3/starch paper	Hold over mouth of test tube	Colourless → blue	
$Ni(OH)_2$[a] paste or on paper	Place sample in stoppered tube. Add 1–2 drops of 6 M HCl and close tube with stopper coated with reagent. Warm gently. Alternatively, hold coated filter paper in gas evolved	Green → grey/black. For small amounts, treat product with benzidine acetate,[b] when blue colour is formed	$0.4 \,\mu g \; SO_2$
Zinc nitroprusside[c] paste	Same procedure as above, but colour intensified when the reaction product is held over ammonia to decolorise the unused reagent	Salmon pink → red	$3.5 \,\mu g \; SO_2$

[a] Nickel(II) hydroxide may be prepared by adding sodium hydroxide to nickel(II) chloride and washing the precipitate free from alkali. Freshly prepared nickel(II) hydroxide should be used.
[b] Benzidine acetate may be prepared by dissolving benzidine or benzidine hydrochloride (50 mg) in glacial acetic acid (10 ml), diluting to 100 ml and filtering.
[c] Zinc nitroprusside may be prepared by the addition of zinc sulphate to a solution of sodium nitroprusside, boiling for a few minutes, filtering the precipitate and washing with water. It should be stored in a dark bottle.

TABLE 2.2

Spot tests for sulphur(IV) oxoanions (Vogel, 1954; Feigl and Anger, 1972)

Reagent	Test	Result	Sensitivity
Acidified $K_2Cr_2O_7$	In acidified solution	Orange → green	
Acidified $KMnO_4$	In acidified solution	Decolorised	
Iodine solution	In acid or neutral solution	Decolorised	
$BaCl_2/SrCl_2$	In neutral solution. Test for solubility of precipitate in hydrochloric acid and the effect of oxidation of precipitate (Br_2/H_2O; H_2O_2) on its solubility in acid	White precipitate soluble in acid—becomes insoluble on oxidation	
$AgNO_3$	In neutral solution. Test for solubility of precipitate in dilute nitric acid and in ammonia	White precipitate soluble in dilute nitric acid and in ammonia	
Lead acetate	In neutral solution. Test for solubility of precipitate in cold dilute nitric acid and the effect of boiling	White precipitate soluble in dilute nitric acid. $PbSO_4$ formed on boiling	

(continued)

TABLE 2.2—contd.

Reagent	Test	Result	Sensitivity
Zinc/dilute H_2SO_4	Reaction carried out in test tube. Test gas with lead acetate paper	H_2S evolved	
Sodium nitroprusside/ zinc sulphate/potassium ferrocyanide	Add 1 drop potassium ferrocyanide (1 M) to 1 drop cold saturated zinc sulphate followed by 1 drop of sodium nitroprusside (1%). Then add 1 drop of test solution	Red colour formed	$3\cdot2\,\mu g$ Na_2SO_3
Malachite Green	1 drop Malachite Green (2·5%) is mixed with 1 drop of neutral test solution	Decolorised	$1\,\mu g$ SO_2
Cobalt(II) azide	Place a drop of a slightly acid solution of a cobalt salt containing ~0·5 mg Co on filter paper followed by 1 drop saturated sodium azide. Add 1 drop test solution (pH 5–6). Test product with o-tolidine (2%) in acetic acid	Violet → yellow. Colour changes to blue with o-tolidine	$0\cdot5\,\mu g$ $NaHSO_3$

induced oxidation using oxygen as the oxidising agent. In the case of the nickel compound, the reaction is particularly remarkable since it is otherwise accomplished only by the use of strong oxidising agents such as free halogen or peroxodisulphate ion, hydrogen peroxide being totally ineffective (Feigl and Anger, 1972). The product is black nickel(IV) hydroxide, which can oxidise benzidine to benzidine blue, the latter being a 1:1 compound of benzidine and the imine formed on oxidation, which, thereby, forms the basis of a more sensitive method of detection of formation of the nickel(IV) product. Cobalt(II) azide will undergo air oxidation to cobalt(III) complexes, but this reaction is greatly accelerated by the presence of sulphur(IV) oxoanions. Since the colour change (violet → yellow) is not very sensitive, an alternative test for the formation of the cobalt(III) product is its ability to oxidise o-tolidine to yield a blue quinone-type product.

The main source of interference is the presence of sulphides or compounds which evolve hydrogen sulphide when acidified. Thus, hydrogen sulphide, which is a good reducing agent, will decolorise acidified potassium permanganate and potassium dichromate solutions, and iodine solution. Black sulphides of silver, lead, cobalt(II) and nickel(II) are readily formed, and the nitroprusside reaction is characteristic for sulphide ion giving a transient purple coloration in neutral or, preferably, alkaline solution. One simple treatment is the addition of precipitating agents such as bismuth nitrate or mercury(II) chloride. Interference from thiosulphates, which decompose in acid solution to sulphur dioxide and elemental sulphur, may be eliminated by the addition of mercury(II) chloride, when the insoluble mercury(II) sulphide and undetectable sulphate ion are formed (Vogel, 1954):

$$Hg^{2+} + S_2O_3^{2-} + H_2O \rightarrow HgS + SO_4^{2-} + 2H^+$$

Triphenylmethane dyes, such as Malachite Green, react in a similar manner with sulphides and with sulphite ion. Since zinc, lead and cadmium salts reduce the sensitivity of this test, the interference due to sulphide cannot be removed by the addition of salts of these elements.

A consequence of the simultaneous liberation of sulphur dioxide and hydrogen sulphide is the possibility of reaction between them, leading to the formation of sulphur, thiosulphates and polythionates:

$$SO_2 + 2H_2S \rightarrow 2H_2O + 3S$$

Subsequent reaction between sulphur and any hydrogen sulphite ion will give rise to thiosulphate ion (Heunisch, 1977) in the first instance, and to polythionates in later reactions:

$$S + HSO_3^- \rightarrow S_2O_3^{2-} + H^+$$

2.3   PRINCIPLES OF QUANTITATIVE ANALYSIS

2.3.1 Iodimetric methods

A prerequisite to almost any investigation of the reactions of sulphur(IV) oxoanions is a knowledge of the concentrations of solutions of these species. Since the composition of solid salts of sulphur(IV) oxoanions is uncertain with respect to oxidation and solutions undergo aerobic oxidation, all solutions must be standardised. The method of iodimetric titration is widely used and forms the standard technique against which all other methods of analysis are compared.

The stoichiometric equation for the titration is:

$$S(IV) + I_2 + H_2O \rightarrow S(VI) + 2H^+ + 2I^-$$

Therefore, one mole of iodine reacts with one mole of sulphur(IV) compound. The reaction is normally carried out in acid solution, the main sources of error apparently lying in the oxidation of sulphur(IV) by air, and in the loss of sulphur dioxide gas from the solution during titration when iodine is added from a burette (Mason and Walsh, 1928a, b). These difficulties may be overcome if the unknown solution of sulphur(IV) oxoanions is added to an excess of iodine, thereby leading to rapid oxidation of the whole of the sample, the excess iodine being determined by titration with standard thiosulphate solution. This procedure constitutes the normally accepted technique for the standardisation of solutions of sulphur(IV) oxoanions (Vogel, 1961), with the additional recommendation that dilute solutions are used. The incorporation of antioxidants is desirable if the standardisation is to be meaningful and if the solutions are to be stable for a significant period of time. Possible additives include mannitol, sucrose and alcohol (Kolthoff et al., 1957), glycerol (Urone and Boggs, 1951) or a small quantity of EDTA (Humphrey et al., 1970). The use of EDTA to complex metal ions capable of catalysing oxidation appears to be particularly effective; when a solution of sodium sulphite (0·001 M) containing EDTA (0·001 M) is left for two, three and four weeks, the observed losses of sulphur(IV) are 5, 10 and 15 % respectively.

If other reducing agents are present in solution with sulphur(IV) oxoanions, it is still possible to determine the amount of sulphur(IV) in the mixture, providing the reducing power of these interfering substances is known. A simple method, which allows for the presence of other reducing agents, is the conversion of the sulphur(IV) oxoanions to the hydroxysulphonate of formaldehyde or acetone, both of which, at acid pH, are relatively stable and do not react with iodine. Thus, the sulphur(IV) content may be found by a comparison of the reducing power towards iodine before and after treatment with the carbonyl compound (Kolthoff *et al.*, 1957).

The apparent lack of reaction between the hydroxysulphonate adduct and iodine has led to the use of iodimetric methods for the study of carbonyl–sulphur(IV) oxoanion equilibria and the rates of formation or dissociation of hydroxysulphonates. For equilibrium studies, reaction mixtures are 'quenched' by the rapid addition of acid. At low pH the rate of decomposition of the adducts is sufficiently slow, on the timescale of the titration, to permit reliable determination of the composition of equilibrium mixtures and the results are consistent with those obtained by spectrophotometric determination of the aldehyde compound in reaction mixtures (Green and Hine, 1974). If necessary, the sulphur(IV) combined in the form of hydroxysulphonate may be estimated by subsequently raising the pH of the reaction mixture to approximately 10, allowing the adducts to decompose, and titrating the mixture after 'quenching' in acid. The kinetics of adduct decomposition may be readily observed by following the decomposition reaction, in an excess of iodine, by spectrophotometric measurement of residual triiodide ion (Green and Hine, 1974). Alternatively, if the amount of iodine present in such reaction mixtures is small compared with that required to react with all the combined sulphur(IV) oxoanion, the progress of decomposition may be conveniently followed by means of an iodine 'clock' reaction (Blackadder and Hinshelwood, 1958).

An instrumental development of titrimetric analysis is coulometric titration with electrogenerated iodine, the titrant being produced at the anode of an electrolysis cell:

$$2I^- \rightarrow I_2 + 2e^-$$

A typical experimental arrangement for coulometric titration is shown in Fig. 2.1, in which the generator and auxiliary electrode are of platinum foil (Vogel, 1961). Current is allowed to flow until there is an excess of iodine, at which point the quantity of electricity, Q, in coulombs, passed through the cell is calculated (current × time) and, hence, the amount of iodine equivalent to the sulphur(IV) oxoanions may be determined, since:

$$\text{Amount } I_2 \text{ formed} = \frac{Q}{2 \times 96\,493} \text{ mol}$$

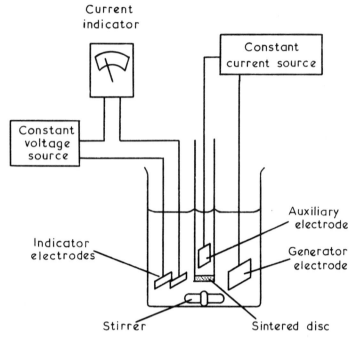

Fig. 2.1.   Arrangement for coulometric titration of sulphur(IV) oxoanions with electrogenerated iodine (Vogel, 1961). Reproduced with permission from A. I. Vogel, *Textbook of Quantitative Inorganic Analysis Including Elementary Instrumental Analysis*, 3rd edition, page 680. © 1961 Longman Group Ltd.

An advantage of this method of analysis is that the standardisation of reagents may be dispensed with since the standard is the electron and accuracy, therefore, lies in the careful measurement of current and time. In addition, no dilution of the sample occurs during analysis, and the method lends itself to automation. An important factor with regard to the latter is the detection of end-point. This may satisfactorily be determined by observation of the first formation of a starch–iodine colour in the cell, but even more convenient is the use of either an amperometric or dead-stop procedure. This arrangement is shown in the figure and involves the use of two indicator electrodes. In amperometric titrations the end-point is determined from the current in the indicator circuit. The voltage on the indicator electrode is set well below the decomposition potential of the pure supporting electrolyte but close to, or above, the decomposition potential of the supporting electrolyte plus free titrant. The indicator current depends upon the amount of free titrant present and will, therefore, be small so long as the titrant is used up, but will increase beyond the equivalence point. The generating time to the equivalence point may then be found from the observed final value of indicator current at the end of the titration, the known rate of generation and appropriate indicator calibration data. The electrodes may be of platinum foil or one electrode may be platinum whilst the other is a reference electrode. In the case of dead-stop endpoint determination, a small potential is used to polarise platinum or tungsten indicator electrodes—free iodine acts as a depolariser removing hydrogen from the cathode and causing a current to flow.

The use of coulometric procedures for the determination of sulphur(IV) oxoanions has been advocated in applications in which sulphur dioxide is formed as part of the analysis of sulphur compounds, or for atmospheric sulphur dioxide analysis. A typical procedure requires the use of 0.2 M potassium iodide in 0.7% acetic acid and a current of 3 mA. For analysis, 5 ml of the unknown solution containing up to 20 ppm SO_2 is added to 30 ml of the electrolyte, and the concentration is determined with a standard deviation of less than $\pm 1\%$ over this range using a dead-stop endpoint determination (Campbell *et al.*, 1975).

2.3.2 Non-iodimetric titration

Acid–base titration may be used to determine the acid form of the sulphur(IV) oxoanions and may be carried out either on the hydrogen ion released on formation of complex with tetrachloromercurate(II), or through reaction with hydrogen peroxide to form sulphuric acid:

$$HgCl_4^{2-} + SO_2 . H_2O \rightarrow HgCl_2SO_3^{2-} + 2Cl^- + 2H^+$$

$$HgCl_4^{2-} + HSO_3^- \rightarrow HgCl_2SO_3^{2-} + 2Cl^- + H^+$$

$$H_2O_2 + SO_2 . H_2O \rightarrow H_2SO_4 + H_2O$$

$$H_2O_2 + HSO_3^- \rightarrow HSO_4^- + H_2O$$

In both cases, sulphite ion does not lead to the production of hydrogen ion. For titration with sodium hydroxide, methyl orange could be used as indicator and these titrations are applied to the measurement of gaseous or liberated sulphur dioxide. The method may also be used for the determination of hydrogen sulphite in the presence of sulphite ion.

Common oxidising agents which could be considered for the titration of sulphur(IV) oxoanions are permanganate, dichromate and iodate ions and ceric sulphate. Although there are numerous reports of non-stoichiometric reactions between sulphur(IV) oxoanions and all these reagents with the exception of iodate ion, the optimum conditions for their use in the estimation of sulphur(IV) were investigated by Rao and Rao (1955). Assuming that the dithionate responsible for the non-stoichiometric behaviour could be oxidised, in the presence of a suitable catalyst, to sulphate, the effectiveness of copper(II) ions and iodine monochloride as catalysts was considered. Both proved useful for permanganate and dichromate titrations, but only iodine monochloride is suitable for ceric sulphate and vanadate titrations, and in all cases a high concentration of acid was required. The titrations were sensitive to concentration of both acid and sulphur(IV) oxoanions, the errors increasing with increasing concentration of sulphur(IV) at a given acid concentration. Careful adjustment of acid concentration leads to conditions (1 M acetic acid) in which no catalyst is required for the dichromate titration. In all cases the titration procedure involved addition of the solution for analysis to an excess of oxidising agent, and estimation of the excess by titration with ferrous ammonium sulphate. The reactions for reduction of the oxidising agents in acid solution are:

$$Cr_2O_7^{2-} + 14H^+ + 6e^- \rightleftharpoons 2Cr^{3+} + 7H_2O$$

$$MnO_4^- + 8H^+ + 5e^- \rightleftharpoons Mn^{2+} + 4H_2O$$

$$Ce^{4+} + e^- \rightleftharpoons Ce^{3+}$$

$$VO_4^{3-} + 6H^+ + e^- \rightleftharpoons VO^{2+} + 3H_2O$$

The stoichiometry of the reaction with sulphur(IV) oxoanions may be calculated, assuming a two-electron transfer to sulphur(IV) in the reduction of the reagent.

Sulphur(IV) oxoanions react quantitatively with periodate ion according to the following equations:

$$IO_4^- + SO_3^{2-} \rightarrow SO_4^{2-} + IO_3^-$$

$$IO_4^- + HSO_3^- \rightarrow SO_4^{2-} + IO_3^- + H^+$$

The analysis may be carried out by the addition of the sample to an excess of periodate in saturated borax followed by the iodometric determination of unreacted periodate (Kaushik and Prosad, 1969). The amount of hydrogen sulphite ion present may be found by adding the sample to the reagent in the presence of a known quantity of borax, and determining the residual borax with hydrochloric acid.

Issa and coworkers (1974) report the ability of lead(IV) acetate to oxidise sulphur(IV) oxoanions to sulphate ion:

$$Pb(CH_3COO)_4 + 2e^- + 2H^+ \rightleftharpoons Pb(CH_3COO)_2 + 2CH_3COOH$$

An advantage of this reaction is that it can be carried out in pH 4 buffer, thereby reducing the loss of gaseous sulphur dioxide normally occurring if titrant is added to the reducing agent in strongly acidic media. The reaction may be followed potentiometrically using platinum and calomel electrodes. When lead(IV) acetate is added to the solution of hydrogen sulphite ion, the endpoint is observed as a single large inflection, the jump in potential being in the region of $50 \, mV/0.02 \, ml$ of titrant. The apparent disadvantages are a somewhat slow reaction around the endpoint, requiring two to three minutes to attain equilibrium, and the need to maintain a pH of approximately 4 for a quantitative endpoint. Therefore, if the effective capacity of the buffer used is inadequate, some form of pH-stat arrangement seems desirable.

Chloramine-T reacts with hydrogen sulphite ion rapidly and quantitatively in neutral or weakly acidic solutions:

$$CH_3C_6H_4SO_2NClH + HSO_3^- + OH^-$$
$$\rightarrow CH_3C_6H_4SO_2NH_2 + HSO_4^- + Cl^-$$

When solutions of sulphur(IV) oxoanions are added to an excess of chloramine-T, the unreacted reagent may be determined iodometrically by the addition of an excess of potassium iodide to the acidified mixture and titration of the liberated iodine with standard sodium thiosulphate solution:

$$CH_3C_6H_4SO_2NClH + 2I^- + H^+ \rightarrow CH_3C_6H_4SO_2NH_2 + I_2 + Cl^-$$

An alternative and convenient method of following this reaction is that of amperometric titration using either a dropping mercury or a rotating platinum electrode (Matsuda, 1979). Good results are obtained with the rotating platinum electrode when chloramine-T is titrated with hydrogen sulphite ion in weakly acidic solution (pH 3–6) at $0 \, V$ relative to the saturated calomel electrode. In the case of the dropping mercury electrode, good results are obtained for both the titration of chloramine-T with hydrogen sulphite ion, and vice-versa, at pH 7 and at $0.2 \, V$ with respect to the saturated calomel electrode. The titration curves are L-shaped for addition of hydrogen sulphite ion to chloramine-T, and reverse L-shaped for the reverse titration, since, under the conditions of the experiment, no response is observed for hydrogen sulphite ion. Relative standard deviations of $\pm 0.5\%$ may be obtained using either electrode for the determination of hydrogen sulphite ion concentration in the range 0.002–0.05 M, with relative errors not exceeding $\pm 1\%$.

2.3.3 Colorimetric methods

2.3.3.1 Use of pararosaniline and triphenylmethyl dyes

The most widely used colorimetric method for the determination of sulphur(IV) oxoanions is based on the Schiff reaction. The reaction, as practised in histochemistry, involves the treatment of a sample suspected of containing aldehyde groups with the Schiff reagent, which consists of a solution of pararosaniline bleached with sulphur(IV) oxoanions. On reaction with aldehyde, a red–violet colour develops. The reverse of this process, that is the reaction of sulphur(IV) oxoanions with acidified pararosaniline in the presence of formaldehyde, constitutes the quantitative analytical reaction. Studies of its stoichiometry, by means of the method of continuous variation, give an apparent reaction of 1.7 moles of sulphur(IV) oxospecies with one mole of pararosaniline (Scaringelli *et al.*, 1967), indicating that it proceeds beyond the 1:1 stage.

When pararosaniline ($\lambda_{\max} = 540$ nm, pH 5) is acidified to below pH 3, it is decolorised as a result of the following reaction:

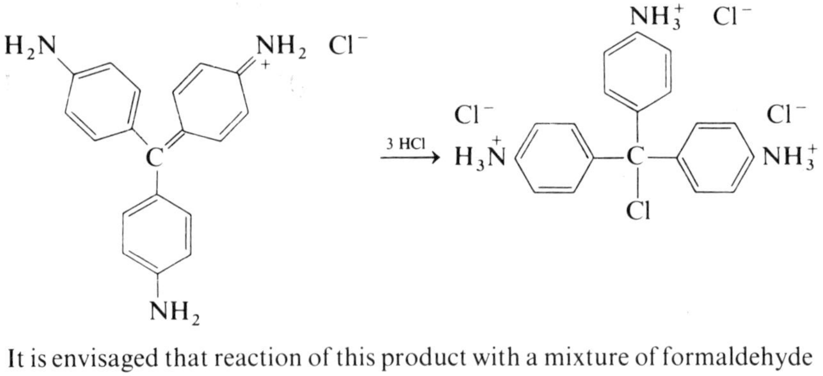

It is envisaged that reaction of this product with a mixture of formaldehyde and sulphur(IV) oxoanion leads to the formation of up to three adducts, the first addition compound being:

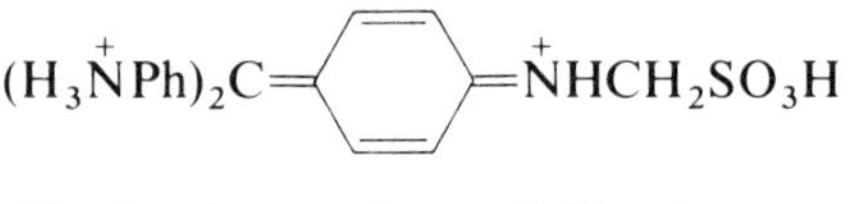

$$(H_3\overset{+}{N}Ph)_2C=\!\!\!\left\langle\;\;\right\rangle\!\!\!=\overset{+}{N}HCH_2SO_3H$$

Further addition will take place at the available nitrogen atoms. Assuming that these three complexes are formed in a stepwise manner, Huitt and

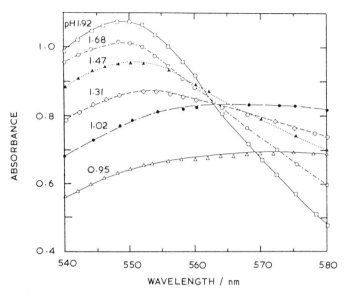

Fig. 2.2. Effect of pH on the absorption spectrum of the pararosaniline–formaldehyde–sulphur(IV) oxoanion complex (Scaringelli *et al.*, 1967). Reproduced with permission from *Anal. Chem.*, 1967, **39**, page 1714. © 1967 American Chemical Society.

Lodge (1964) calculated the equilibrium constants K_1, K_2 and K_3 respectively for the formation of 1:1, 1:2 and 1:3 (pararosaniline: sulphur(IV)) complexes, and also their molar extinction coefficients at 560 nm, ε_1, ε_2 and ε_3, at pH 1·2, as follows:

$$K_1 = 2·2 \times 10^5 \, \text{M}^{-1} \qquad \varepsilon_1 = 14\,900 \, \text{M}^{-1} \, \text{cm}^{-1}$$
$$K_2 = 1·7 \times 10^5 \, \text{M}^{-1} \qquad \varepsilon_2 = 44\,800 \, \text{M}^{-1} \, \text{cm}^{-1}$$
$$K_3 = 0·2 \times 10^5 \, \text{M}^{-1} \qquad \varepsilon_3 = 65\,600 \, \text{M}^{-1} \, \text{cm}^{-1}$$

Measurement of absorbance resulting from the complex is made normally at low pH when pararosaniline is colourless. The absorption spectrum of the product is, however, sensitive to pH under these conditions, as illustrated in Fig. 2.2 between pH 0·95–1·92, the data being characteristic of an acid–base indicator, with an isosbestic point (Scaringelli *et al.*, 1967). Therefore, in acid solution careful control of pH is necessary.

The results also show that the reaction products are stronger acids than pararosaniline and this explains why they may be extracted preferentially into butanol or pentanol in order to enhance the sensitivity of the analysis (Dasgupta *et al.*, 1980).

An important observation is that the addition of sodium hydroxy-methanesulphonate to acid-bleached pararosaniline does not lead to the formation of the expected product, whereas prior decomposition by alkali followed by addition to acidified reagent gives a good yield of coloured product. Dasgupta and coworkers (1980) consider that the reaction of interest takes place through adduct formation between formaldehyde and pararosaniline, and is followed by reaction with sulphur(IV) oxoanions:

$$(NH_2Ph)_3C^+ + HCHO \rightarrow (NH_2Ph)_2\overset{+}{C}PhNHCH_2OH \xrightarrow[-OH]{HSO_3^-}$$

$$(NH_2Ph)_2\overset{+}{C}PhNHCH_2SO_3H \xrightarrow{HCHO} etc.$$

The final reaction product has the same spectral characteristics as that obtained in the Schiff staining procedure. In the latter case, the bleaching of pararosaniline with sulphur(IV) oxoanions is expected to yield the sulphonate:

$$(NH_2Ph)_3C—SO_3^-$$

Subsequent reactions with aldehyde are expected to be analogous to those shown above but with concomitant decomposition of the sulphonate.

The presence of dimethylformamide (3–5 M) leads to considerable enhancement of the extinction coefficients of the possible products, values of 31 500, 62 300 and 119 200 $M^{-1} cm^{-1}$ at 555 nm being reported respectively for 1:1, 1:2 and 1:3 (pararosaniline:sulphur(IV) oxoanion) adducts, with equilibrium constants being only slightly affected.

The practical value of molar extinction coefficient for this analysis may be as high as 48 000 $M^{-1} cm^{-1}$, calculated on the basis of the amount of sulphur(IV) analysed. Impurities in the pararosaniline do, however, present problems (Scaringelli et al., 1967). In particular, an undesirable violet impurity is detectable by paper chromatography and may be removed from the reagent by partitioning between the acidified solution and butanol, when the impurity transfers into the organic phase.

An important development of this reaction has been the use of tetrachloromercurate(II) ion for the trapping of sulphur dioxide (with particular reference to atmospheric analysis) and constitutes the basis of the method of West and Gaeke (1956). Tetrachloromercurate(II) ion forms, on reaction with sulphur(IV) oxoanions, a particularly stable product which is also very resistant to oxidation:

$$HgCl_4^{2-} + SO_2 + H_2O \rightarrow HgCl_2SO_3^{2-} + 2Cl^- + 2H^+$$

Other triphenylmethyl dyes which are potentially suitable for the determination of sulphur(IV) oxoanions are Malachite Green and Crystal Violet:

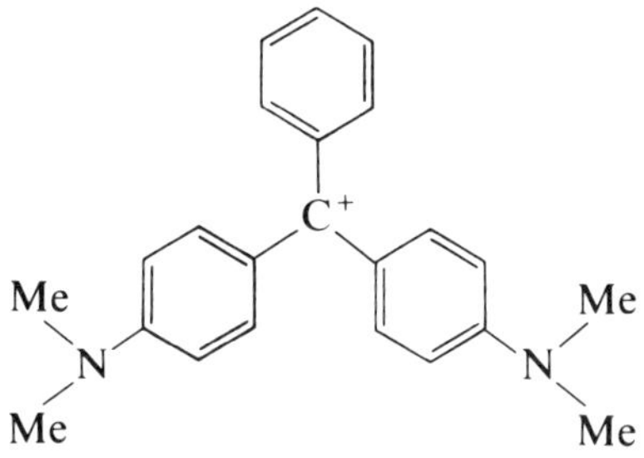

Malachite Green 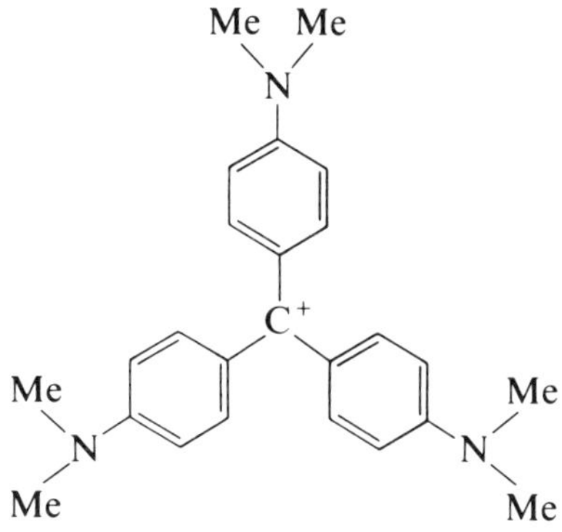   Crystal Violet

Both dyes react readily with sulphur(IV) oxoanions at the pH of acetic acid/acetate buffers (pH 3·7–5·7), giving a colourless sulphonate:

$$(Ar)_3C^+ + HSO_3^- \rightleftharpoons (Ar)_3C - SO_3^- + H^+$$

At an ionic strength of 1×10^{-3} M and 23 °C, the equilibrium constants for the reactions of Malachite Green and Crystal Violet with HSO_3^- are given respectively as $> 10^6$ and 8×10^3, whilst the second order rate constants for the formation of product are respectively 4×10^3 and $4·7 \times 10^2$ $M^{-1} s^{-1}$ (Ritchie and Virtanen, 1973).

2.3.3.2 Reaction with disulphide

One of the most promising alternative analytical reactions to the use of triphenylmethyl cations is that with the disulphides, 5,5′-dithiobis(2-nitrobenzoic acid) and 4,4′-dithiopyridine (Humphrey et al., 1970). The reaction, which is essentially quantitative at pH 7, leads to the formation of thiols, present as the thiolate anion. The extensive delocalisation of negative charge is probably an important contributory factor to the quantitative nature of the reaction, and also renders the products highly coloured. The important spectral characteristics at pH 7 are:

5-mercapto-2-nitrobenzoic acid: $\lambda_{max} = 412$ nm
$\varepsilon_{max} = 15\,500$ $M^{-1} cm^{-1}$

4-mercaptopyridine: $\lambda_{max} = 324$ nm
$\varepsilon_{max} = 21\,500$ $M^{-1} cm^{-1}$

The disulphide reagent shows negligible absorbance at the absorbance maximum of the reduced form and Beer's law is obeyed for both compounds to absorbances in excess of 1 (1 cm cells). Absorbance readings are stable for 1–2 h, after which they decrease slowly, probably as a result of air oxidation of the thiol. The main source of interference is expected to be the presence of SH-groups, since both disulphides are commonly used for the determination of these groups. Cyanide ion will react slowly with these disulphides, and thiosulphate is expected to produce some colour. The reaction will not take place at all in the presence of mercury(II) ions, probably as a result of the high stability of the mercury(II)–sulphite complex. The ability of transition metals to catalyse oxidation of the thiol may be checked using EDTA.

Although the sensitivity of this method is only half that of the pararosaniline reaction, it is sufficiently high for many applications, and compensating factors are its overall simplicity and need of less critical conditions. The procedure involves direct addition of the unknown sample to the reagent $(1 \times 10^{-3}$ M$)$ in pH 7 phosphate buffer, and making up to volume with buffer. The reaction is quantitative over a wide pH range (pH 6–9 for 5,5′-dithiobis(2-nitrobenzoic acid), and pH 4–7 for 4,4′-dithiopyridine), and colour is independent of pH. This may be compared with the need for strict pH control in the pararosaniline determination, in which colour development requires some 30 min, whereas the reaction between disulphide and sulphite is complete in approximately 2 min. These advantages are reinforced by an unambiguous 1:1 stoichiometric ratio, whereas the pararosaniline reaction involves stepwise addition to an unknown extent in any particular case.

2.3.3.3 Oxidation with iron(III)

Despite the well-accepted non-stoichiometric nature of the reduction of iron(III) to iron(II) by sulphur(IV) oxoanions, this reaction, combined with the spectrophotometric determination of the iron(II) product, has been frequently advocated for use in quantitative analysis. The colour-producing reactions are those of iron(II) with 1,10-phenanthroline (Stephens and Lindstrom, 1964), 2,4,6-tri(2-pyridyl)-1,3,5-triazine (TPTZ) (Stephens and Suddeth, 1970) and 3-(2-pyridyl)-5,6-bis(4-phenylsulphonic acid)-1,2,4-triazine sodium salt (ferrozine) (Attari and Jaselskis, 1972). In addition to the effect of composition of the reaction mixture, other factors which affect the yield of coloured product in these analyses are pH and temperature. In all the reported analytical experiments the conversion of iron(III) to iron(II) was carried out in the presence of chelating agent, therefore it is not possible to separate the component reactions and, indeed, it is likely that the chromotropic reagent influences the course of the reduction by forming a complex with iron(III).

Considering first the reaction in the presence of 1,10-phenanthroline, the most striking observation is that of the effect of temperature on the development of colour, the absorbance at 45 °C being exactly twice the absorbance at 20 °C. Since, in both cases, the results represent final absorbance values, the simplest explanation must be that of a changing stoichiometry. Attari and coworkers (1970), therefore, considered that, at the lower temperature, the reaction involved the one-electron oxidation of sulphur(IV) to give dithionate, whereas, at the higher temperature, the stoichiometry was consistent with a two-electron oxidation giving sulphate. The presence of acetate ion in the form of a buffer was also critical to the reduction of iron(III) by sulphur(IV) oxoanions, and the effect was attributed to the formation of a labile complex of iron(III) involving both the phenanthroline and the acetate ions. It is significant that ions such as phosphate, which form stable iron(III) complexes, hinder the reduction and are, therefore, unsuitable in this application. The reaction is relatively insensitive to pH, a broad maximum in the value of absorbance as a function of pH being observed in the range pH 5–6 (Stephens and Lindstrom, 1964), and the specification for pH control is, thus, pH $5 \cdot 5 \pm 0 \cdot 5$. It is widely thought that the ratio of concentrations of iron(III) to sulphur(IV) affects the relative amounts of dithionate and sulphate formed in the reaction, an excess of iron(III) favouring sulphate production (Carlyle and Zeck, 1973). Thus, the molar ratio of these two reactants is likely to affect the analytical result. The data for the determination of sulphur(IV) oxoanions by the 1,10-phenanthroline method at 20 °C (Attari et al., 1970) are shown in Fig. 2.3, from which it is evident that Beer's law was not obeyed over the range of concentrations studied. The initial iron(III) concentration was 200 μM, representing a molar excess over sulphur(IV) oxoanions of $2 \cdot 3$ at the highest concentration and 23 at the lowest. The extinction coefficient of the complex between iron(II) and 1,10-phenanthroline is $1 \cdot 1 \times 10^4 \, \mathrm{M^{-1} \, cm^{-1}}$ at 510 nm, and a line with this slope has been drawn through the data points at the lowest concentrations. Since these data correspond well to this line, the implication is that one mole of sulphite gives rise to one mole of complex and, therefore, the reaction involves the formation of dithionate, despite the high molar ratio of oxidising agent to reducing agent. At these low concentrations of sulphur(IV) oxoanions the extinction coefficient for the overall reaction is equal to that of the expected reaction product, and

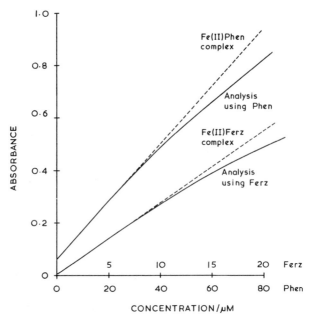

Fig. 2.3. Calibration curve for the determination of sulphur(IV) oxoanion with iron(III) in the presence of 1,10-phenanthroline (phen) and ferrozine (ferz). Continuous lines show experimental data, whilst broken lines show the calculated absorbance of the respective iron(II)-complexes assuming that the concentration of complex is equal to the concentration of sulphur(IV) oxoanion originally present.
Data from Attari and coworkers (1970) and Attari and Jaselskis (1972).

the assay may be regarded as quantitative. When the concentration ratio of 1,10-phenanthroline to iron(III) is less than 10:1, this also affects the analysis. Since the iron(III)/phenanthroline blank for these determinations gives a significant time dependent absorbance, it is necessary to limit the amount of this reagent which, in turn, limits the maximum concentration of iron(III).

The reduction of iron(III) to iron(II) by sulphur(IV) oxoanions in the presence of ferrozine differs from that carried out in the presence of 1,10-phenanthroline in three respects (Attari and Jaselskis, 1972):

1.   The ferrozine reaction has a much lower temperature coefficient, the absorbance increasing by less than 10% over the range 20–45 °C. By comparing the colour produced in the analytical reaction with the absorbance of the complex between ferrozine and iron(II) (Fig. 2.3), it is clear that, at low concentrations of

sulphur(IV) oxoanions (with initial iron(III) concentration of 100 μM), the reaction involves one-electron oxidation of sulphur(IV) to dithionate.

2. The pH dependence is much greater with ferrozine than for the 1,10-phenanthroline procedure, a sharp maximum being observed at pH 3·7–4·0. Thus, the specification for pH control is pH 3·85 ± 0·15.

3. The extinction coefficient of the reaction product is $2·8 \times 10^4 \, M^{-1} \, cm^{-1}$ at 562 nm and, at low concentrations, this value represents the sensitivity of the overall analysis. The use of ferrozine leads to an analysis which has more than twice the sensitivity of that using 1,10-phenanthroline.

As in the previous case, the analysis is dependent upon the presence of acetate ion, and blanks containing the iron(III) complex of ferrozine are unstable, with a progressive increase in absorbance with time.

The iron(II) complex of TPTZ has an extinction coefficient of $2·2 \times 10^4 \, M^{-1} \, cm^{-1}$ at 593 nm, and the reduction of iron(III) to iron(II) in the presence of this reagent is similar to that in the presence of ferrozine (Stephens and Suddeth, 1970). The temperature dependence of the reaction is an increase of 18 % in absorbance when the temperature is increased from 20–40 °C, but the stoichiometry was not investigated in detail.

In all cases colour development is adequate after 10 min and it is evident that, providing the experimental conditions are carefully controlled, the production of iron(II) by reduction of iron(III) and subsequent colorimetric determination offers a good method for the determination of sulphur(IV) oxoanions. When ferrozine is used as the reagent, the analysis is almost twice as sensitive as with the method using 5,5′-dithiobis(2-nitrobenzoic acid), but less sensitive than that using pararosaniline. The spectrophotometric determination of iron(II), however, suffers from a number of interference factors, an important contributor being EDTA, which competes effectively with the reagent for iron.

2.3.3.4 Reaction with mercury(II) chloranilate and mercury(II) thiocyanate
The sparingly soluble metal ion salts of chloranilic acid (2,5-dichloro-3,6-dihydroxy-p-benzoquinone) are important reagents for the analysis of anions, which combine with the metal in question to form a much less soluble compound, liberating the chloranilate ion, which is highly coloured,

$$Y^- + MCh \, (solid) \rightarrow Ch^- + MY \, (solid)$$

where Y^- is the anion to be determined. This reaction may be illustrated by the determination of sulphate ion using barium chloranilate (Vogel, 1961),

$$SO_4^{2-} + BaC_6Cl_2O_4 + H^+ \rightarrow BaSO_4 + HC_6Cl_2O_4^-$$

The chloranilate ion absorbs in the visible region at 530 nm and much more strongly in the ultraviolet at 332 nm.

The general procedure for these analyses is to shake the test solution with an excess of the solid chloranilate, remove the precipitate and measure the absorbance of the solution at the appropriate wavelength. The blank is prepared by equilibrating the solid chloranilate with the background medium, and the absorbance of the reference, therefore, depends upon the solubility of the solid compound. This procedure was applied to the determination of sulphur(IV) oxoanions by Humphrey and Hinze (1971) using mercuric chloranilate as the sparingly soluble salt, when the reaction is presumably:

$$HgCh + SO_3^{2-} \rightarrow HgSO_3 + Ch^{2-}$$

The product, $HgSO_3$, appears to be undissociated. The highest sensitivity may be obtained when the solvent is ethanol:water (1:1), with apparent extinction coefficients of 1000 M^{-1} cm^{-1} at 525 nm and 16 800 M^{-1} cm^{-1} at 330 nm, based upon the concentration of sulphite ion used in the analysis. The useful concentration range is approximately 5–100 μg/ml SO$_2$ when absorbances are measured at 525 nm, and approximately 0·5–8·0 μg/ml when measurements are made at 330 nm. Beer's law is obeyed in this range. Since chloranilic acid is a p-benzoquinone, it can undergo addition reactions with sulphur(IV) oxoanions. When the amount of sulphite ion exceeds the amount of available mercuric chloranilate, the absorbance of the solution is decreased.

Soluble, slightly dissociated mercury(II) thiocyanate reacts with certain anions which form less dissociated or insoluble mercury(II) compounds, releasing thiocyanate ion. This is true of sulphite ion and, if the free thiocyanate is estimated as ferric thiocyanate (FeSCN^{2+}) by the addition of iron(III) ions, the reaction forms a convenient basis for the quantitative analysis of sulphur(IV) oxoanions (Hinze et al., 1972). When analyses are carried out in acidified ethanol or methanol, the extinction coefficient, calculated on the basis of the amount of sulphur(IV) oxoanion present, is very similar to that of ferric thiocyanate at 470 and 485 nm respectively. However, the most sensitive determinations are observed when the solvent is ethanol:water (3:1) acidified with perchloric acid, when the apparent extinction coefficient is 7700 M^{-1} cm^{-1} at 470 nm, compared with a value of

$4300\,\text{M}^{-1}\text{cm}^{-1}$ for ferric thiocyanate under these conditions. If the reaction between mercuric thiocyanate and sulphite ion is,

$$Hg(SCN)_2 + SO_3^{2-} \rightleftharpoons HgSO_3 + 2SCN^-$$

the yield of product is only 50 % in acidified ethanol or methanol and close to 90 % in aqueous ethanol. However, if only one thiocyanate is released, the reaction in the alcohols is quantitative, whilst the yield of colour in aqueous ethanol is excessive. Although experimental data are few, it appears that a useful concentration range is $1–20\,\mu\text{g/ml}$ SO_2.

Both the mercuric chloranilate and mercuric thiocyanate methods are subject to serious interference from anions which form stable complexes or insoluble products with mercury(II). Thus, interference is expected from sulphide, cyanide, chloride, iodide, nitrite, etc. and, therefore, these methods are suitable for the analysis of solutions in which sulphur(IV) oxoanions are the only anions present. In both cases, the reaction time is of the order of 15 min.

2.3.3.5 Formation of Meisenheimer complex

A highly coloured Meisenheimer complex is formed between 2,4,6-trinitrobenzoic acid and sulphur(IV) oxoanions at pH 6·5 (Blasius and Ziegler, 1974). The complex absorbs at around 450 nm and in the range 500–550 nm but, unfortunately, the absorbance is strongly pH dependent up to at least pH 7, and close control of pH is essential. The choice of pH 6·5 is based upon the pH dependence of the colour, measured at 436 nm, resulting from reaction between trinitrobenzoic acid and sulphide ion, this pH value representing a maximum absorbance at 436 nm due to the sulphite adduct, before the absorbance of the sulphide adduct becomes significant. Beer's law is very closely obeyed when reaction mixtures contain $5 \times 10^{-3}\,\text{M}$ trinitrobenzoic acid and $0–7 \times 10^{-4}\,\text{M}$ sulphur(IV) oxoanion, using a citrate buffer at pH 6·5, and giving an extinction coefficient of $617\,\text{M}^{-1}\text{cm}^{-1}$ at 436 nm. The time required for reaction is 2 min and measurements are not affected by the presence of sulphate, thiosulphate, sulphide and polythionate ions at concentrations up to $0·01\,\text{M}$.

2.3.3.6 Reaction with sodium nitroprusside

The reaction between sulphur(IV) oxoanions, intensified by the presence of zinc ions and pyridine, and sodium nitroprusside has been adapted for the quantitative analysis of sulphur(IV) oxoanions (Fogg et al., 1966). Continuous variation curves indicate that, with a slight excess of

TABLE 2.3

A comparison of the relative sensitivities of various methods available for the colorimetric determination of sulphur(IV) oxoanions. The apparent extinction coefficients are in terms of the sulphur(IV) oxoanion concentration used in the determinations

Reagent	Apparent extinction coefficient ($M^{-1}cm^{-1}$)	λ_{max} (nm)
Pararosaniline	48 000 max	560
5,5'-Dithiobis(2-nitrobenzoic acid)	15 500	412
4,4'-Dithiopyridine	21 500	324
Fe(III)/1,10-phenanthroline	11 000	510
Fe(III)/ferrozine	28 000	562
Mercuric chloranilate	1 000	525
	16 800	330
Mercuric thiocyanate/Fe(III)	7 700	470
2,4,6-Trinitrobenzoic acid	617	436

nitroprusside, about 93% of the sulphur(IV) oxoanions added may be converted to sulphitonitroprusside, with an extinction coefficient of $3800\,M^{-1}\,cm^{-1}$ for the complex-product. The main problem associated with this experiment is the gradual formation of a white crystalline precipitate, which can be delayed but not eliminated by the addition of ethanol. Reaction mixtures are also required to contain gelatin in order to retain the coloured species in solution. Nevertheless, graphs of optical density as a function of concentration are linear to a concentration of 2.7×10^{-4} M of sulphur(IV) oxoanion in the final solution. The principal source of interference arises from the presence of sulphide ion, but the extent of this interference may be estimated by the degree to which the test solution reacts with the reagent before addition of zinc. The latter only enhances the colour of the sulphitonitroprusside product. Thus, interference by sulphide may be eliminated by reversing the order in which the reagents are added.

A list of the methods discussed here is shown in Table 2.3, together with a comparison of their relative sensitivities. Since the method involving the reaction with sodium nitroprusside cannot be recommended owing to the uncertainty of obtaining clear, measurable solutions, it has been omitted from the list.

2.3.4 Fluorimetric analysis

In contrast to the large volume of published information on the colorimetric determination of sulphur(IV) oxoanions, that concerning the use of fluorimetry is sparse. There exist two types of reaction: those which lead to the quenching of the fluorescence of a fluorescent compound, and those in which non-fluorescent reactants lead to the formation of fluorescent products.

The first type of reaction is well illustrated by the quenching of the fluorescence of 5-aminofluorescein (Axelrod *et al.*, 1970). In acid solution, the reagent owes its fluorescence to resonance within the dibenzopyran system:

$$HO \quad O \quad \overset{+}{C}OH \qquad \leftrightarrow \qquad HO \quad \overset{+}{O} \quad OH$$

$$^{+}NH_3 \qquad\qquad\qquad\qquad ^{+}NH_3$$

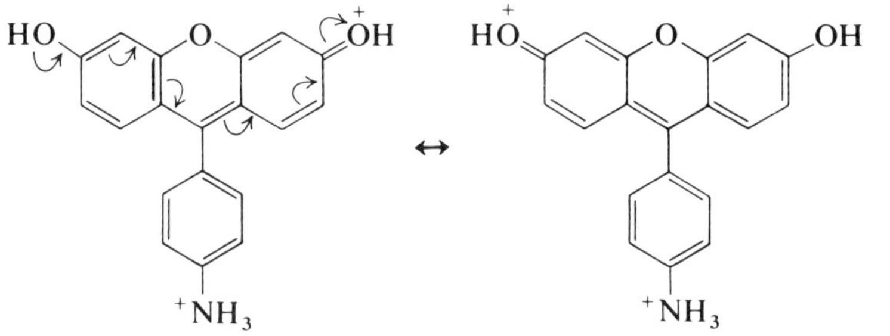

Any reaction which will break the conjugation in this part of the molecule would be expected to suppress the fluorescence. It is, perhaps, not surprising that the additional stabilisation conferred on this molecule by conversion of the phenylammonium moiety to an imino-group, with reduction of the quinoid type structure, causes it to react with formaldehyde and sulphur(IV) oxoanions in a reaction reminiscent of that with pararosaniline. Although the reaction product has not been characterised, it is expected to have the structure:

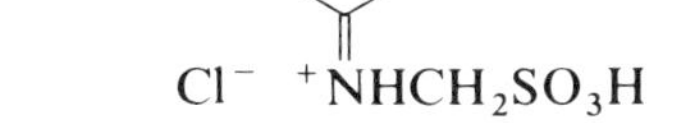

$$HO \quad O \quad OH$$

$$Cl^- \quad ^{+}NHCH_2SO_3H$$

The reaction is accompanied by a bathochromic shift in absorbance, as deduced from the description, and the product is either weakly fluorescent or non-fluorescent. The structure showing loss of resonance within the dibenzopyran system is consistent with the latter observation.

Pure 5-aminofluorescein shows a concentration dependent excitation spectrum, with a maximum at 440 nm at concentrations of the order of 10^{-6} M, but giving two maxima, at 405 and 460 nm, at higher concentrations (10^{-4} M) with concomitant loss of the peak at 440 nm. Since a typical convenient working concentration of this reagent is of the order of 10^{-4} M, it is expected that its removal by the reaction with formaldehyde and sulphur(IV) oxoanions will lead to a change in spectroscopic behaviour. Although a detailed study of the effect of medium on this property was not carried out, the practical result, in the form of a calibration curve for the determination of sulphur(IV) oxoanions, is very encouraging, for excitation at the two maxima of 405 and 460 nm. Irrespective of the excitation wavelength, and independent of concentration, emission takes place at 515 nm. In addition to 5-amino-fluorescein (10^{-4} M) and formaldehyde (0·2–0·7 %), mixtures contained tetrachloromercurate(II) ion (0·1 M) since, as in the case of the pararosaniline reaction (Scaringelli et al., 1967), its presence increased the sensitivity. A reaction time of the order of 1 h is desirable.

The analysis showed no interference from nitrate, sulphate, iron(II) and ammonium ions and hydrogen peroxide at concentrations of 10^{-2} M, and the presence of potassium, calcium, magnesium, copper, acetate and nitrite ions at the same concentration caused an error of $\pm 5\%$. Iodide ion at a concentration of 10^{-3} M caused a similar error.

A reaction of sulphur(IV) oxoanions which leads to the production of a fluorescent product is that with indophenol blue (Nakamura and Tamura, 1975a):

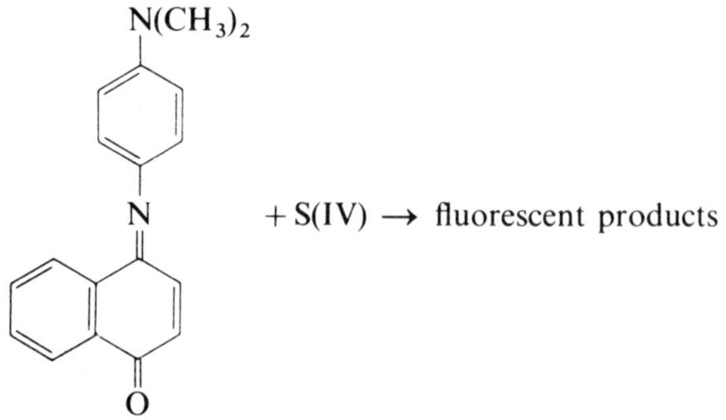

$+ \text{S(IV)} \rightarrow$ fluorescent products

The products are fluorescent in alkaline media, with an excitation wavelength of 340 nm, and emission takes place at 435 nm. The reaction itself is very sensitive to pH, a maximum yield of fluorescent product being obtained in the range pH 3·5–4·0, and falling off sharply outside this range. Since the pH optimum for the reaction is different from that for the emission of fluorescence (pH > 11), it is necessary to introduce a further experimental step by adding sodium hydroxide before measurement. A typical analytical procedure, therefore, involves the mixing of the test solution with pH 3·62 citrate buffer and ethanolic indophenol blue (10^{-4} M), such that the final mixture contains 25 % ethanol. After allowing the reaction to proceed for 30 min at 37 °C, 1 ml sodium hydroxide (2 M) is added and the fluorescence determined. Under these conditions, the analysis of 5×10^{-8} mol sulphur(IV) oxoanion (present at a final concentration equivalent to a 1×10^{-5} M sulphur(IV) oxoanion solution) gives $5·00 \pm 0·07 \times 10^{-8}$ mol and analysis of a solution at twice this concentration gives $9·99 \pm 0·01 \times 10^{-8}$ mol. Calibration graphs of fluorescence intensity, plotted as a function of the amount of sulphur(IV) oxoanion used, are linear up to final equivalent concentrations in the region of 4×10^{-5} M of the species in the solution for fluorimetry. This work is also supported by very extensive data on interference from anions, cations and a range of organic molecules. The effects may be divided into two categories: the inhibition of fluorescence by cations, and the fluorescence yield of anions and organic compounds.

Interference from cations decreases in the order,

$$\text{Mn}^{2+} > \text{Hg}^{+} > \text{Ag}^{+} > \text{Sb}^{3+} > \text{Hg}^{2+} > \text{Bi}^{3+} > \text{Fe}^{3+} > \text{Sn}^{2+} > \text{As}^{3+}$$

at concentrations in the region of 8×10^{-5} M in the final solution, the effect of manganese(II) ion being to reduce the fluorescence by 27 % and that of mercury(II) ion by 6 %. When the concentration of cation is increased by a factor of 100, most metal ions, with the exception of cadmium, show an effect, the suppression of fluorescence by sodium and potassium ions being 0·6 % and 1·6 % respectively.

Dithionite and sulphide ions lead to the formation of fluorescent products, the yield of fluorescence being 40 % and 17 % respectively, calculated on a molar basis with respect to the salt, $NaHSO_3$. Other sulphur oxoanions, including thiosulphate, and the inorganic compounds potassium cyanide and potassium thiocyanate, give little or no fluorescence, the maximum yield being 0·7 % for potassium disulphate ($K_2S_2O_8$). The only significant interference from organic compounds is observed with the reduced form of cysteine, the greatest effect being that of reduced glutathione with a relative fluorescence yield of 8 %. Mercaptoethanol shows a yield of 1 %. Otherwise, with final solutions containing an equivalent of $1·7 \times 10^{-5}$ M organic compound, no interference can be detected from, for example, organic thiosulphonates, disulphides, sulphite esters, formaldehyde hydroxysulphonate, cysteic acid, sulphamates, thiocyanates and isothiocyanates. There is no reported study of the stoichiometry of this reaction but this aspect is of interest since it appears that a mixture of fluorescent products, one with a blue and one with a violet fluorescence, is formed.

2.3.5 Molecular spectroscopic analysis

The use of molecular spectroscopy, electronic and infrared, for the identification and quantitative analysis of sulphur(IV) oxoanions in solution is, perhaps, the simplest available technique for their determination, particularly when no sample work-up is necessary. Colorimetric and fluorimetric techniques involve the conversion of all the sulphur(IV) oxospecies into one form as a final product and, since the sulphur(IV) species in aqueous solution are easily and rapidly interconvertible, the analysis is that of total sulphur(IV) oxospecies. On the other hand, the spectroscopic methods to be described here are carried out on aqueous (or solid) systems which are either in equilibrium or in which the position of equilibrium is only slightly perturbed. They are, therefore, capable of providing data on the concentrations of individual sulphur(IV) oxospecies.

Ultraviolet absorption spectroscopy may be used to detect headspace sulphur dioxide over solutions containing sulphur(IV) oxoanions (Cresser and Isaacson, 1976). In preliminary experiments it was shown that sufficient gas for spectrophotometric determination may be produced when 5 ml hydrochloric acid (6 M) are added to 100 mg sodium sulphite in a 100 ml volumetric flask fitted with a rubber septum, and 0·05–0·2 ml of the gas are withdrawn and placed in a 1 cm silica cell also fitted with a rubber septum. A spectrum of sulphur dioxide obtained by this method is shown in Fig. 2.4. In adapting this principle to the quantitative analysis of sulphur(IV) oxoanions, it must be remembered that the method gives an indication of the activity of $SO_2 . H_2O$ in solution and, since it is well known that sulphur dioxide and oxoanions interact extensively with ionic species in solution, the result of any such analysis would be dependent upon the composition of the medium. Thus, the calibration data would only be valid for the system for which they were obtained. Other important considerations include satisfactory temperature control and, if measurements are to be used for studies of the distribution of activities of sulphur(IV) oxoanions among various species, the amount of sulphur dioxide transferred to the gas phase should be small in comparison with the amount present in solution. Assuming an extinction coefficient of $500 \text{ M}^{-1} \text{cm}^{-1}$ for sulphur dioxide at 276 nm, the amount required to give

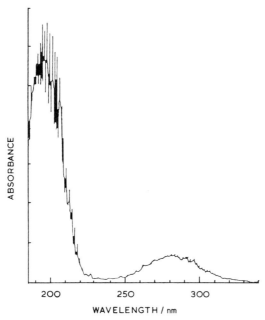

Fig. 2.4. Ultraviolet spectrum of sulphur dioxide gas taken from the headspace above an acidified solution of sulphur(IV) oxoanion (Cresser and Isaacson, 1976). Reproduced with permission from *Talanta*, 1976, **23**, page 886. © 1976 Pergamon Press Ltd.

an absorbance of 0.01 in a 1 cm cell at this wavelength is 2×10^{-5} mol litre^{-1}, corresponding to 1.3 ppm of the gas. If this absorbance can be easily measured, the sensitivity of this method is of the order of 1 ppm in the gas phase and, therefore, the method is potentially very useful. Assuming that Henry's law is obeyed at these concentrations, the concentration required in aqueous solution to exert sufficient vapour pressure is in the region of $4-5 \times 10^{-4}$ M. An enhancement of sensitivity is possible by measuring the absorbance of the gas at 198 nm, where it has an extinction coefficient of 2820 M^{-1} cm^{-1} (reported by Bhatty and Townshend, 1971, for solutions in sulphuric acid, and presumed here to be similar in the gas phase on account of the very weak interaction existing between sulphur dioxide and water).

The direct spectroscopic determination of sulphur(IV) oxoanions in aqueous solutions has been widely used for the study of equilibria between the species and is applicable to simple solutions. Measurement must, however, be made in the range, 240–320 nm, in which a very large number of chromophores absorb. There are, however, large differences in reported values of extinction coefficients for the various sulphur(IV) oxospecies. The extinction coefficient of hydrated sulphur dioxide at 276 nm was reported as $498 \pm 6 \, M^{-1} cm^{-1}$ (Huss and Eckert, 1977) and is preferred to the previously reported values of $608 \, M^{-1} cm^{-1}$ (Ratkowsky and McCarthy, 1962) and $610 \, M^{-1} cm^{-1}$ (Bhatty and Townshend, 1971), and is in good agreement with a value of $500 \, M^{-1} cm^{-1}$ (Scoggins, 1970). The disulphite ion absorbs at 255 nm ($E_{max} = 1980 \, M^{-1} cm^{-1}$) and at this wavelength the contribution from hydrogen sulphite ion ($E = 1 \, M^{-1} cm^{-1}$) may be neglected, that from sulphite ion ($E = 9 \cdot 4 \, M^{-1} cm^{-1}$) is small, whilst dissolved sulphur dioxide ($E = 450 \, M^{-1} cm^{-1}$) interferes significantly (Bourne et al., 1974). In principle, for a mixture containing n components with different (albeit slightly different) spectral properties, it is a simple task to calculate the concentration of each component by simultaneous solution of n linear equations of the form:

$$A_{\lambda_j} = \sum_{i=1}^{n} \varepsilon_{i(\lambda_j)} c_i$$

where an equation is written, in each case, for $j = 1$ to n. A_{λ_j} is the absorbance of the mixture at a wavelength, λ_j, $\varepsilon_{i(\lambda_j)}$ is the extinction coefficient of the pure component, i, at that wavelength and c_i is the concentration of component i in the mixture. The technique is subject to three limitations: first, the precision of the data is stretched as the number of unknowns in the system is increased, a likely practical proposition being the analysis of three components; secondly, the accuracy depends upon that with which extinction coefficients of the individual species are known, and thirdly, the extent to which the component spectra differ will be reflected in the uncertainty of the final result. Eriksen and Lind (1972) demonstrate this technique, with satisfactory results considering the uncertainties in the data, for the analysis of a four-component mixture containing sulphur dioxide, hydrogen sulphite, disulphite and thiosulphate ions.

The most reliable spectrophotometric method for the determination of sulphur(IV) oxoanions in solution and one which gives the greatest resolution is that of Raman spectroscopy, as illustrated by the work of Meyer and his group (Meyer et al., 1980). A Raman spectrum of sulphur dioxide, hydrogen sulphite, disulphite and sulphate ions is shown in Fig. 2.5. Hydrogen sulphate ion shows scattering at $1050 \, cm^{-1}$, whilst

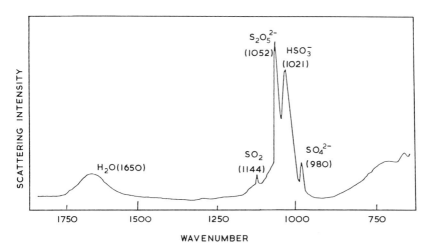

Fig. 2.5. Raman spectrum of a mixture of sulphur oxoanions (Meyer *et al.*, 1980).
Reproduced with permission from *Anal. Chim. Acta*, 1980, **117**, page 310. © 1980
Elsevier Scientific Publishing Co.

sulphite ion may be observed at 933 and 966 cm^{-1}. Detection limits for a
number of sulphur(IV) oxoanions, determined from single scans recorded
at a scan speed of 100 cm^{-1} min^{-1}, are shown in Table 2.4, although
multiple scanning and slower recording would lower the detection limit by a
factor of about ten. It is, of course, not possible to calibrate directly the
procedure for sulphur (IV) oxoanions by dissolving the appropriate
compounds in water since interconversion between the various species
takes place and additional information, in the form of the respective

TABLE 2.4

Detection limits by Raman spectroscopy of sulphur oxospecies (Meyer *et al.*, 1980)

Species	Detection limit (10^{-3} M)	Species	Detection limit (10^{-3} M)
$S_2O_5^{2-}$	0·05	HSO_4^-	4
$S_2O_4^{2-}$	0·8	SO_4^{2-}	7
$S_2O_6^{2-}$	1	$S_4O_6^{2-}$	9
$S_2O_3^{2-}$	3	SO_3^{2-}	10
HSO_3^-	3	SO_2	30

Spectra determined as single scans. Reproduced with permission from *Anal. Chim.
Acta*, 1980, **117**, page 311. © 1980 Elsevier Scientific Publishing Co.

equilibrium constants, under the experimental conditions is required. The relative scattering intensities due to SO_2, HSO_3^-, $S_2O_5^{2-}$ and SO_3^{2-}, at respective wavelengths of 1144, 1021, 1052 and 933/966 cm^{-1}, are estimated as 1·80, 2·00, 11·55 and 2·20 respectively, these values being expressed with respect to the intensity of the water band at 1650 cm^{-1}, which could be used as an internal standard. For thiosulphate, the concentration of which can be reliably measured, a graph of scattered intensity relative to that of water as a function of concentration gave a straight line over the range of concentrations, 0·01 M to saturation.

2.3.6 Molecular emission analysis

When sulphur-containing compounds are burnt in a hydrogen-rich flame, a blue molecular emission takes place from the core of the flame showing a band spectrum originating from S_2 molecules. The usual wavelength for monitoring this emission is chosen as 384 nm. The introduction of too much oxygen leads to the S_2 species being short-lived, presumably as a result of conversion to various oxospecies, and emission is consequently reduced. Because of the important variations in the reducing conditions found in different parts of the flame, the most efficient method of analysing sulphur-containing compounds by molecular emission spectroscopy would require the placing of the sample in the desired part of the flame and the confining of combustion to that region, so that reducing conditions are maintained and the lifetime of the S_2 species is extended. This requirement is well satisfied by a technique named, by Belcher and coworkers (1973), molecular emission cavity analysis, in which the sample is placed in a small cavity cut in a stainless steel rod, which may then be introduced into the flame. The desired emission is observed but the behaviour of the sample in the flame is clearly different from the behaviour of a sample introduced in the conventional way since, although initial addition of oxygen lowers the emission intensity as expected, further addition leads to a restoration of emitted light. As well as acting as a means of measuring the total sulphur content of a sample in the same way as any atomic emission technique is used to determine the total amount of an element present, the cavity technique has one important additional feature. If the cavity is introduced, cold, into the flame, it will proceed to heat up at a reproducible rate and, therefore, if the sample is a mixture of components, these can be selectively volatilised and analysed. The temperatures at which the compounds appear are often lower than their boiling or decomposition points, a result probably due to the reducing nature of the environment. The cavity consists of a hexagonal aperture (5 mm maximum) at the end of a 3 cm long stainless steel Allen screw. One further advantage of this presentation of the sample is the ability to present solid samples without the need for extraction. The instrumental variables in this method of analysis are the composition of the flame with respect to hydrogen, oxygen and a diluent such as nitrogen, the position of the cavity in the flame and the composition of material forming the cavity. Adjustment of these variables allows the technique to be adapted for a wide range of sulphur(IV) oxoanion mixtures, such as SO_3^{2-}/SO_4^{2-}; $SO_3^{2-}/S_2O_8^{2-}$; $S_2O_3^{2-}/SO_4^{2-}$; $S^{2-}/SO_3^{2-}/SO_4^{2-}$ (Al-Abachi et al., 1976) but, although the appearance times of sodium disulphite, sodium sulphite and sulphuric acid are given as 1·7, 2·4 and 8·5 s respectively, indicating that a separation is possible (Belcher et al., 1975), unfortunately no studies of these mixtures have been carried out.

An important consideration is the relationship between the area under the emission curve and the amount of substance analysed. If the rate of production of S_2 emission is determined by the interaction of two sulphur-containing species, then the relationship between intensity and concentration is expected to be of the form,

$$I = kc^2$$

where k is a constant. In practice, the exponent is between 1 and 2, those for sulphite and disulphite being 1·9 and 2·0 respectively, and for sulphuric acid, 1·8 (Belcher et al., 1975). Cations tend to depress the emission but this effect may be removed by the addition, as appropriate, of phosphoric acid, pH 7 phosphate buffer or diammonium hydrogen orthophosphate. The detection limits of sulphite and of sulphate ions are given as 6 and 20 mg/5 μl respectively (Al-Abachi et al., 1976).

Schubert and coworkers (1980) describe the application of this technique to the analysis of solid samples containing the four common forms of inorganic sulphur, that is elemental sulphur, sulphide, sulphite and sulphate ions. A typical result, indicating the high resolution which may be obtained in practice, is shown in Fig. 2.6. The procedure used by these workers is rather more complex than that described previously, requiring a change in the operating conditions during the course of a run. An air supply is turned on, as indicated by the negative spike at 28–29 s, and is required for successful observation of the sulphate component. This report also differs considerably from the previous work in relation to claims of detection limit, which for sulphite ion is stated as 0·2 pg in a 1·5 mg sample. The detection limit for sulphate is 8 ng in a 1·5 mg sample, which is more consistent with the earlier work.

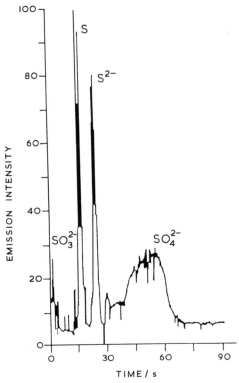

Fig. 2.6.   Separation of four inorganic forms of sulphur by molecular emission cavity analysis (Schubert *et al.*, 1980). Reproduced with permission from *Anal. Chem.*, 1980, **52**, page 964. © 1980 American Chemical Society.

2.3.7 Analysis by atomic absorption spectroscopy

Two indirect methods of analysis will serve to illustrate the use of atomic absorption spectroscopy for the determination of sulphur(IV) oxoanions. Both methods involve the measurement of the concentration of metal ion which, in one case, is rendered soluble in the presence of sulphur(IV) oxoanion, whilst in the other the metal ion is precipitated by sulphate and the residual concentration is determined.

When mercury(II) oxide is shaken with an aqueous solution containing sulphur(IV) oxoanions, some mercury(II) dissolves forming the complex $Hg(SO_3)_2^{2-}$. After removal of excess oxide, the soluble mercury may be determined by atomic absorption at 253·7 nm (Jungreis and Anavi, 1969). If the experiment is carried out using 0·1 g mercury(II) oxide with 10 ml of a solution containing between $1 \cdot 5$–10×10^{-4} M sulphur(IV) oxoanion at pH 11, the calibration graph of absorbance as a function of concentration is linear and passes through the origin.

In contrast to this procedure, the mixing of an oxidised (by means of hydrogen peroxide) solution of sulphur(IV) oxoanions with a standard solution of lead(II) perchlorate gives a good yield of lead(II) sulphate precipitate in the presence of ethanol. If the precipitate is removed by centrifuging, the residual lead may be determined by atomic absorption at $283 \cdot 3$ nm (Rose and Boltz, 1969).

2.3.8 Polarographic analysis

The reduction of sulphur dioxide at the dropping mercury electrode is described by Kolthoff and Miller (1941) and the reaction may quantitatively be accounted for by the transfer of two electrons:

$$SO_2 + 2H^+ + 2e^- \rightarrow H_2SO_2$$

In $0 \cdot 1$ M nitric acid, the half-wave potential is given as $-0 \cdot 37$ V relative to the saturated calomel electrode, and the diffusion current is found to be proportional to the concentration of the sulphur(IV) species in the concentration range 0–$2 \cdot 3 \times 10^{-3}$ M. As pH is raised the value of the half wave potential changes at a rate of $-0 \cdot 0625$ V per unit of pH in the range of pH 3–7 (Cermak and Smutek, 1975). The situation around neutrality appears to be complex since Kolthoff and Miller (1941) observed under these conditions a polarogram with two waves. The first had a half wave potential in the region of $-0 \cdot 67$ V relative to the saturated calomel electrode and, presumably, attributable to the primary reduction process, whilst the second wave, at $-1 \cdot 23$ V, was considered to be due to a further reduction of the product.

Since the wave height is proportional to the amount of sulphur dioxide present, it will rapidly diminish in intensity with increasing pH and, therefore, the most sensitive analysis may be carried out in strongly acidic media. One of the obvious difficulties is that solutions need to be purged of oxygen and, since the normal procedure involves passing nitrogen through the solution, loss of sulphur dioxide is anticipated. A very significant improvement on the use of aqueous solutions is the taking of measurements in dimethylsulphoxide (DMSO) in which the gas has a very high solubility. This solvent has been found to be so effective that it may be used directly for the scrubbing of air samples containing sulphur dioxide. A differential pulse polarographic peak of 16 μA is possible when the concentration of the gas is 1 μM, with a background electrolyte of $0 \cdot 1$ M lithium chloride (Garber and Wilson, 1972). Bruno and coworkers (1979) note that up to 10 % water in the organic solvent does not interfere and they have studied the analysis of samples of aqueous sulphur(IV) oxoanion solutions by this method. The height of the differential pulse polarographic peak is found to be independent of apparent pH in the range pH 2–3, and suitable conditions for the analysis are obtained when the solution in the polarographic cell is acidified with sulphuric acid so that it contains between 10–50 mM acid. Under these conditions the calibration curve is found to be perfectly linear for concentrations in the range 0–20 μM SO_2 in the polarographic cell, and a useful working range is 3–20 μM. Thus, if the polarographic cell contains 10 ml DMSO (0·1 M with respect to background electrolyte, lithium chloride) and 200 μl of 0·5 M sulphuric acid are added, the maximum amount of aqueous sample which may be added is of the order of 900 μl if the 10 % limit of water content is not to be exceeded. Given a practical lower limit of 3 μM sulphur(IV) oxospecies in the polarographic cell, the limit of the analysis is of the order of 40 μM sulphur(IV) oxospecies in the sample.

Indirect polarographic methods for the determination of sulphur(IV) oxoanions have also been reported. The use of mercury(II) chloranilate for the spectrophotometric determination of sulphur(IV) oxoanions has already been considered, but in cases in which solutions contain other absorbing species which interfere with the analysis, an alternative polarographic method of determining the released chloranilic acid is advocated (Humphrey and Laird, 1971). Since the reaction products, chloranilic acid and mercury(II) sulphite, are both reducible, the method has a higher sensitivity than, say, polarographic determination of chloranilic acid released when barium chloranilate is allowed to react with sulphate ion.

Chloride, fluoride and sulphate ions may be determined chemically by reaction with an insoluble metal iodate, releasing IO_3^-, which may be determined either titrimetrically or polarographically. For example, the use of barium iodate for the determination of sulphate ion proceeds according to the following equation:

$$Ba(IO_3)_2 + SO_4^{2-} \rightarrow BaSO_4 + 2IO_3^-$$

The result is that for each monovalent ion the amount of iodate released is equivalent to a reduction involving 6 electrons, and 12 electrons in the case of divalent ions. Thus, the analysis has been through a process of chemical amplification. The analysis of sulphur(IV) oxoanions by such a process depends upon the reaction of mercury(II) iodate with sulphite ion, leading to the formation of elemental mercury:

$$Hg(IO_3)_2 + SO_3^{2-} + H_2O \rightarrow Hg + 2IO_3^- + 2H^+ + SO_4^{2-}$$

The procedure involves shaking mercury(II) iodate with a solution of sulphur(IV) oxoanion in ethanol:water (1:1), filtering and adding concentrated perchloric acid to give an approximate hydrogen ion concentration of 0·12 M. Iodate is then determined by measuring the reduction current at $-0·5$ V relative to the saturated calomel electrode on an instrument set to zero at an applied potential of $+0·1$ V (Humphrey and Sharp, 1976).

2.3.9 Use of ion selective electrodes

There are two types of ion selective electrode which may be used for the direct determination of sulphur(IV) oxoanions in solution. The first is the membrane electrode, which is effectively a gas-sensing probe and relies on the diffusion of the gas through a selectively permeable membrane, followed by a reaction within the electrode to generate a change in pH. The second type of electrode has its active surface in contact with the solution and a chemical reaction takes place at the interface. The characteristics of these sensors will now be considered in detail.

2.3.9.1 Membrane electrode

The essential features of a membrane electrode used to measure sulphur dioxide in solution are shown in Fig. 2.7 (Barnett, 1975). The partial pressure of sulphur dioxide in a dilute solution of the gas is related to its concentration as $SO_2 . H_2O$ through Henry's law:

$$[SO_2 . H_2O] = K_H pSO_2$$

where K_H is Henry's constant. When the membrane electrode is inserted into the solution, molecular sulphur dioxide will pass through the membrane until the partial pressure, or activity, of the gas within the sensor is equal to the partial pressure in the sample. The normal filling solution for the measurement of sulphur dioxide is an aqueous solution of hydrogen sulphite ion and the passage of sulphur dioxide into this solution leads to the formation of an $SO_2 . H_2O/HSO_3^-$ buffer whose pH is given by,

$$pH = pK_a + \log \frac{[HSO_3^-]}{[SO_2 . H_2O]}$$

where pK_a refers to the acid dissociation constant of $SO_2 . H_2O$. The

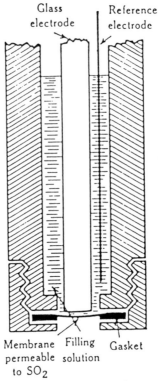

Glass electrode   Reference electrode

Membrane
permeable
to SO$_2$

Filling
solution

Gasket

Fig. 2.7. Essential features of a gas permeable membrane electrode for the determination of sulphur dioxide in aqueous solution (Barnett, 1975). Reproduced from *CSIRO Food Res. Quart.*, 1975, **35**, page 68.

change in pH within the electrode assembly is measured by means of a glass electrode, the response of which is essentially Nernstian, that is,

$$E_m = E_0 + \frac{2 \cdot 303 RT}{F} \log [H^+]$$

where E_m is the measured potential. Therefore, at 25 °C,

$$E_m = E_0 - 0 \cdot 0591 \, pH$$

Therefore, for the combined electrode,

$$E_m = E^* + 0 \cdot 0591 \log [SO_2 . H_2O] - 0 \cdot 0591 \log [HSO_3^-]$$

where E^* is a constant. If the amount of hydrogen sulphite ion present in the filling solution is large and may be regarded as constant, the expression reduces to,

$$E_m = E^{**} + 0{\cdot}0591 \log [SO_2 . H_2O]$$

where E^{**} is another constant. This is the situation within the electrode. Since $[SO_2 . H_2O]$ is linearly related to the pressure of sulphur dioxide inside the cell, and at equilibrium the partial pressure of the gas is equal to that in the sample, then provided that the partial pressure of sulphur dioxide is related to the concentration of dissolved species in the sample as it is in the electrode, the concentration term in the equation is that of $SO_2 . H_2O$ in the sample. If this is not the case, then provided that the ionic environments of the sample and the electrodes are kept constant, the relationship will be obeyed, with any constant terms which arise being incorporated in E^{**}, and the response should still be Nernstian.

Measurements made with a commercial electrode show that the linearity of response is good over a change of three orders of magnitude in sulphur dioxide concentration from 10^{-1} to 10^{-4} M. The detection limit, specified as the sulphur dioxide concentration which increases the potential by 1 mV from that of the electrode immersed in the background electrolyte, can be of the order of 5×10^{-6} M, depending upon the concentration of the filling solution (Bailey and Riley, 1975).

The measurement operation involves a number of mass transfer steps, any of which could be rate determining. If mass transfer of sulphur dioxide within the sample to the membrane can be neglected, the limiting steps are diffusion through the membrane, transport in the internal electrolyte and the response time of the glass electrode. The membrane may be silicone rubber of 0·025 mm thickness, the important parameter for the rate of mass transfer through the membrane being the product, DK, where D is the diffusion coefficient of the diffusing species in the membrane, and K is the partition coefficient between the membrane and the sample (Ross et al., 1973). Very thin electrolyte films may be obtained by stretching the membrane by means of pressure from the glass electrode, thereby reducing transport time. The response time of the glass electrode is generally relatively fast, provided that the electrode is in good condition. Thus, for a decadic increase in concentration from 10^{-5} to 10^{-2} M the response times are of the order of 400, 36 and 30 s for the three steps respectively and, when the concentration of sulphur dioxide is similarly decreased from 10^{-3} to 10^{-5} M, they are of the order of 50 and 150 s respectively (Bailey and Riley, 1975). These response times, though relatively long by some electrochemical standards, are, nevertheless, short when compared with other analytical methods. The only serious interference in the use of this electrode arises from the presence of acetic acid at concentrations greater than 10^{-3} M, which causes an increase in potential. Osmotic effects resulting from large concentration differences between the internal electrolyte and the sample are negligible.

Růžička and Hansen (1974) constructed an electrode which works on the same principle, but used an air gap in place of a membrane. They argue that, 'since the diffusion of gases in air is much more rapid than in solid, aqueous or even porous media, the air gap electrode will exhibit a faster response than any porous membrane electrode'. For sulphur dioxide analysis the electrolyte is 1×10^{-3} M sodium hydrogen sulphite in 20% aqueous ethanol containing a wetting agent, with its pH adjusted to 5·0 with sodium hydroxide (Hansen *et al.*, 1974). The film is applied by allowing the electrode to rest on a sponge soaked with the electrolyte and is renewed between measurements. Although the electrode appeared to function well, the reported slope of the calibration curve is 0·50 pH units per decade change in concentration instead of the expected value of 1 pH unit per decade, and no specific analysis times are stated except that they are of a few minutes. Other developments include modification of the pH-sensing element, such as the use of a moulded antimony in polythene electrode in place of the usual glass one. The advantages of the former include the possibility of the manufacture of 'microelectrodes' and greater control of the electrode surface (Mascini and Cremisini, 1977).

The sulphur dioxide gas-sensing membrane electrode may be used to carry out two types of analysis. First, it may be used at the pH of the sample to determine the amount of free $SO_2 . H_2O$ present. Secondly, pre-treatment of the sample with acid to pH 1 will convert most of the sulphur(IV) oxoanion to sulphur dioxide, which may be determined giving in theory the total sulphur(IV) content of the sample. This, however, is not strictly accurate since the measurement is essentially that of the activity of dissolved sulphur dioxide and need not necessarily correspond to the total sulphur(IV) oxoanion concentration as determined by, say, iodimetric or spectrophotometric methods. Calibration of the electrode is carried out in terms of concentration using dilute solutions of salts of sulphur(IV) oxoanions, and it is important that such a calibration medium is made as similar as possible, with respect to ionic strength, to the medium on which analyses are to be carried out. The presence of certain salts which are known to associate with sulphur dioxide forming weak complexes should also be avoided. The response of the electrode will also be affected by temperature, for two reasons: first, the value of the constant term, $2·303RT/F$, is a function of absolute temperature, its value increasing by over 10 % as the temperature is raised from 10–40 °C; secondly, the effect of temperature on Henry's constant may be significant. In the temperature range 26–43 °C, the measured temperature coefficient for a membrane electrode is of the order of $0.5\,mV\,°C^{-1}$ at a concentration of $10^{-3}\,M\,SO_2$, rendering it relatively insensitive to temperature in normal use (Bailey and Riley, 1975).

2.3.9.2 Solid state electrode
A solid state electrode sensitive to sulphur(IV) oxoanions may be prepared by compressing a mixture of mercury(II) sulphide and mercury(I) chloride in a potassium bromide press. The measuring system comprises the resulting disc mounted at the end of a glass tube, an internal contact to its surface being made with a small amount of mercury, and measurements of emf are made with reference to a saturated calomel electrode (Tseng and Gutknecht, 1976). The response is independent of pH over the range pH 3·3–8·5 at a sulphur(IV) oxoanion concentration of 10^{-3} M, and at pH 5–6 the measured response is linear when emf is plotted as a function of the logarithm of concentration over the range 0.05–10^{-5} M, with a slope of 50 mV per decade. The reaction taking place at the surface of the electrode is,

$$Hg_2Cl_2 + 2SO_3^{2-} \rightleftharpoons Hg + Hg(SO_3)_2^{2-} + 2Cl^-$$

It is estimated that the equilibrium constant for this reaction is in the region of 10^4 M. With such a large driving force, each sulphite ion will displace one chloride ion. The composition of this electrode is the same as that of the one which may be used for the determination of chloride ion, its operation being dependent on the solubility of mercury(I) chloride,

$$Hg_2Cl_2 \rightleftharpoons Hg_2^{2+} + 2Cl^-$$

Thus, the sulphur(IV) oxoanion sensitive application operates by the production of a quantitative amount of chloride ion which is then determined by the electrode, and the idea is confirmed by its identical response towards sulphur(IV) oxoanion and chloride ion over the range of concentrations 10^{-2}–10^{-5} M. The active component is, therefore, mercury(I) chloride, the mercury(II) sulphide serving as a binder. Interference from other anions has been observed, the selectivity coefficients, K_{sel}, relative to sulphur(IV) oxoanion being:

	Cl^-	Br^-	I^-	SCN^-	NO_3^-	ClO_4^-
K_{sel}	1	$10^{2·1}$	$10^{5·0}$	$10^{0·8}$	$\sim 10^{-3}$	$\sim 10^{-3}$

Response times are of the order of 30 s.

Mohan and Rechnitz (1973) describe a solid state electrode which is almost equally sensitive to sulphite and sulphate ions. The active surface is prepared by pressing a disc from a mixture of equimolar amounts of silver(II) sulphide, lead(II) sulphide and lead(II) sulphate, containing 5 mol % copper(I) sulphide. For sulphate ion, graphs of emf as a function of the logarithm of concentration are linear over at least three decades of concentration, the slope being 28 mV/decade. For sulphite ion, the slope is 26 mV/decade. The sensitivity with respect to other anions parallels, in general, the solubility product of the lead salt of the anion in question. Anions which form lead salts which are less soluble than lead(II) sulphate interfere, whereas those which form salts which are more soluble do not. Interference from iodide is often significant but not reproducible.

2.3.9.3 Indirect potentiometric measurement

The principle of indirect potentiometric measurement of sulphur(IV) oxoanions is the reaction of the sample with an excess of a reagent and subsequent use of an ion selective electrode to determine either the residual reagent or a reaction product. Such a method has been devised for the determination of atmospheric sulphur dioxide by means of an iodide selective membrane electrode (Mascini and Muratori, 1975). The reaction involves the absorption of sulphur dioxide in a solution containing 10^{-4} M potassium iodide in 0·1 M acetate buffer at pH 4·7, saturated with iodine. The potassium iodide is required for long term stability. An initial potential of 150 mV increases to around 180 mV when the concentration of liberated iodide ion increases to $1·4 \times 10^{-2}$ M.

So far, all the methods of electrochemical analysis have involved the establishment of equilibrium conditions, and the measurement is that of a steady potential. When measurements are made under non-equilibrium conditions in circumstances in which the rate of change in potential of the sensing electrode is related to the concentration of the species under analysis, the technique is described as that of chronopotentiometry. In the case of gas-sensing membrane electrodes, the rate of diffusion of gas across the membrane will be determined by the concentration gradient between the external and internal solutions. Thus, for a given internal solution, the initial rate of change of potential will be proportional to the logarithm of the concentration of the gas in the external medium. This idea could be applied to the use of an electrode in which the internal electrolyte is renewed for each measurement, as in the case of the 'air-gap' probe described above. Sekerka and Lechner (1978) describe an indirect method of sulphur dioxide analysis based upon this principle. A silver(I) selective electrode is fitted with a suitable gas-permeable membrane, the sensing element being immersed in an internal electrolyte containing silver nitrate and potassium bromate. When diffusion of sulphur dioxide between the external solution and the internal electrolyte takes place, the following reactions occur:

$$3SO_3^{2-} + BrO_3^- \rightarrow Br^- + 3SO_4^{2-}$$

$$AgNO_3 + Br^- \rightarrow AgBr\downarrow + NO_3^-$$

The change in silver ion concentration as it is precipitated as silver bromide is, thus, measured. Calibration curves of initial rate of change of potential as a function of concentration are, as expected, non-linear, but apparently offer a reliable means of determination of sulphur dioxide in aqueous solutions to a detection limit of 5×10^{-7} M, when an initial rate of change of potential of 2 mV in 30 min may be obtained.

2.3.10 Piezoelectric effect

A large number of ingenious specific detectors have been devised for the instrumental measurement of atmospheric sulphur dioxide levels and, whereas most have only novelty value in the context of this book, one particular method, that of the use of piezoelectric detectors, will be considered here.

The piezoelectric material used for these purposes is quartz, which acts as a mechanical resonator of very high stability and of very low loss at frequencies of the order of MHz. In a practical application the crystal forms part of an oscillator circuit which allows the vibrations to be sustained and the frequency of the oscillation may be determined with sufficient accuracy by means of a digital frequency meter. The frequency of oscillation is set by the thickness and 'type of cut' of the quartz crystal but will be affected by temperature, the loading of the crystal surfaces and the electrical loading by the oscillator. If the temperature and the circuit conditions are kept constant, the change in frequency of oscillation will be proportional to the weight of substance adsorbed at the surfaces of the crystal (Sauerbrey, 1959). By coating the electrode surface of the crystal with a substance which will selectively adsorb a particular gas, the concentration of the latter may be determined (King, 1964).

For the analysis of sulphur dioxide in gas mixtures suitable coatings are amines, the compound being applied to the surfaces in a volatile solvent, which is allowed to evaporate leaving a deposit. The loading due to this amine gives a reduction in resonant frequency of about 5 kHz for an oscillator frequency in the region of 9 MHz. The detection system is sensitive to sulphur dioxide in air in amounts of the order of 1 part in 10^9 (Karmarkar and Guilbault, 1974). The detector response time is stated to be a few seconds, complete reversibility of response being observed over approximately 5 min. Although there is no interference from air or carbon dioxide, the greatest problem associated with this procedure is its response to water vapour. Water is physically adsorbed onto the surfaces of the crystal and gives rise to large frequency shifts and the detector is, therefore, unsuitable for measurements of equilibrium vapour pressures of sulphur dioxide over aqueous solutions. In its basic form it may, however, be useful for the detection of the gas where prior separation from water, for example by gas chromatography, is carried out. However, it is feasible to use such a detector in conjunction with a gas-permeable, hydrophobic membrane in much the same way as a membrane is used to isolate the internal electrolyte from the external medium in analyses using gas-membrane electrodes (Webber and Guilbault, 1977).

2.4   SEPARATION OF SULPHUR(IV) OXOSPECIES

The majority of chromatographic separations involving sulphur(IV) oxospecies have been reported for mixtures with other inorganic species. These will be reviewed here in order to illustrate the variety of chromatographic behaviour shown by sulphur(IV) oxospecies and simple inorganic products formed on oxidation of the anion. The separation of sulphur(IV) oxoanions from related organic compounds such as sulphonates will be considered in Section 2.7.3.

2.4.1 Paper and thin layer chromatography
In any paper or thin layer chromatographic separation a considerable area of the sample is exposed to gases present in the atmosphere of the chromatography tank and dissolved in the chromatography solvent. This exposure is likely to be for times in excess of 1 h and, if the atmosphere contains oxygen, there is a risk of oxidation occurring unless the composition of the solvent is such that it acts as an antioxidant. This would be true for mixtures containing large quantities of alcohols or ketones. Once the chromatogram has been developed, it is usually left to dry before spraying to identify spots. The extent to which sulphur(IV) oxoanions undergo oxidation when adsorbed onto paper in the dry or partially dry state is not known. However, the specific tests for sulphur(IV) oxoanions are found to be positive under these conditions and, according to published

work, the sulphur(IV) state may be maintained without special precautions. However, in the case of an unknown separation, it would be wise to check for the presence of sulphur(V) and sulphur(VI) in the position expected for the sulphur(IV) species, if the latter does not respond to a specific test.

A single paper chromatographic separation is capable of resolving the common inorganic anions, sulphite ion being particularly well resolved from the others. The best solvent is a mixture of ethanol:pyridine:water: 0·880 ammonia solution (15:5:4:1) and results in the sulphite ion running at R_f 0·20, with its nearest neighbours, arsenite and bromate, at R_f 0·12 and 0·41 respectively. Sulphate is found at R_f 0·07 (Elbeih and Abou-Elnaga, 1960). A general spray reagent for inorganic anions consists of a mixture of ammoniacal silver nitrate, naphthylamine-5-sulphonic acid and fluorescein which gives, with sulphite ion, a pale brown coloration, appearing dark grey under ultraviolet light. Confirmation of the presence of this ion may be obtained by spraying with a solution containing kojic acid and o-coumaric acid in ethanol, giving a medium grey spot under ultraviolet light. Once dry, the paper may be sprayed with ammoniacal silver nitrate, which gives a yellowish-white, slightly fluorescent spot under ultraviolet light whilst the paper is still wet and a brown spot in daylight when dry. An alternative colour reaction has been described by Mitchell and Waring (1978) and involves the spraying of dry chromatograms with a saturated solution of barium chloride and potassium permanganate. Paper chromatograms are oversprayed with 4 M hydrochloric acid and left to dry at room temperature. The colours which are produced are stable for up to 12 months. Sulphite ion appears as a khaki coloured spot on a dark purple background after the first spraying, and as a pale violet spot on a white background after the acid spray. Other spray reagents include tetrazotised o-dianisidine, stabilised with zinc chloride, which gives a bright yellow dye (Gringras and Sjoestedt, 1961), and dichlorofluorescein which gives a pink coloration (Haworth and Ziegert, 1968). Sulphite, sulphate and dithionate ions and also polythionate ions may be separated by means of paper chromatography using $tert$-butanol:acetone:water (3:21:5) containing 0·5% potassium acetate (Pollard $et\ al.$, 1964). When a paper chromatogram is eluted with this solvent, two phases are formed on the paper by frontal analysis. The leading phase contains free acetic acid, whilst the rear phase contains a high concentration of potassium ions. If a sample is applied to the paper whilst dry and then eluted, thionates are converted to the acid form when the acid front reaches the spot. However, if the amount of compound in the spot exceeds the amount of hydrogen ion available, then only partial conversion of the acid form will take place and the component will run as two spots, one as the acid form and one as the salt form. The recommended procedure for the use of this solvent system is to apply the mixture to the wet paper such that it is introduced into the rear phase where the buffer capacity is high. The phase boundary may be shown up by incorporating 1 % phenol red in the solvent. Under these conditions sulphate, dithionate and sulphite ions have mobilities of 0·02, 0·38 and 0·77 respectively, relative to tetrathionate ion which is taken as a standard (mobility = 1·00). The hydroxysulphonate of formaldehyde runs with a mobility of 0·71 in this solvent system.

Inorganic anions may be separated by means of thin layer chromatography on alumina, silica and cellulose. Although many anions run well on alumina, the mobility of sulphite ion is low and silica gel is recommended with butanol:water:pyridine:ammonia:acetone (8:12:4:1:8) as solvent, when the anion runs at the solvent front (Hashmi and Chughtai, 1968). Circular thin layer chromatography, using this system and visualising the sulphite ring by means of a mixture of sodium nitroprusside, zinc sulphate and potassium ferrocyanide, offers a semiquantitative technique for sulphur(IV) oxoanions with a sensitivity of 1 μg and an accuracy of well in excess of $\pm 5 \%$ when the developed colour is compared with standards. On microcrystalline cellulose the mobility of sulphite ion is low (R_f 0·05) in acetone:water:ethylacetoacetate (6:1:3), and the ion runs at R_f 0·42 in butanol:2-propanol:1·5 M ammonia solution (1:2:3) (Haworth and Ziegert, 1968).

Sulphate ion shows up as a bright yellow spot in daylight and ultraviolet light with the general reagent suggested by Elbeih and Abou-Elnaga (1960), and may be confirmed by a violet colour under ultraviolet light when chromatograms are treated with naphthylamine-5-sulphonic acid in 50 % ethanol. Sulphate and dithionate both give a blue to violet-blue coloration on a brown background when treated with benzidine followed by potassium permanganate (Garnier and Duval, 1959). It is interesting that this colour reaction is also stated for disulphite ion, whereas sulphite ion gives a white or yellowish-white colour under the same conditions.

Paper and cellulose thin layer chromatography of inorganic anions have also been reported by Okumura and Nishikawa (1976). An interesting feature of this work is the detection method using a spray reagent containing aluminium(III)–morin complex. This compound is fluorescent, the components on the chromatograms quenching the fluorescence if they can form complexes themselves with aluminium(III), thereby displacing the morin. The quenching is described as 'strong' for sulphite and sulphate ions but, overall, the test is not specific, a positive reaction being obtained with most anions.

2.4.2 Electrophoretic separations

It is expected that sulphur(IV) oxoanions will be amenable to electrophoretic separation, the mobility being a function of pH, and a small number of investigations involving these ions have been reported. As in the case of paper chromatographic separations, oxidation of the sample is theoretically possible and, perhaps, more likely, since aqueous solutions containing only inorganic ions may be used as the electrolyte. Sulphur(IV) oxoanion migration at pH 7 (Wood, 1955) and in 0·1 M sodium hydroxide (Grassini and Lederer, 1959) has been described, but the most comprehensive data are provided by Vepřek-Šiška and Eckschlager (1965), who measured the mobilities of sulphur(IV), sulphur(V) and sulphur(VI) oxoanions as a function of pH. Their results are summarised in Table 2.5, from which it is evident that the oxoanions may be separated at any of the pH values studied, although a large excess of sulphite ion interferes with the detection of sulphate ion. The data, however, show an unexplained feature in the low, pH independent mobility of disulphite ion when compared with that of sulphite ion. In alkaline solution, the former is expected to be hydrolysed and converted to the latter. A possible reason for this effect is that the pH at the disulphite spot on the electrophoretogram was modified by the presence of the component. The pH of an aqueous solution containing hydrogen sulphite ion may be in the region pH 4–5 and the sample will act as a buffer, tending to prevent pH rise during the

TABLE 2.5
Electrophoretic mobility of sulphur(IV), sulphur(V) and sulphur(VI) oxoanions as a function of pH in Britton–Robinson buffer on Whatman No. 1 paper

Ion	Electrophoretic mobility at given pH					
	4·1	6·1	7·2	8·0	9·1	11·6
SO_3^{2-}			32·5	34·0	35·8	40·0
SO_4^{2-}	33·7	36·2	38·8	39·6	41·2	45·0
$S_2O_5^{2-}$		26·6	26·6	26·5	26·8	
$S_2O_6^{2-}$	42·8	43·0	45·7	47·2	54·0	56·5

Mobilities expressed as mm/100 V/h for a paper length of 30 cm (Vepřek-Šiška and Eckschlager, 1965). Reproduced with permission from *Coll. Czech. Chem. Comm.*, 1965, **30**, page 2547. © 1965 Academia.

experiment. At pH > 8 sulphite, sulphate and dithionate ions will have no buffering capacity and, for practical purposes, the actual pH will be that of the buffer.

2.4.3 Ion exchange separations

Inorganic anions may be separated by means of anion exchange chromatography, the separations being based upon differences in ion exchange affinities or in acid strength. The ion exchange affinity of an anion is defined by considering the exchange of a species, B^{n-}, present in solution, with the species A^{m-}, which is bound to the resin, thus:

$$R(A)_n + mB^{n-} \rightleftharpoons R(B)_m + nA^{m-}$$

If it is assumed that the process is in equilibrium, the equilibrium constant for the reaction, which is also referred to as the selectivity coefficient, is defined by,

$$K_A^B = \frac{[R-B]^m[A^{m-}]^n}{[R-A]^n[B^{n-}]^m}$$

If the anions, A and B, have equal charge, a selectivity coefficient $K_A^B > 1$ implies that the resin has a greater affinity for ion B than for ion A, a value of 1 denotes no selectivity, whilst for $K_A^B < 1$ the order of affinities is reversed (Samuelson, 1963). In exchanges involving anions of different charges the situation is more complicated, since the units of K are those of concentration, and it is necessary to specify actual concentrations, particularly the concentration in the resin phase. This is conveniently expressed as the mole fraction or in terms of equivalents of the ion in question. The relationship between this selectivity coefficient and the thermodynamic equilibrium constant, K_{therm},

$$K_{therm} = \frac{(a_{R\ B})^m (a_{A^m})^n}{(a_{R\ A})^n (a_{B^n})^m}$$

is the product of the activity coefficients of the species concerned. Since it is not possible to predict the activity coefficients of resin-bound anions which are present in this phase at very high concentrations, the selectivity coefficients are taken as empirical parameters which, even for the exchange of ions of equal charge, are expected to be concentration dependent. Since ion adsorption and desorption requires some solvation and desolvation to occur, the size and charge on the ion will help to determine the tendency of an ion to exchange. Large ions, which are relatively poorly solvated, are expected to show a higher affinity for the resin than small, more highly solvated ions. Van der Waals' interaction forces between certain ions, such as large organic ions, and the resin may contribute considerably to the selectivity coefficient. When this effect operates, ions which contain groups similar to those present in the resin matrix, for example ions containing aromatic groups and polystyrene resins, are preferentially taken up by the resin.

Anions may be separated using strongly basic ion exchange resins, which are fully ionised owing to the presence of the quarternary ammonium functional group either as resin-bonded trimethylamine or N,N-dimethylethanolamine:

$$-CH_2-CH- \qquad\qquad -CH_2-CH-$$

$$CH_2N^+(CH_3)_3Cl^- \qquad\qquad CH_2N^+(CH_3)_2(CH_2CH_2OH)Cl^-$$

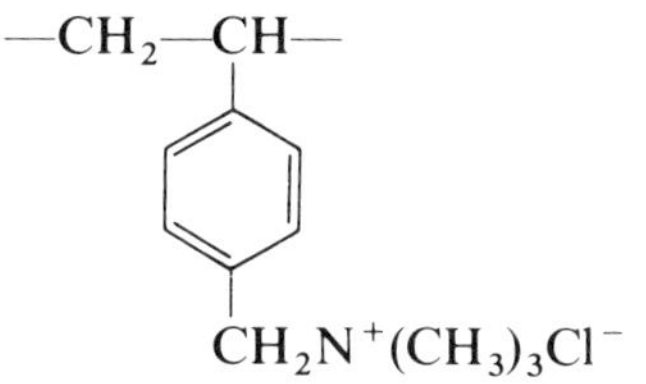

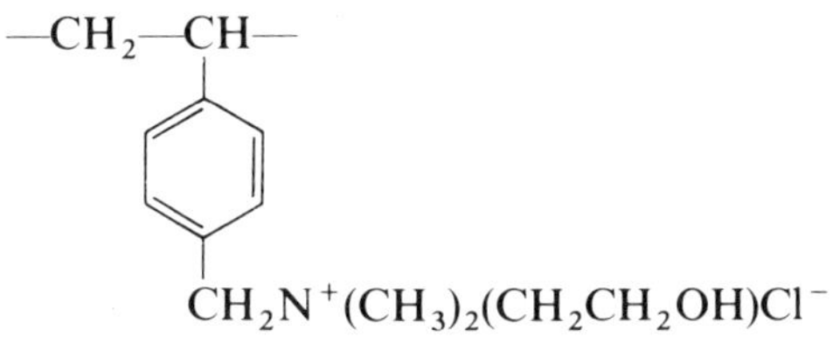

A combination of experimental studies of the affinities of anions for these two types of resin suggests that a series of increasing selectivity coefficient is as follows (Peterson, 1953):

$$NH_2CH_2COO^- < F^- < CH_3COO^- < HCOO^- < IO_3^- < ClCH_2COO^-$$
$$< H_2PO_4^- < HCO_3^- < BrO_3^- < Cl^- < CN^- < HSO_3^- < NO_2^-$$
$$< Cl_2CHCOO^- < F_3CCOO^- < Br^- < NO_3^- < PhSO_3^- < HSO_4^-$$
$$< PhO^- < I^- < p\text{-}CH_3PhSO_3^- < Cl_3CCOO^- < SCN^- < ClO_4^-$$
$$< salicylate < C_6H_3Cl_2O^- < \beta\text{-naphthalenesulphonate}$$

When a strongly basic ion exchanger is exposed to a medium containing a mixture of sulphite and hydrogen sulphite ions, the sulphite ion is apparently bound preferentially (Shikanova *et al.*, 1977). The selectivity coefficient of sulphite ion places it around hydrogen sulphate ion in the order of selectivity shown above (Gürtler *et al.*, 1968). The position of hydroxide ion in this series depends upon the type of resin used, this ion having the lowest selectivity coefficient for the trimethylammonium type resin and a value between those of hydrogen carbonate and bromate ions in the case of the N,N-dimethylethanolamine resin.

Clearly, there are a very large number of counter ion/eluant systems which are possible for the desorption of sulphur(IV) oxoanions from strongly basic anion exchangers, the choice being affected by the composition of the mixture and also by the detection system. Examples of

TABLE 2.6

Systems for ion exchange separations of sulphur oxoanion mixtures which contain sulphur(IV) oxoanion

Resin	Eluant	Mixture	Reference
Dowex 1 2% crosslinked	0·1 M NaNO$_3$ + NH$_3$ (pH adjusted to 9·7)	SO$_3^{2-}$/SO$_4^{2-}$	Iguchi, 1958b
	As above plus 30% (v/v) acetone	SO$_3^{2-}$/S^{2-}	
	As for SO$_3^{2-}$/S$_2^-$ followed by 0·1 M NaNO$_3$(SO$_4^{2-}$ eluted) and by 1 M NaNO$_3$ for S$_2$O$_3^{2-}$	SO$_3^{2-}$/S^{2-}/SO$_4^{2-}$/S$_2$O$_3^{2-}$	
De-Acidite FF 2% crosslinked	2 M potassium hydrogen phthalate	SO$_3^{2-}$, S$_2$O$_3^{2-}$	Pollard et al., 1964
Dowex 2, low surface capacity specially prepared; followed by Dowex 50W-X8 H$^+$	0·005, 0·01, 0·015 M sodium phenate	Inorganic anions including SO$_3^{2-}$, SO$_4^{2-}$ and also a range of carboxylic acids	Small et al., 1975

ion exchange systems which allow separation of sulphur(IV) oxoanions from other anions are shown in Table 2.6. Strongly basic ion exchange resins also have a very high affinity for polythionates, the conditions required for elution of dithionate, trithionate, tetrathionate and penta- thionate from a Dowex 1 (2% crosslinked) column being 1 M, 3 M, 6 M and 9 M hydrochloric acid respectively (Iguchi, 1958a).

The last entry in Table 2.6 is interesting since the solvent system was designed for conductimetric detection of column effluent. The main separation is carried out by anion exchange, the choice of sodium phenate as the eluant being to allow the possibility of subsequent conversion to phenol in a 'stripper' column containing a cation exchange resin in the H^+ form. Thus, the effluent will have a low conductivity, since phenol is a weak acid, making possible the detection of the more strongly acidic anions by conductivity measurements. Although sodium hydroxide would appear to be an equally good eluant, it has a very low displacement potential, indicated by its selectivity coefficient, and leads to extensive tailing of the elution curves of the more tightly bound ions (Small et al., 1975). Although the order in which ions are eluted is generally that expected for anion exchange separations, some anomalies arise, particularly for anions which are partially dissociated in the acid form. This is attributable to an ion exclusion effect based upon the Donnan membrane equilibrium between the hydrogen ions of the solute and the fixed hydrogen ions of the resin. Whilst the presence of this effect leads to the recommendation that the volume of the stripper bed should be as small as possible, ion exclusion may be used as a chromatographic technique in its own right (Wheaton and Bauman, 1953). Thus, it is possible to separate acids on a cation exchange resin, the pK_a, solubility and molecular weight being important factors in chromatographic behaviour under these conditions (Harlow and Morman, 1964). The retention data correspond to the equation:

$$V_R = V_o + K_d V_i$$

in which V_R is the retention volume, V_o is the column void volume, V_i is the inner volume and K_d is the distribution coefficient of the acid in question (Tanaka et al., 1979). This equation is the same as that for gel chromatography based upon the steric exclusion effect (Saunders and Pecsok, 1968).

Ion exchangers are also available with the functional group, diethyl- aminoethyl (DEAE), attached to cellulose (powder or in paper form), a crosslinked dextran network or porous silica for HPLC. Sulphur(IV) oxoanions are readily adsorbed on to columns of DEAE media and may be eluted and separated from sulphate ion by means of a linear 0–0·05 M sulphuric acid gradient (Panahi, 1982). As an alternative to commercially manufactured ion exchange papers, suitable media may be prepared by depositing high molecular weight amines, such as tri-*n*-octylamine, onto Whatman No. 1 paper (Przeszlakowski and Kocjan, 1977). Such papers are capable of separating a wide range of inorganic anions when eluted with either dilute acids (hydrochloric or nitric) or their sodium salts, but the only relevant system to the applications described here is the separation of sulphite, sulphide and thiosulphate ions using 0·5 M sodium chloride as solvent, when the respective R_f values are 0·04, 0·26 and 0·40. Although this technique is that of ion exchange with the stationary phase being a film-like exchanger, it is strictly a development of liquid–liquid extraction processes involving acids and high molecular weight amines and should, perhaps, be better termed 'extraction chromatography'.

2.4.4 Analysis by gas chromatography

Gas chromatography may be used to determine sulphur dioxide in the headspace above solutions of sulphur(IV) oxospecies, thereby allowing direct measurement of the activity of the gas in solution. If solutions are acidified to pH < 1, the method allows, at least in principle, the analysis of total sulphur(IV) oxospecies to the same extent as do other headspace methods described previously, but using a very specific detection technique. This requires a chromatographic procedure capable of separating sulphur dioxide from at least the components of air and from water vapour, if the gas chromatography detector is one of low specificity. Extensive data on the separation of mixtures containing carbon dioxide, nitrous oxide, nitrogen dioxide, hydrogen sulphide, sulphur dioxide, chlorine, hydrogen chloride and ammonia have been compiled by Bethea and Meador (1969) and the phases, with their respective supports, capable of separating sulphur dioxide from air and carbon dioxide with good peak shape are shown in Table 2.7.

Whilst thermal conductivity detectors are excellent non-specific detectors for the analysis of gas mixtures, they lack sensitivity. Unfortunately, the flame ionisation detector, used very extensively for the detection of organic compounds in gas chromatography effluent, has a negligible response to inorganic gases under normal operating conditions. Alternative means of detection of sulphur dioxide include the use of the electron capture detector which, though more sensitive than thermal conductivity detectors, still leaves much to be desired and there is the risk of interference from strongly capturing impurities. The gas may be effectively

TABLE 2.7

Gas chromatographic phases and supports useful for separating sulphur dioxide from air and carbon dioxide

Liquid phase	Solid support
Silicone oil SF-96 Triacetin Di-n-decyl phthalate Arochlor 1232 QF-1 (FS-1265) Fluoro Silicone XE60	Chromosorb T
Fluorolube Grease HG 1200 Kel F Grease F 90 Halocarbon 11-14 Silicone oil DC 200	Chromosorb W, AW, DMCS
Orthophosphoric acid	Porapak Q
None	Porapak I
Di-n-ethyl-hexyl adipate	Chromosorb W, AW

Loading of liquid phase is 10 % for all except o-phosphoric acid, which has a loading of 3 %. Temperature is 32 °C except for columns with o-phosphoric acid and with di-n-ethyl-hexyl adipate, when it is 100 °C and 50 °C respectively (Bethea and Meador, 1969). Reproduced from the *Journal of Chromatographic Science* by permission of Preston Publications, Inc.

detected using a dielectric constant measuring cell (Winefordner *et al.*, 1965) which has a sensitivity of 3×10^{-10} g SO_2 s^{-1}, or by using a flame photometric detector (Hamano *et al.*, 1979). The piezoelectric detector described in Section 2.3.10 also offers a convenient solution, but perhaps the simplest approach is to consider the modifications to the well known flame ionisation detector which enhance its sensitivity towards inorganic gases. This may be achieved by adding into the flame a continuous supply of methane or a fluorinated hydrocarbon. In the case of methane addition, a detection limit of 7×10^{-8} g SO_2 s^{-1} is possible, but the main limitation to improvement is the high noise level of the system (Russev *et al.*, 1976). An interesting development, reported by the Government Chemist's Group (Russev *et al.*, 1976), is the use of a hydrogen-rich flame in an atmosphere of oxygen instead of air. The data shown plotted in Fig. 2.8 illustrate the large changes in the response for nitric oxide when the flow rate of carrier gas (nitrogen) and the hydrogen and oxygen are altered, the response towards sulphur dioxide being reported as the same. The response is independent of nitrogen flow rate at low flow rates and the detection limit towards sulphur

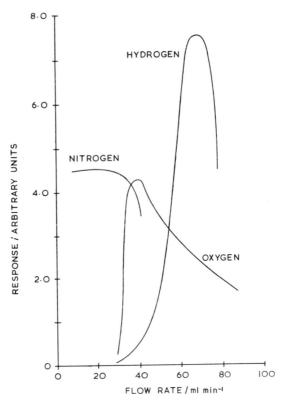

Fig. 2.8. Response of the flame ionisation detector towards nitric oxide and sulphur dioxide as a function of the flow rate of nitrogen, hydrogen and oxygen (Russev *et al.*, 1976). Reproduced with permission from *J. Chromatog.*, 1976, **119**, page 463. © 1976 Elsevier Scientific Publishing Co.

dioxide under optimum conditions is $4 \times 10^{-10} \mathrm{g\,s^{-1}}$, two orders of magnitude lower than the figure previously reported. The response to oxygen and carbon dioxide is two orders of magnitude lower than for sulphur dioxide. The technique is convenient but limited to some extent by the need for careful control of gas flow rate in order to ensure that the instrument is operating at the point of zero slope on each of the curves shown in Fig. 2.8. Otherwise the response is likely to become unstable.

The analysis of a simple inorganic mixture consisting of a solidified melt of a small quantity of sodium sulphide and sulphite in a medium of alkali metal carbonate has been described by Birk and coworkers (1970). The sample is placed in a reaction vessel of 80 ml capacity, the vessel is evacuated and 1 ml sulphuric acid (75 %) is added. The mixture is heated for 1 min at 20 °C before sampling 5 ml of the headspace for analysis. Thermal conductivity detector response is found to be a linear function of the weight of alkali metal sulphite present over the range 0–20 mg.

2.5 QUALITATIVE ANALYSIS IN FOODS

The method of qualitative analysis of sulphur dioxide in foods which is recommended by the Association of Official Analytical Chemists (AOAC) involves reduction of the sample with sulphur-free zinc in hydrochloric acid and detection, using lead acetate paper, of the hydrogen sulphide evolved by decomposition of the dithionite ion so formed (Horowitz, 1975). Traces of metallic sulphides occasionally present in vegetables also give positive results. Nishima and Matsumoto (1969) suggest the use of potassium iodate/starch paper in preference to zinc, although compounds such as methylthiol, hydrolysates of sinigrin, as well as hydrogen sulphide reduce iodate ion and, therefore, interfere.

A specific qualitative test based on the decolorisation of Malachite Green by sulphur(IV) oxoanions is recommended by AOAC for identifying the additive in meats. The procedure simply involves the mixing of 3·5 g of ground meat, on a white surface impervious to moisture, with 0·5 ml of 0·02 % aqueous Malachite Green. After a few minutes normal meat appears blue-green in colour whilst the reagent is decolorised in sulphited meat.

2.6 QUANTITATIVE ANALYSIS IN FOODS

2.6.1 Sample size
The quality of any analysis of sulphur dioxide in food is limited by the chemistry of the method and by the nature of the sampling operation. Assuming that the chemical deficiencies of any method may be assessed by studying well defined systems, the second limitation becomes dominant for well founded analytical methods. One important factor which may contribute significantly to the quality of the result is the size of sample taken in relation to the heterogeneity of the product. This is particularly true if two methods, one involving a large sample and another, a more sensitive method, requiring a small sample are compared. Thus, if one method of analysis which is relatively insensitive requires a given weight of a food and

another method which is more sensitive requires only one hundredth of this amount, the standard deviation calculated from repeated determinations in the latter case will be ten times greater than that of the former, assuming that both sets of data obey a normal distribution.

When it is desired to know accurately the amount of the additive present in the food, the choice of approach is between repeated analyses using a small scale sampling method and the technique requiring a larger sample, the final decision depending upon the relative merits of the two methods with regard to time, resources and the requirement of certain skills. However, an obvious solution which eliminates, at least in theory, a large part of the small scale sampling problem is to homogenise a larger amount of the product and to use a representative sample of this homogenate. Unfortunately, some foods, such as dehydrated fruits, do not lend themselves to satisfactory homogenisation on account of texture. It is possible to grind dehydrated vegetables in, say, a mortar and pestle. Although this is probably the best way of obtaining as uniform a sample as possible, it is not wholly satisfactory since it will lead to the production of a range of particle sizes and may lead to the detachment of crystals of salts of sulphurous acid as a finely divided solid.

It will now be seen that there are available numerous methods for the analysis of the additive in any particular food, requiring a variety of sample sizes ranging over at least an order of magnitude. The interchange of these methods for quality control purposes must, therefore, be subject to an awareness of the sampling problems. There will, of course, come a point beyond which the sample size will become effectively unimportant in determining the standard deviation of the result, and the limiting value so reached will then reflect the variability of the analytical procedure itself.

There are no reported distributions of sulphur dioxide within commercially prepared foods and it is, therefore, not possible to shed light on the type of distribution function expected in practice. Any attempt to analyse the effect of varying sample size on the experimental data is critically dependent upon the form of the distribution curve, which may be skewed, and its constancy within a process over a period of time is essential for continued application. The use of sampling for purposes of process control is beyond the scope of this book and is discussed fully by Weatherill (1969).

2.6.2 Sample stability
A well known effect in the sampling of foods for sulphur dioxide is the time dependent loss of the additive observed when samples are stored. These

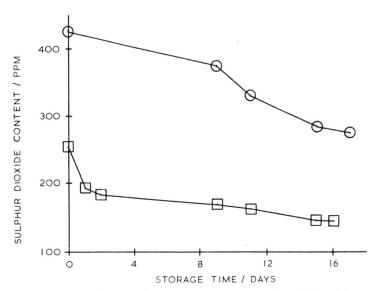

Fig. 2.9. The effect of storage time, at room temperature in sealed containers, on the analysis of sulphur dioxide in dried peas ⊙ and in lemon juice ⊡ (Jennings *et al.*, 1978).

changes take place over time periods of the order of days and are expected to be a result of gaseous loss of sulphur dioxide and chemical reaction through the action of sulphur(IV) oxoanions as a food preservative leading to the irreversible combination of the additive with food components and through oxidation to sulphate ion. Once samples are exposed to air, the rate of loss is often greater, as may be illustrated by the change in analysis with time of storage after first opening of samples of dried peas and of lemon juice, as shown in Fig. 2.9 (Jennings *et al.*, 1978). It is essential that analyses are always carried out without delay.

2.6.3 Analytical methods

Over the pH range of foods (pH 2–7) and at the concentrations of sulphur(IV) oxoanions present in foods, it is expected that the additive will be found as one or more of the species $SO_2 . H_2O$, HSO_3^-, $S_2O_5^{2-}$ and SO_3^{2-} in any particular case, and all these species will be encountered when the diversity of products to which sulphur(IV) oxoanions may be added are considered. In order that the specification may be uniform and applicable in all cases, legislation concerning the use of this additive refers to the amount present in terms of the compound, sulphur dioxide, limits by weight of this substance being given as parts per million. Thus, analytical procedures involve either conversion of total sulphur(IV) oxoanion present in the sample to the single product, sulphur dioxide, or if the analysis leads to the determination of the amount of the additive in gram ions per kilogram of sample, the accepted result is obtained by multiplication by 64 000.

In analysing to legal requirements for sulphur(IV) oxoanions in food it is not necessary to specify the ionic form of the additive in the original material. Thus, when the additive is being referred to in this context, the normally used name, *sulphur dioxide*, will be adopted. It must be stressed that, under these circumstances, this name is being used to comply with current analytical practice by expressing the amounts of the additive as parts per million of *sulphur dioxide*, but it would be equally or even more appropriate, in other circumstances, to state the amount of the additive in terms of, say, equivalents of sulphur(IV) per unit weight of food. The latter is, of course, independent of the gram ionic weight of the species present in the mixture.

When present in food, the additive may be found in three chemical forms which are distinguishable by the analytical methods available for the determination of *sulphur dioxide*:

1. free sulphur(IV) oxoanion
2. reversibly bound sulphur(IV) oxoanion
3. irreversibly bound sulphur(IV) oxoanion.

Current legislation requires determination of only the free and reversibly bound *sulphur dioxide*, the latter being referred to simply as bound *sulphur dioxide* to distinguish it from that which is free. Irreversible combination is the formation of organic and inorganic products which do not decompose to sulphur(IV) oxospecies under the conditions of analysis. The analytical distinction between free and reversibly bound *sulphur dioxide* is based upon the difference between the result of analysis when the amount of *sulphur dioxide* is measured in an acidified sample, and that found for a sample which has been treated with alkali prior to acidifying. These measurements give the free and the total (free + reversibly bound) *sulphur dioxide* respectively. The most important contribution to reversible binding of the additive is from the presence of carbonyl compounds in foods and the success of the distinction between the two forms depends upon the stability of the hydroxysulphonate adduct during the analysis of the free additive, and its tendency not to re-form during analysis of total additive after alkaline decomposition. It is shown in Section 1.11.1 that equilibrium constants for the formation of adduct with a large number of aldehydes and ketones show almost pH independent behaviour between pH 2–6, but the rate constants for formation and decomposition of the adducts pass through a minimum at around pH 2. Providing the hydroxysulphonates of food components behave similarly with respect to pH, an ideal medium for the determination of the free *sulphur dioxide* would be at pH 2.

The choice of analytical method for a particular food is made after consideration of whether free or total *sulphur dioxide* is to be estimated, the physical state of the food, the amount of additive present and the presence of interfering substances (Carswell, 1977). The methods which are available are broadly classified according to whether separation of sulphur(IV) oxospecies from the food is required prior to its determination; if this is not the case, the method is referred to as direct. The most commonly used separation method in indirect analysis is that of distillation of sulphur dioxide from acidified samples.

2.6.4 Direct analysis of *sulphur dioxide*
2.6.4.1 Iodimetric methods
The simplest method available for the analysis of sulphur(IV) oxoanions is iodimetric titration and, in some cases, it is possible to use such a titration for the determination of *sulphur dioxide* in food, provided the reducing power of the other components of the food is known. This method was originally introduced by Ripper (1892) and is still recommended by the AOAC as an Official Final Action (Horowitz, 1975). In its simplest form, the procedure is applicable to uncoloured or pale foods and lends itself particularly to the determination of both free and reversibly bound *sulphur dioxide* since the titrations may be carried out rapidly after pH adjustment. A blank titration, to determine the reducing power of components other than sulphur(IV) oxoanions present in the sample, is carried out by first adding an excess of acetone (Downer, 1943; Bennet and Donovan, 1943; Prater *et al.*, 1944; Potter and Hendel, 1951), formaldehyde (Ponting and Johnson, 1945; Reifer and Mangan, 1945; Potter and Hendel, 1951; Joslyn, 1955; Ross and Treadway, 1972) or glyoxal (Tanner and Sandoz, 1972; Potter and Hendel, 1951). The original choice of acetone for this purpose was based upon its use by Mapson (1941) for the removal of sulphur(IV) oxoanions from solutions to be analysed for ascorbic acid. The endpoints with formaldehyde are, however, much more stable and the greater desirability of this reagent over the others was demonstrated by Potter and Hendel (1951) for analyses on extracts of dehydrated white potato, when the results from seven analysts, with no previous experience

of the technique, gave the following deviations from an indirect reference procedure:

formaldehyde	-17% to $+4\%$
acetone	-39% to -6%
glyoxal	-22% to -6%

The experiments involving the use of acetone also showed the poorest agreement between analysts. The only preference for the use of acetone is stated by Prater and coworkers (1944), who found that the addition of formaldehyde led to *sulphur dioxide* levels consistently 5–10% greater than those obtained by an indirect reference method. Reifer and Mangan (1945), however, suggest that the quantity of formaldehyde used by Prater and coworkers was too small, leading to the observed discrepancy. The iodimetric end-point for the blank, using starch as indicator, is indistinct, particularly in the presence of formaldehyde. This is illustrated by blank titrations of 0·75, 0·35, 0·21 and 0·14 ml of 0·05 N iodine for an addition of 0·1, 0·2, 1·0 and 5·0 ml of 1% starch solution respectively (Joslyn, 1955). A method of reducing this difficulty, and one which also leads to stable end-points, is the removal of the free sulphur(IV) oxoanions by oxidation with hydrogen peroxide prior to determination of the blank. Such a procedure was advocated by Potter (1954) who treated an alkaline extract of dehydrated cabbage with acid and an excess of hydrogen peroxide. A comparison of blank titrations obtained in this way with those obtained using acetone gives the following results:

Sample	1	2	3	4	5	6
Acetone	3·80	4·65	5·10	6·56	5·04	4·52
H_2O_2	3·86	4·60	5·26	6·38	4·99	4·38

where titres are in ml 0·05 N iodine solution. The two methods show agreement in general but the 'compromise' nature of the titration in the presence of acetone is eliminated. The blank in the determination of *sulphur dioxide* in sausage meat is determined with the aid of hydrogen peroxide (Pearson and Wong, 1971). The difficulties encountered in visually detecting the end-point may be largely overcome by adopting an electrometric procedure, such methods also making possible the analysis of highly coloured samples such as red wines or blackcurrant juice. Ingram (1947a, b) advocates the use of a bright platinum electrode with a silver/silver chloride reference electrode. This shows a change in potential of 10–30 mV ml^{-1} 0·05 N iodine solution at the endpoint for the titration of citrus juices. Joslyn (1955), however, found the method less satisfactory for white wines which showed changes of the order of $1\,mV\,ml^{-1}$ titrant and an indistinct inflexion point. Amperometric methods are now frequently reported for the detection of this iodine endpoint as, for example, adopted for the determination of *sulphur dioxide* in wine by Brun and coworkers (1961).

Contrary to the normally accepted recommendation that sulphur(IV) oxoanion solutions be added to an excess of iodine for standardisation, the titration of foods involves addition of iodine from a burette, the only suggestion to combat losses being that of rapid addition. For the determination of total *sulphur dioxide* in dehydrated fruit and vegetables and in sausage the reversibly bound additive is liberated during an alkaline extraction procedure, which typically involves suspending the ground sample in sodium hydroxide for 20 min (Prater *et al.*, 1944; Potter and Hendel, 1951; Potter, 1954; Pearson and Wong, 1971). The sample is titrated immediately after acidifying. The procedures used for liquid samples generally give the free and the combined *sulphur dioxide* levels separately, and may be divided into two groups according to whether both determinations are carried out on a single sample or whether two samples are used. In the case of a single sample, this is first acidified to allow titration of free *sulphur dioxide*. Raising the pH to above pH 9, holding for 5–10 min and re-acidifying allows titration of the reversibly bound *sulphur dioxide*. In the two-sample procedure, one aliquot is acidified and titrated to give the free *sulphur dioxide* as above, but the other sample is treated with alkali and titrated for total *sulphur dioxide* after subsequent addition of acid. The accuracy of the determination of the reversibly bound *sulphur dioxide* depends upon the quantitative decomposition of any hydroxy-sulphonate-type adducts, these tending to reform to some extent when the pH is lowered. In order to improve the yield of this reversibly bound *sulphur dioxide*, a two-step decomposition procedure is used in which alkaline decomposition and determination of liberated *sulphur dioxide* is repeated after the normal determination of reversibly bound *sulphur dioxide*. The improvement in analysis of total *sulphur dioxide* by the introduction of the second decomposition step is illustrated by Jaulmes and Hamelle (1961), for the analysis of wine samples, as follows:

Simple decomposition	Two-step decomposition
437	470
381	422
355	445
367	516
390	430

where amounts of *sulphur dioxide* are given in units of parts per million in the original wine sample and each pair of results refers to a separate sample. Prolonged alkaline treatment may, however, lead to changes in the components of wines which change their sulphur(IV) oxoanion binding capacity (Lay, 1970, 1971).

2.6.4.2 Spectrophotometric methods

The majority of direct spectrophotometric methods for the determination of *sulphur dioxide* in foods are based upon the reaction between pararosaniline, formaldehyde and sulphur(IV) oxoanions discussed in detail in Section 2.3.3.1. Such methods may also be used to differentiate between free and reversibly bound *sulphur dioxide*, and a typical procedure for such a spectrophotometric analysis is shown by the determination of the additive in beer (Stone and Laschiver, 1957). This procedure for the measurement of total *sulphur dioxide* is also the Official Final Action recommended by the AOAC for beer (Horowitz, 1975). In order to measure free *sulphur dioxide*, the sample, after appropriate dilution, is mixed with 5 ml of the colour reagent (0·01 % pararosaniline in 2 % hydrochloric acid) and with 5 ml formaldehyde solution (0·2 % in water). The mixture is diluted to 50 ml with water and its absorbance measured at 550 nm after 30 min. For measurement of total *sulphur dioxide*, advantage is taken of the ability of tetrachloromercurate(II) ion to trap the liberated sulphur(IV) oxoanions when the mixture is treated with alkali. Thus, the sample is added to a mixture of 2 ml tetrachloromercurate(II) reagent (27·2 g mercuric chloride + 11·7 g sodium chloride in 1 litre water) and 5 ml 0·05 M sulphuric acid, the mixture is made alkaline by the addition of 0·1 M sodium hydroxide and re-acidified after 15 s with sulphuric acid. The colour reaction is then carried out as for free *sulphur dioxide*. A blank is provided by oxidising the sulphur(IV) oxoanion in the sample by titration with iodine and carrying out the determination as described for free *sulphur dioxide*. For photometer readings in the region of 200 when beer is used, blanks with values close to 80 are expected and, therefore, a considerable correction needs to be made. However, it is found that this blank is highly reproducible and is the same for a wide range of beers and, therefore, does not need to be checked on each occasion. Concentrations of *sulphur dioxide* are read from calibration curves prepared by standard addition of solutions containing sulphur(IV) oxoanions. When compared with a recognised indirect reference analysis procedure, this spectrophotometric assay gives recoveries in the range 94–105 % for total *sulphur dioxide* contents in the range 5·5–30·4 ppm.

TABLE 2.8

Comparison of results obtained by spectrophotometric and iodimetric determinations of free and total *sulphur dioxide* in five white wines (Joslyn, 1955)

Wine	Free SO$_2$ (ppm)		Total SO$_2$ (ppm)	
	Spectrophotometric	Iodimetric	Spectrophotometric	Iodimetric
A	17·0	9·5	100·0	130·0
B	95·0	66·5	208·0	189·0
C	5·0	6·0	97·5	67·0
D	66·7	30·0	183·5	371·2
E	14·3	3·8	41·6	103·8

Reproduced with permission from *J. Agric. Food Chem.*, 1955, **3**, page 693. © 1955 American Chemical Society.

Joslyn (1955) compared direct iodimetric and spectrophotometric determinations of both free and reversibly bound *sulphur dioxide* in white wine and the results, which demonstrate the very considerable differences which may be observed, are shown in Table 2.8.

The expected amount of *sulphur dioxide* in white sugar is in the range < 1–20 ppm and the sensitivity of the spectrophotometric analysis is appropriate for its determination. The application of the pararosaniline reaction to the analysis of white sugar is described by Carruthers and coworkers (1965), this method now being recommended by the International Commission on Uniform Methods of Sugar Analysis (ICUMSA). A feature of the procedure is the apparent need for alkaline pre-treatment of samples containing only sucrose and sulphur(IV) oxoanions. This is clearly illustrated in Fig. 2.10 which shows, as the lower curve, the time dependent change in analysis result when sucrose and sulphur(IV) oxoanions are left to stand for given lengths of time before addition of spectrophotometric reagent. In each case, treatment of such samples with sodium hydroxide increases the absorbance to a constant value. Although it is tempting to attribute the cause of this change to the presence of small amounts of glucose in the sample, it is not the predominant effect because addition of as much as 1 % hydrolysed sucrose to the samples does not increase the difference between the colour yields. Recoveries of *sulphur dioxide* when sulphur(IV) oxoanions are added to white sugar are close to 100 %, the results being, on average, some 9 % lower than those obtained by direct iodine titration, which is reputed to give slightly excessive results for the estimation.

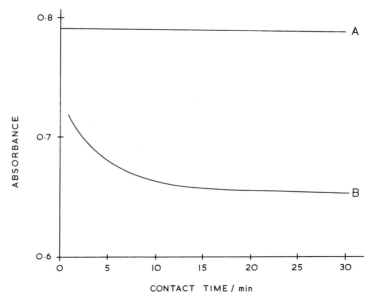

Fig. 2.10. Effect of alkaline pre-treatment on the analysis of *sulphur dioxide* in sucrose solution (400 g litre^{-1}) by the pararosaniline method. A, alkali treated; B, no alkali. Contact time refers to the length of time for which the sucrose solution and sulphur(IV) oxoanion were left to stand before analysis (Carruthers *et al.*, 1965). Figure reproduced from *International Sugar Journal*, 1965, **67**, page 365. © 1965 The International Sugar Journal Ltd.

The spectrophotometric procedure may be used for the determination of the additive in solid foods such as dehydrated fruits (apples, pears, peaches, apricots and golden raisins) following an appropriate extraction reminiscent of the alkaline treatment preceding titration of fruits and vegetables with iodine. Samples of the alkaline extract are acidified and mixed with a solution containing tetrachloromercurate(II) ions before reaction with pararosaniline reagent and formaldehyde (Nury *et al.*, 1959; Taylor *et al.*, 1961; Brekke, 1963; Nury and Bolin, 1965). Colour development in this system appears to be very sensitive to temperature, the maximum yield being obtainable at 22 °C, and the recommended development time is 30 min (Taylor *et al.*, 1961).

One obvious possibility offered by spectrophotometric analysis is that of an automated process, an example of an 'autoanalyser' arrangement for the analysis of simple liquid samples such as beer being shown in Fig. 2.11 (Saletan and Scharoun, 1965). The procedure, which is designed to

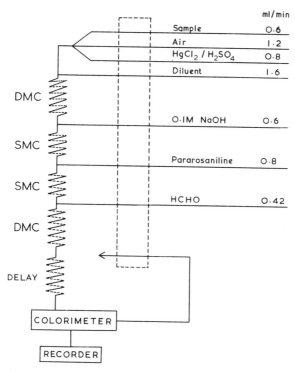

Fig. 2.11. Manifold diagram showing autoanalyser arrangement for analysis of *sulphur dioxide* using pararosaniline reagent (Saletan and Scharoun, 1965). DMC, double mixing coil; SMC, single mixing coil. Reagents: 0·015 M Na$_2$HgCl$_4$ in 0·05 M H$_2$SO$_4$; 0·02% pararosaniline in 8% HCl; 0·4% HCHO containing surfactant; solution of surfactant as diluent. After *Proc. Am. Soc. Brew. Chem.*, 1965, page 203.

measure the total *sulphur dioxide*, follows the method recommended by AOAC for beer analysis described at the beginning of this section, and is reported to give excellent results. The arrangement is similar to one reported later by Logsdon and Carter (1975) for the automated analysis of atmospheric sulphur dioxide, the main differences being the introduction of a sample in sodium tetrachloromercurate(II), the addition of sulphamic acid to reduce interference from oxides of nitrogen in atmospheric samples and the reverse order of addition of pararosaniline and formaldehyde. Since aldehydes and ketones are not expected in such atmospheric samples, no treatment with sodium hydroxide prior to development of colour is necessary.

2.6.4.3 Polarographic analysis

The most intense polarographic waves resulting from the reduction of sulphur(IV) oxoanions are observed in strong acid, any determination of the additive under these conditions being susceptible to losses during the removal of oxygen by bubbling nitrogen through the solution. Diemair and coworkers (1961) describe the polarographic determination of free *sulphur dioxide* in wine without acidification, that is in the pH range 3–3·5 typical of most wines. Calibration curves, prepared by measuring the wave height when sulphited model wines are analysed, are linear to concentrations of the order of 70 ppm with only a small pH dependence in slope over the pH range. The analytical procedure involves, in addition to recording the polarogram of the sample, the measurement of pH in order that the appropriate calibration data may be used. When the free *sulphur dioxide* content of glucose solutions (20 g/litre) to which sulphur(IV) oxoanions have been added is determined, iodimetric and polarographic methods give identical results to concentrations of 65 ppm free *sulphur dioxide*. However, when wines containing this level of additive are analysed by both methods, the polarographic analysis gives results which are lower by about 10 ppm. The correspondence is better as the concentration is reduced.

Differential pulse polarograms of wine samples, 50 μl, added to 10 ml of a solution of 0·1 M lithium chloride in dimethylsulphoxide with 200 μl of 0·5 M sulphuric acid are reported by Bruno and coworkers (1979). The sample may be that of the original wine or that obtained after treatment of the wine with sodium hydroxide giving respectively the responses for free and for

TABLE 2.9

Comparison of direct iodimetric titration and of polaro-graphic analysis of alkaline extracts of dehydrated vegetables (Prater *et al.*, 1944)

Commodity	Amount of *sulphur dioxide* found (ppm)	
	Iodimetric titration	Polarographic analysis
Cabbage	1 104; 1 070	1 100; 1 050; 1 060
Carrot	489; 499	483; 465
Potato A	687; 653	675; 690
Potato B	2 160; 2 225	2 140; 2 150

Reproduced with permission from *Ind. Eng. Chem.* (*Anal. Edn*), 1944, **16**, page 156. © 1944 American Chemical Society.

total *sulphur dioxide*. The calibration method advocated was one of standard addition, in which case known amounts of *sulphur dioxide* are added to the wine sample and polarograms recorded. A graph of wave height on the differential pulse polarogram is plotted as a function of the amount of *sulphur dioxide* added, the negative intercept on the concentration axis being equal to minus the concentration of *sulphur dioxide* in the sample. Five replicate analyses on a sample of commercial wine gave a value for free *sulphur dioxide* of 16.6 ± 0.37 ppm. In the determination of total *sulphur dioxide* by the polarographic and iodimetric methods the latter gave values 7–10 ppm higher when eight samples of white and red wines were analysed.

The correspondence between polarographic analysis and iodimetric titration appears highly acceptable when alkaline extracts of dehydrated cabbage, carrot and potato are analysed, a set of results published by Prater and coworkers (1944) being shown in Table 2.9.

2.6.4.4 Use of selective electrodes

The only serious work with sulphur(IV) oxospecies selective electrodes in food analysis appears to be that using the sulphur dioxide gas-sensing membrane probe described in Section 2.3.9.1. The response of a commercial model is conveniently Nernstian over the range 3–3000 ppm, which is suitable for direct measurement on a number of food products. A typical procedure for the determination of free *sulphur dioxide* involves the insertion of the probe into a continuously stirred solution, acidified to pH 1·0. Total *sulphur dioxide* is determined on a separate aliquot, to which 2 M sodium hydroxide has been added in order to raise the pH above 13·0. After 15 min the solution is acidified with 2 M sulphuric acid, which lowers the pH to 1·0 (Jennings *et al.*, 1978). Since the method is dependent upon the concentration of the liberated sulphur dioxide rather than on the total amount present, corrections due to volume changes are required. The ideal working range appears to be a concentration of 20–100 ppm *sulphur dioxide*, when successive analyses for the free additive at a variety of dilutions have a standard deviation in the region of 3–5 % of the mean value, and samples originally containing some 30 ppm of *sulphur dioxide* are taken (wine and orange squash). In the case of a fruit syrup containing in the region of 250 ppm *sulphur dioxide*, the standard deviation of the data obtained from several analyses of varying sample size is 8 % of the mean value. Alcoholic samples, especially red wines, are found to have an adverse effect upon the membrane, which requires replacement after each series of measurements on any one type of sample containing alcohol. The life of the membrane in use with non-alcoholic samples is about two weeks, the internal filling solution requiring replacement weekly.

The 'air-gap' sulphur dioxide sensor, described by Hansen and coworkers (1974) and, briefly, in Section 2.3.9.1, has also been used for the determination of the additive in wine samples acidified with 20% (v/v) orthophosphoric acid in 20% aqueous ethanol. The accuracy of the determinations is limited by the standard deviation of the slope of the calibration curve obtained by analysing standard sulphur(IV) oxoanion solutions, this slope being ± 0.04 pH unit for the measurement of the response of the glass electrode, corresponding to an error in the region of $\pm 16\%$ in *sulphur dioxide* concentration. A blank, in which the sample is oxidised with dichromate ion, gives an apparent *sulphur dioxide* concentration of 4×10^{-7} M, corresponding to the actual limit of the electrode response.

2.6.5 Indirect methods of analysis

Indirect methods of *sulphur dioxide* analysis involve some method whereby the additive is separated from the other components of the food and determined subsequently by a more or less specific technique according to the extent to which interfering components are carried over during the separation. The normally used methods involve mass transfer through the gas phase above the sample. If these are assisted, for example by heat or by the use of an inert carrier gas, the methods will be referred to as distillation techniques, as distinct from processes such as equilibration with the headspace for subsequent headspace analysis or microdiffusion techniques, the rates of which are controlled by the rates of gaseous diffusion of sulphur dioxide in the atmosphere above the sample.

2.6.5.1 Distillation methods

Distillation is always carried out from strongly acid solution and involves four steps:

(a)   Desorption of sulphur dioxide.
(b)   Optional separation of sulphur dioxide from water vapour and some other volatile components.
(c)   Mass transfer to the receiver.
(d)   Detection of sulphur(IV) oxoanion in the distillate by absorption and reaction.

Although the solubility of sulphur dioxide in water is usually much greater, even at 100 °C, than the amount of the gas dissolved in an acidified solution or suspension of a food, some will pass into the gas phase since the solution of sulphur dioxide will exert a significant vapour pressure due to the gas. If left undisturbed the system will come to equilibrium, the rate of attainment of equilibrium being increased by the passage of a gas, such as water vapour or an inert gas, through the solution during boiling. The rate of mass transfer of sulphur dioxide from solution to the gas phase will depend upon the difference between the equilibrium vapour pressure of the gas and that at the moment in question. Therefore, in order to ensure rapid and complete mass transfer of sulphur dioxide out of solution, it is necessary to maintain the vapour pressure of the gas above the solution as low as possible. It is not strictly necessary to convert all sulphur(IV) oxoanions in the food to $SO_2 . H_2O$ before distillation is commenced since as the gas, which is in dynamic equilibrium with oxoanions, is removed, protonation of oxoanions will take place to maintain equilibrium in solution, thus forming more $SO_2 . H_2O$ until complete conversion has taken place. The rate of desorption will, however, also depend upon the concentration of $SO_2 . H_2O$ in solution which should, therefore, be maintained at as high a level as possible by adequate acidification.

The simplest desorption procedure which could be envisaged is the passage of an inert gas, oxygen-free nitrogen, through an acidified sample at room temperature. It is claimed that, at pH 1–2, with a carrier gas flow rate of $300\,ml\,min^{-1}$, this experiment gives complete desorption of free sulphur dioxide, when results are compared with direct iodimetric titration, from solutions of 10% glucose, fructose and sucrose in approximately 20 min (Lloyd and Cowle, 1963). In order to test the extent of decomposition of hydroxysulphonate-type adducts at ambient and sub-ambient temperatures during such a desorption experiment, Fujita and coworkers (1979) measured the amount of sulphur dioxide produced when the hydroxysulphonates of acetaldehyde, pyruvic acid and D-mannose were acidified with phosphoric acid (10 ml of 25% acid to 20 ml of sample) and air was bubbled at a rate of 1 litre min^{-1} for up to 30 min. The use of air in place of an inert gas has very little effect on recovery. None of the adducts are decomposed when the mixture is held at $0\,°C$ with bubbling continuing for 30 min, but at $20\,°C$ the extent of decomposition of acetaldehyde, pyruvic acid and mannose hydroxysulphonates is 0·4, 10·5 and 1·8 % after 10 min and 0·6, 17·9 and 2·1 % after 30 min respectively. The instability of hydroxysulphonates under distillation conditions at $20\,°C$ had been noted earlier by Burroughs and Sparks (1964) and invalidates the original assumption that the time required for desorption of free sulphur dioxide at room temperature is sufficiently short to give a good estimation of the free *sulphur dioxide* content of soft drinks (Lloyd and Cowle, 1963). More appropriate conditions for such a desorption experiment, therefore, appear to be those of temperature at $0\,°C$, when the yield of sulphur dioxide from solutions of pure sulphur(IV) oxoanions treated with acid is 98 % after 15 min and 100 % after 30 min of passage of air (Fujita *et al.*, 1979).

The most widespread use of distillation methods is for the determination of total *sulphur dioxide*, which requires decomposition of all the adducts. This is conventionally carried out by boiling acidified solutions of the food, the available methods being adaptations of the procedure suggested by Monier-Williams in 1927. The important modifications can be divided into those which involve changes to the distillation medium and arrangement of apparatus, and those which involve modification of the analysis of distillate. A summary of the essential details of the Monier-Williams method and those of a representative selection of distillation analyses is given in Table 2.10. The first point concerns the choice of acid in the distillation medium; this can be orthophosphoric, sulphuric or hydrochloric acid. The choice of orthophosphoric acid by several workers is, perhaps, rather surprising since it is a much weaker acid than the others, the pH of distillation media used by Rothenfusser (1929) and Tanner (1963) being 1·5 and 2·0 respectively, whereas those containing hydrochloric acid, for example in the methods of Reith and Willems (1958) or Zonneveld and Meyer (1960), were in the region of pH 0·3 (Heintze, 1970). Thus, the use of hydrochloric acid leads to almost complete conversion of sulphur(IV) oxoanion to $SO_2 . H_2O$, whereas, at the higher pH, only some 50 % conversion takes place. Heintze (1962) argues that when such 'high pH' media are employed with samples containing substrates for non-enzymic browning, the relatively slow liberation of sulphur dioxide from such solutions may lead to significant irreversible combination of the additive. This possibility is supported by the comments of a number of investigators that hydrochloric acid leads to a more rapid and complete liberation of the additive than does phosphoric acid (for example, Nichols and Reed, 1932; Zonneveld and Meyer, 1960). In support of the use of phosphoric acid is its lower volatility, which renders less likely the possibility of carry-over and interference in acid–base titrations. This potential difficulty may be completely overcome by the use of a more specific detection method but, in the author's experience, it is very rare to have significant blanks when hydrochloric acid is used in the distillation medium for Monier-Williams analysis with alkalimetric determination of distilled sulphur dioxide. The second variable in the composition of the distillation medium is the addition of methanol. The main reason for this addition is the lowering of the reflux temperature, which tends to reduce interference from the transfer of volatile acids other than sulphur dioxide to the receiver (Zonneveld and Meyer, 1960). Other advantages of adding methanol include increased protection against oxidation and, in some cases, such as the analysis of pectin samples, when aqueous media give solutions of high viscosity, the use of a solvent with a higher methanol content is beneficial.

One variation of the distillation method involves the boiling of the acidified solution under reflux, the liberated sulphur dioxide being removed through the top of the condenser by the passage of a slow stream of inert gas through the boiling solution. Under these conditions the condenser appears to give a good separation of sulphur dioxide from other reducing agents or from acids, rendering non-specific means of detection, such as alkalimetric analysis following oxidation with hydrogen peroxide, satisfactory. The use of a downward condenser, implying co-distillation or steam distillation of volatile components, cannot be universally applicable unless the methods of analysis of sulphur(IV) oxoanions in the distillate are made more specific. The use of a downward condenser does, however, lead to much faster transfer of sulphur dioxide to the receiver and, indeed, some 'rapid' methods of *sulphur dioxide* analysis, with distillation times in the range 3–5 min, involve such an arrangement. Whilst the presence of reducing agents other than sulphur dioxide in the distillate will tend to increase the titre in iodimetric determination and, likewise, the transfer of volatile acidity will tend to increase alkalimetric endpoints, the transfer of aldehydes will have the opposite effect. It is established that acetaldehyde hydroxysulphonate may be transferred into the receiver on distillation (Diemair *et al.*, 1961), a process which, presumably, involves dissociation of the adduct, transfer of components separately in the gas phase followed by recombination in the receiver. If this is suspected, then appropriate conditions necessary for the decomposition of such hydroxysulphonate adducts need to be used before estimation of distilled sulphur dioxide.

Since distillation methods are the most widely employed standard procedures for the analysis of *sulphur dioxide* in foods, and are often regarded as reference methods against which other analytical techniques are assessed, it would be appropriate to conclude this section with the experimental details of well established procedures. Those chosen are the modified Monier-Williams distillation method recommended by AOAC (Horowitz, 1975), the Tanner (1963) method, which is also the EEC standard procedure for the analysis of sulphur dioxide in wines, and a rapid distillation technique appropriate to beverages and dehydrated vegetables (Wedzicha and Johnson, 1979; Wedzicha and Bindra, 1980). Finally, an

TABLE 2.10

Comparison of some features of reported distillation methods for the analysis of *sulphur dioxide* in foods

Distillation medium	Condenser arrangement	Boiling time	Detection method	Application	Reference
200 ml HCl + 500 ml H_2O	Reflux	1 h (dried fruit $1\frac{1}{2}$ h)	H_2O_2/NaOH H_2O_2/BaCl$_2$	Corn syrup, sugar, jam, port wine, dried fruit, gelatin, sausage, jam, mustard, onion	Monier-Williams, 1927
10 ml 25% H_3PO_4 + 300 ml H_2O	Co-distillation	Until distillate shows negative reaction for SO_2	H_2O_2/benzidine	Wine, beer, vinegar, chopped meat, fats and oils, dried fruit, glycerine, soap	Rothenfusser, 1929
20 ml 10 M HCl + 400 ml H_2O	Reflux	30 min	H_2O_2/NaOH	Finely divided dried foods	Shipton, 1954
5 ml 25% H_3PO_4 + 10–20 ml sample	Reflux	15 min	H_2O_2/NaOH	Wine	Paul, 1954
20 ml HCl + 300 g sample	Reflux	45 min	Sodium tetrachloromercurate/pararosaniline	Malt, beer	Beetch and Oetzel, 1957
40 ml 15% HCl + 280 ml H_2O	Reflux	1 h	H_2O_2/BaCl$_2$–EDTA	Most foods	Reith and Willems, 1958
50 ml 4 N HCl + 225 ml CH_3OH + 50 ml H_2O	Reflux	1 h	H_2O_2/NaOH	Dried vegetables	Zonneveld and Meyer, 1960

Sample preparation	Method	Time	Determination	Application	Reference
5 ml 85% H_3PO_4 + 50 ml H_2O + 25 ml sample	Co-distillation	Until 50 ml remains	pH 3–4 buffer/I_2	Wine	Diemair et al., 1961
40 ml 15% HCl + 300 ml H_2O	Reflux	45 min	H_2O_2/NaOH	Preserves, jam	Nehring, 1961
10 ml 25% H_3PO_4 + 20 ml sample	Reflux	15 min	H_2O_2/NaOH	Wine	Rankine, 1962
5 ml 25% H_3PO_4 + 10–20 ml sample	Reflux	15 min	I_2/$Na_2S_2O_3$	Cider	Burroughs and Sparks, 1963
10 ml 25% H_3PO_4 + 150 ml CH_3OH + 150 ml H_2O	Reflux	15 min	H_2O_2/NaOH	Drinks, concentrates, vinegars	Tanner, 1963
10 ml 1 M HCl + 40 ml H_2O	Co-distillation	15 min	NaOH/peroxodisulphatotitanic acid	Dried fruit, fruit syrups, beverages, concentrates	Nagaraja and Manjrekar, 1971
10 ml 16% H_2SO_4 + 10% sample + 2 ml CH_3OH	Reflux	3 min	IO_3^-/$Na_2S_2O_3$	Wine, fruit juices	Rebelein, 1973
50 ml 2 M H_2SO_4 + 5–10 ml sample	Co-distillation	4 min	5,5'-dithiobis-(2-nitrobenzoic acid)	Ginger ale	Wedzicha and Johnson, 1979
50 ml 2 M H_2SO_4	Co-distillation	5 min	5,5'-dithiobis-(2-nitrobenzoic acid)	Dehydrated cabbage, carrot, pea, potato	Wedzicha and Bindra, 1980

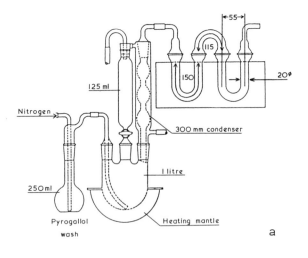

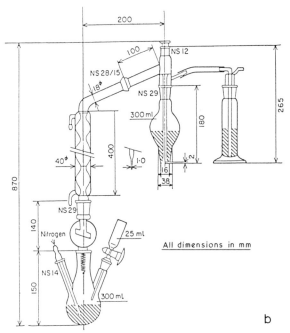

Fig. 2.12. Distillation apparatus for the determination of *sulphur dioxide*. (a) Recommended by AOAC (Horowitz, 1975). (Reproduced with permission from *Official Methods of Analysis of the Association of Official Analytical Chemists*, 12th edition, page 367. © 1975 Association of Official Analytical Chemists.) (b) Recommended by Tanner (1963). (Reproduced with permission from *J. Ass. Public Analysts*, 1978, **16**, page 68. © 1978 Association of Public Analysts.)

automated distillation, originally developed for the determination of alcohol in beers and wines (Lidzey *et al.*, 1971) in conjunction with an autoanalyser arranged for determination of *sulphur dioxide* by means of the pararosaniline–formaldehyde reaction (Jennings *et al.*, 1978), will be described.

The apparatus used for the AOAC and Tanner distillation methods is shown in Fig. 2.12. In both cases the sample is weighed or pipetted into the distillation flask and diluted with solvent, which in the AOAC method is water (400 ml), and in the Tanner method is water (20–40 ml) followed by methanol (50 ml). The trapping solution is, in both cases, hydrogen peroxide neutralised with respect to the acid–base indicator used for titration, which is methyl red in the case of the AOAC method and a mixed indicator consisting of methyl red and methylene blue for the Tanner method. The analysis is started by adding the appropriate acid (90 ml hydrochloric acid diluted 1 + 2 for the AOAC method, and 15 ml of 88 % orthophosphoric acid for the Tanner method), and distillation carried out for 105 and for 30 min for the two methods respectively. The procedure is complete for the Tanner method with only alkalimetric determination of sulphuric acid in the receiver to follow, whereas the AOAC distillation is completed by turning off the water in the condenser and heating until the trap begins to warm up, that is, until steam passes over. Titration of the contents of the receiver follows or, alternatively, sulphuric acid may be determined gravimetrically as sulphate. The carrier gas flow rate is usually more a matter of experience and should be consistent with a good recovery of the additive in the distillation period but not so high as to cause large amounts of sulphuric acid to be formed in the guard bottle. A reasonable flow rate is in the region of one bubble per second, which usually gives negligible amounts of acidity in the guard bottle. The AOAC method is described as unsuitable for dried onions, leeks and cabbage since, in these cases, the unsulphited vegetables lead to the formation of significant amounts of acidity in the receiver. For example, the pH of the distillate from freeze dried fresh cabbage is close to pH 3·5 and a significant titre, using sodium hydroxide, may be obtained if an indicator such as methyl red, with a pH range of 4·2–6·3 is used. However, the titration of sulphuric acid is almost complete at pH 3·5 and it is, therefore, appropriate to use an indicator with a lower pK_a to reduce this interference. The best choice would seem to be bromophenol blue (pH 3·0–4·6) which was, in fact, the indicator originally used by Monier-Williams, the main practical criticism being the difficulty in reproducing the colour at the endpoint and requiring a trained eye. A more satisfactory procedure, which has proved very

TABLE 2.11
Comparison of analysis for *sulphur dioxide* in some dehydrated vegetables by the Monier-Williams and rapid distillation methods

Dehydrated vegetable	Conventional Monier-Williams analysis	Rapid method
	$SO_2{}^a$ (ppm)	$SO_2{}^b$ (ppm)
Cabbage	547 ± 39	601 ± 15
Carrot	556 ± 48	565 ± 46
Pea	818 ± 27	857 ± 22
Potato	323 ± 9	348 ± 10

All errors shown are standard deviations. Sample weight: ^{a}10 g, ^{b}1 g.

successful in the author's laboratory, is to titrate the distillate to pH 3·5 with the aid of a pH meter.

In a rapid method of *sulphur dioxide* analysis, the distillation flask is of 100 ml capacity connected through a sloping condenser to a receiver adaptor whose outlet is below the surface of a solution of 5,5'-dithiobis(2-nitrobenzoic acid) buffered to pH 8·0. Alternatively, if only liquid samples are to be analysed, a semi-micro Kjeldahl assembly (Quickfit 21/100 MC) also with a capacity of 100 ml may be used for the distillation apparatus, the outlet of the condenser being placed below the surface of the solution of reagent. Oxygen-free nitrogen is passed through the distillation medium (2 M sulphuric acid) and the sample is either added through the third neck of the distillation flask or through the air lock of the semi-micro Kjeldahl apparatus after the acid has been heated to boiling to expel air. Considering in detail the procedure employing a three-necked distillation flask, a distillation time of 5 min with standard additions of *sulphur dioxide* in the range $1·4–6·8 \times 10^{-6}$ mol gives an overall recovery of 96 %. In experiments on dehydrated vegetables (cabbage, carrot, peas and potato) a constant recovery of *sulphur dioxide* is obtained for distillation times in excess of 5 min, and a time of 5 min is, therefore, adopted for analysis. The results of analyses for the four vegetables, summarised from eight measurements in each case and compared with the conventional Monier-Williams analysis, modified insofar as titration of distillate was to pH 3·5 using a pH meter, are shown in Table 2.11. Unsulphited cabbage, carrot, peas and potato give results indistinguishable from the blank. Statistical analysis showed the variances for the Monier-Williams and the rapid method to be homogeneous. Comparison of the two techniques by means of a Student's

t-test for each vegetable shows that, with the exception of dehydrated carrot, the differences in the means are significant at the 95 % level and, in all cases, it may be seen that the result for the rapid method is consistently higher. If a paired t-test is carried out, it is found that this conclusion is true, as a general statement, only at the 90 % level but, allowing for the small number of pairs available, it is possible to say that the technique gives a yield of sulphur dioxide, from dehydrated vegetables, which is at least as high as that which may be attained with the conventional Monier-Williams procedure. This may be a little surprising since the short distillation time may not be sufficient to allow complete decomposition of hydroxy-sulphonate-type adducts. Expressed as percentages of means, standard deviations are: cabbage, 2·5 %; carrot, 8 %; peas, 5 %; potato, 2·7 %. The single most important factor in reproducibility was the homogeneity of the sample. For example, whole dehydrated peas when analysed as 0·5 g samples showed a spread of as much as 50 % about the mean, the values shown in the table being for samples ground with a pestle and mortar before analysis. Dehydrated onion is often present in dehydrated formulations such as soups, and possible interferences arising from this vegetable may be important. The conventional Monier-Williams analysis gives, for unsulphited onion, an apparent *sulphur dioxide* level of 30–40 ppm, the rapid method giving a result of 30–60 ppm. In the latter case the size of the effect depends upon the length of distillation, the apparent yield of sulphur dioxide increasing to 217 ppm after 15 min. It is unlikely that the presence of onion as a minor constituent of dehydrated formulations will give rise to a significant error.

A flash distillation unit which may be incorporated into an autoanalyser system for sulphur dioxide, and a manifold diagram for the supporting connections to reagents by way of a proportionating pump are shown in Fig. 2.13. The principle of operation of this unit is straightforward. The acidified sample is passed to the top of a ten-turn helix where it is mixed with a fast stream of nitrogen, causing the sample to be distributed as a thin film along the bottom surface of the coil. Sulphur dioxide is then released into the rapidly moving gas phase. Excess liquid remains in the trap at the bottom of the coil, from which it is removed to waste by way of the pump. The gas, on the other hand, is washed with a solution containing tetrachloromercurate(II) ion and the latter passed on for spectrophoto-metric analysis. The sample capacity of this unit, when connected in series with an autoanalyser system for spectrophotometric analysis using pararosaniline, is given as 12 samples per hour, with an overall linear response over the range 2–14 ppm sulphur dioxide.

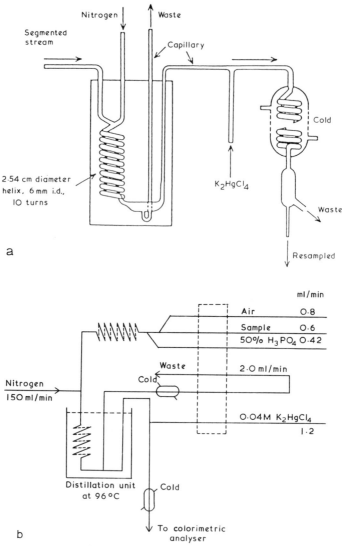

Fig. 2.13. Automated distillation analysis of *sulphur dioxide* in liquid samples. (a) Distillation unit for automatic determination of sulphur dioxide using an autoanalyser arrangement (Lidzey *et al.*, 1971; Jennings *et al.*, 1978). (b) Manifold diagram showing connections to flash distillation unit for analysis of sulphur dioxide in beverages and other solutions (Jennings *et al.*, 1978). Both reproduced with permission from *J. Ass. Public Analysts*, 1978, **16**, page 69. © 1978 Association of Public Analysts.

2.6.5.2 Headspace analysis

A scheme for the analysis of free and combined *sulphur dioxide* in a wide range of foods by gas chromatographic measurement of headspace sulphur dioxide above acidified samples is described by Hamano and coworkers (1979). The essential operation is the shaking of 1 ml samples of appropriate extracts of the food with 5 ml of 10% phosphoric acid in a Nessler tube fitted with a silicone rubber septum, and the temperature of the system is maintained at 0 °C by means of a water bath. After shaking, 1 ml samples of the headspace are withdrawn and analysed using a 3 mm × 5 m column filled with APS-1000 (40/60 mesh) at 70 °C. The detector employed by these workers was the flame photometric detector. The method is highly specific and not subject to interference from foods which evolve hydrogen sulphide, alkyl isothiocyanate, methyl mercaptan or dimethyl sulphide on acidifying (Mitsuhashi *et al.*, 1979).

2.6.5.3 Analysis by microdiffusion

The general principle of the microdiffusion method of analysis consists of the absorption, by simple gaseous diffusion, of a volatile substance dissolved in one liquid, in a trapping medium where it may be analysed. The method may be applied to a wide range of analyses described by Conway (1962) but, unfortunately, the analysis of sulphur(IV) oxoanions by this technique was not known at that time. An example of a microdiffusion cell, that originally described by Conway and Byrne (1933), is shown in Fig. 2.14. The sample to be investigated is measured into the outer chamber with the trapping reagent in the central compartment. An example of the possible speed of gas transfer between the solutions is the quantitative absorption of ammonia from alkaline media into acid in times as short as 10 min. The solution in the inner compartment may then be analysed by conventional means. A procedure for the determination of volatile acids and *sulphur dioxide* in wines using the Conway cell has been described by Owades and Dono (1967); in this the sample is acidified with phosphoric acid, and for the determination of total volatile acidity the absorbing medium is a solution of sodium citrate. The acidity is determined titrimetrically with barium hydroxide. Despite the addition of anhydrous sodium sulphate to the sample compartment to dehydrate the sample and, thus, speed up release of volatile components, the experiment still requires 16–19 h at 55–60 °C. To determine the volatile acidity exclusive of sulphur dioxide, the procedure is repeated after adding to the sample a solution of mercuric oxide in sulphuric acid. When calculated by difference, *sulphur dioxide* levels in a domestic red wine and a domestic white wine are

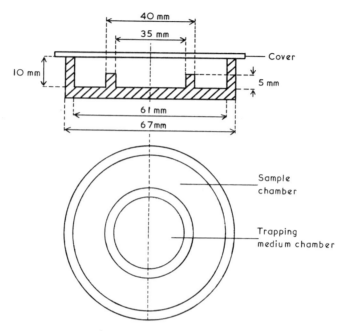

Fig. 2.14. Conway microdiffusion cell (Conway and Byrne, 1933). Reproduced with permission from *Biochem. J.*, 1933, **27**, page 420. © 1933 The Biochemical Society.

respectively 74·5 and 132·0 ppm compared with values of 75·1 and 122·5 ppm obtained by Monier-Williams analysis. Although, for a small number of analyses, the method is time-consuming, it permits a large number of simultaneous determinations to be carried out. If only *sulphur dioxide* is to be determined, the method could be modified to advantage using a specific, spectrophotometric method for analysis of the contents of the inner chamber. Trublin and coworkers (1976) describe a method based upon this approach. For absorption of sulphur dioxide, from a sample acidified with phosphoric acid, into a solution of zinc acetate in 2% ethanol, for spectrophotometric determination with sodium nitroprusside/pyridine, a diffusion time of $2\frac{1}{2}$ h at 37 °C is required. Despite problems associated with the use of sodium nitroprusside for the spectrophotometric analysis of sulphur(IV) oxoanions, results for the analysis of wines agree well with distillation analysis and further studies of microdiffusion methods for the analysis of *sulphur dioxide* in foods are, therefore, warranted.

2.6.6 Comparison of direct and indirect methods of analysis

Whenever a new method or a variant of an established method for the analysis of *sulphur dioxide* in food is devised, it is normal practice to compare its results with those obtained using a reference procedure and, therefore, such comparisons in the literature are numerous. To keep the coverage to reasonable proportion, one example, recently published by the Government Chemist's Group (Jennings *et al.*, 1978), of the comparison between three very different methods of *sulphur dioxide* analysis will be considered here. The methods compared are:

1. The Tanner distillation method
2. The use of a gas-sensing membrane electrode
3. An autoanalyser procedure involving the flash distillation apparatus shown in Fig. 2.13, followed by spectrophotometric determination with pararosaniline, but also with added sulphamic acid, as described in Section 2.6.4.2.

The results of comparative determinations of total *sulphur dioxide* in a wide range of beverages are shown in Table 2.12. In general, there is satisfactory

TABLE 2.12

Comparison of three methods for the determination of *sulphur dioxide* in liquid foods (Jennings *et al.*, 1978)

Sample	Total SO$_2$ concentration (ppm)		
	Tanner's method	Probe method	Automatic method
White wine A	205	180	220
White wine B	130	125	130
Red wine A	85	115	75
Red wine B	155	65	165
Lemon juice A	215	250	175
Lemon juice B	85	115	85
Beer A	5	10	5
Beer B	<1	5	<1
Cider A	85	90	90
Cider B	115	105	120
Sherry A	125	120	135
Grapefruit juice A	95	130	110
Orange perry	50	45	50
Orange juice A	40	35	45
Fruit syrup	90	100	95

correlation between the results using the Tanner method and those obtained by the autoanalyser. The gas-sensing membrane electrode gives results which tend to be more variable in comparison with other methods, but the simplicity and speed of the measurements renders it, nevertheless, a highly desirable approach, providing the response of the food system being studied is well known.

2.6.7 Conclusion

From the point of view of legislative control of the uses of *sulphur dioxide* in foods, there is no serious problem in meeting the required specification, since this is defined in terms of a recommended method of analysis. Any time-saving or automatic method which may be adopted by the individual for quality control purposes should then be referred to the standard method. In the development of such standard methods, perhaps the three most important criteria have been:

1.   The lack of apparent *sulphur dioxide* in the unsulphited food
2.   The maximum recovery of additive from the food
3.   Quantitative recovery of *sulphur dioxide* when the sample is spiked with known amounts of the additive.

Whereas the testing of the first point leads to an unambiguous result, it does not automatically follow that methods satisfying requirements (2) and (3) are showing total recoverable *sulphur dioxide* in the food. The fact that a number of quite different analytical procedures give rise to similar results is encouraging in this respect, but it is, nevertheless, important that, when specifying the *sulphur dioxide* content of foods, the result be accompanied by a statement of the analytical method.

2.7 ANALYSIS OF PRODUCTS ARISING FROM REACTIONS OF SULPHUR(IV) OXOANIONS IN FOODS

Legislative control over the use of *sulphur dioxide* as a food additive is limited to the amount of the additive which is recoverable, and takes no account of that which is combined irreversibly. This section will survey some of the methods of analysis available for separation and identification of these stable products. The investigator may wish to carry out such analyses in order to understand the mechanism of the preservative action of the additive, to identify compounds of toxicological interest and to be able to account for the total amount of *sulphur dioxide* added to a food in terms of the unreacted and reacted additive. The first, somewhat obvious, question which arises is, how much of the additive is present in the form of organic products and how much in the form of inorganic compounds? One way in which this type of information may be obtained is through the use of combustion analysis or ashing methods when sulphited foods are prepared using ^{35}S-labelled additive.

2.7.1 Combustion analysis

The simplest form of combustion analysis is ignition of the sample in a crucible. Under these conditions, organic sulphur compounds in which the sulphur atom is not part of an acid group will be converted to gaseous products, whereas, in the case of the sodium salt of a sulphonic acid, thermal decomposition in the presence of oxygen leads to:

$$2RSO_3Na \xrightarrow{\;O_2\;} Na_2SO_4 + SO_3\uparrow + \text{other gaseous products}$$

If, on the other hand, the sodium salt is first converted to free sulphonic acid, only gaseous sulphur-containing products will be formed. The determination of residual sulphate after ignition of the salt, therefore, represents the determination of the metallic element, but the amount of sulphur in this form will be half of the amount of sulphonic acid originally present (Cheronis and Ma, 1964).

With regard to inorganic compounds, the heating of sodium sulphite to 1000 °C in the presence of oxygen shows that oxidation starts at 400 °C according to the equation,

$$2Na_2SO_3 + O_2 \rightarrow 2Na_2SO_4$$

reaching a maximum rate at 760 °C; the melting point of the product, sodium sulphate, being reached at 890 °C. Although sodium sulphate is, itself, quite stable with respect to the evolution of sulphur-containing gases over this temperature range, sodium hydrogen sulphate undergoes a two-stage decomposition:

$$2NaHSO_4 . H_2O \rightarrow Na_2S_2O_7 + 3H_2O$$
$$Na_2S_2O_7 \rightarrow Na_2SO_4 + SO_3$$

Evolution of sulphur trioxide takes place up to at least 800 °C and appears to be hindered by the presence of an excess of sodium sulphate (Erdey *et al.*, 1966). The conversion of disulphate to sulphate may be carried out quantitatively with greater ease by the addition of a small amount of powdered ammonium carbonate. This leads to decomposition of disulphate giving, as one of the products, the more volatile ammonium sulphate

$$Na_2S_2O_7 + (NH_4)_2CO_3 \rightarrow Na_2SO_4 + (NH_4)_2SO_4 + CO_2$$

When sodium disulphite is heated, decomposition leads to the loss of sulphur dioxide:

$$Na_2S_2O_5 \rightarrow Na_2SO_3 + SO_2\uparrow$$

The reaction takes place at around 170 °C, the sodium sulphite produced in this way being oxidised to sulphate as the temperature is further increased (Erdey *et al.*, 1966).

A popular combustion technique is that of oxygen-flask combustion in which the sample is burnt in an excess of pure oxygen in a large flask to retain both the gaseous product and the residue. Thus, it permits the determination of total sulphur in the sample without the need for solvent extraction and may be particularly useful when it is desired to determine the amount of a radio label in biological material with the minimum of solvent quenching. An example of a typical apparatus for this purpose, in which the sample is held on a platinum gauze and is ignited electrically, has been described by Dobbs (1966), radioactivity in the combustion products being determined by adding a known volume of scintillation fluid directly to the contents of the flask after cooling, and counting an aliquot once absorption of gases is complete. If, after combustion, the gases are separated from the solid residue, it is found that organic and inorganic compounds are decomposed on combustion, in a manner similar to that on ignition in a crucible, despite the much higher sample temperature in oxygen-flask combustion. Thus, the sodium salts of sulphonic acids and sodium disulphite give equal amounts of their sulphur in gaseous and solid products, whilst the amount of sulphur-containing gas, from sodium sulphite and sodium sulphate, is negligible. In addition, it is found that the sodium salts of hydroxysulphonates behave as do other sodium sulphonates, with similar behaviour expected of *S*-sulphonate compounds, when only the sulphur in the sulphonate group is taken into account (Herrera and Wedzicha, 1981).

When sulphur(IV) oxoanions have undergone partial reaction with components of foods which are subsequently dehydrated, it is expected that the sulphur(IV)-derived products mixture will contain the following organic and inorganic species:

Organic	*Inorganic*
hydroxysulphonates	sulphate
sulphonates	sulphite
	disulphite

It is assumed that at the pH of food all these components will exist as metal ion salts and that no significant amount of hydrogen sulphate is present. Therefore, the gaseous products will contain sulphur according to:

$$S_{gas} = S_{disulphite} + \tfrac{1}{2}S_{hydroxysulphonate} + \tfrac{1}{2}S_{sulphonate}$$

and the residue will contain sulphur according to:

$$S_{residue} = S_{sulphate} + S_{sulphite} + S_{disulphite} + \tfrac{1}{2}S_{hydroxysulphonate} + \tfrac{1}{2}S_{sulphonate}$$

where each quantity refers to molar amounts of the respective sulphur-containing compound. One additional measurement which is easily carried out is the analysis of recoverable sulphur dioxide, which would normally be carried out by a distillation technique such as the Monier-Williams method. Thus:

$$S_{Monier\text{-}Williams} = S_{sulphite} + 2S_{disulphite} + S_{hydroxysulphonate}$$

Thus, using the three results, it is possible to write,

$$S_{residue} - S_{gas} = S_{sulphate} + S_{sulphite}$$

and,

$$S_{total} - S_{Monier\text{-}Williams} = S_{sulphate} + S_{sulphonate}$$

where S_{total} is the sum of gaseous product and residue. If the conditions are such that the amount of sulphite present can be considered as negligible, the method gives directly the amount of sulphate in the mixture, from which the amount of stable sulphonate may easily be found. The decision as to whether the sulphite content may be neglected can, of course, be checked by means of an independent sulphate analysis. The sum of the amounts of sulphur present in the gaseous product and in sulphate should equal that in the residue. If $S_{residue} > S_{sulphate} + S_{gas}$, then sulphite is also present.

Both disulphite ion and hydroxysulphonate salts yield equal amounts of sulphur in gaseous product and in residue on combustion, whilst all the sulphur is recoverable by Monier-Williams distillation. Thus, the two components are indistinguishable by this technique. It does, however, offer a relatively simple method by which the conversion of sulphur(IV) oxoanions to stable sulphonates may be determined for solid samples, such as dehydrated foods, without the need for extraction methods and, thereby, minimises subsequent changes, although the extent to which the combustion itself leads to the formation of organo-sulphur compounds is not known. The speed of the oxygen-flask combustion method, giving complete combustion in a few seconds, is, perhaps, more desirable than the longer crucible ignition method in which the partly decomposed organic compound is held, at high temperature, in contact with sulphur(IV) and sulphur(VI) oxospecies. It must, however, be remembered that the technique is only one which essentially distinguishes ions bearing a single charge per sulphur atom from those bearing a double negative charge per sulphur atom. Naturally occurring sulphur-containing compounds will interfere with this analysis and it is, therefore, applicable only when specific determinations of sulphur originating from the additive can be carried out. This may conveniently be done with the help of ^{35}S-labelled additive, the amount of sulphur in each of the analyses being expressed in terms of ^{35}S activity.

One further question is whether the composition of gaseous combustion products with respect to oxides of sulphur can provide a clue as to the composition of the mixture analysed. Any such analysis is complicated by the equilibrium between sulphur dioxide, sulphur trioxide and oxygen:

$$SO_2 + \tfrac{1}{2}O_2 \rightleftharpoons SO_3$$

whose equilibrium constant has a value of unity in the region of 1000 K, sulphur trioxide being favoured at lower temperatures and dioxide at higher temperatures (Lovejoy et al., 1962). Further work is required to explore the potential of such measurement.

2.7.2 Identification and quantitative analysis of organic products derived from sulphur(IV) oxoanions

2.7.2.1 Labile adducts

Ample evidence is available for the formation of labile adducts of the hydroxysulphonate type, resulting from reactions between carbonyl compounds and sulphur(IV) oxoanions, from measurements of free and loosely bound *sulphur dioxide* in food. This result, however, only gives the total extent of binding, and does not provide information regarding the distribution of the additive among the carbonyl components. If the food in question is a liquid, such as wine, the amounts of individual hydroxy-sulphonates may be determined by calculation on the basis of the respective dissociation constants. An example of such an investigation applied to wine is provided by Burroughs and Sparks (1973*a*, *b*, *c*). Having established the identity and concentration of components expected to contribute to the binding of the additive, it may be necessary to determine dissociation constants of the adducts in question at the pH and in the ionic environment of the food. Unfortunately, unlike the materials available for fundamental studies of interactions between sulphur(IV) oxoanions and carbonyl compounds, it is often not possible to isolate the natural products in a pure dry state and frequently, therefore, the concentrations of solutions are uncertain. One simple method of overcoming this difficulty is seen on re-writing the law of mass action expression for decomposition of the adduct in the following form:

$$[\text{bound } sulphur\ dioxide] = [\text{free carbonyl}] - K\ \frac{[\text{bound } sulphur\ dioxide]}{[\text{free } sulphur\ dioxide]}$$

where K is the dissociation constant. If the free and bound *sulphur dioxide* content of equilibrium mixtures of carbonyl compound and sulphur(IV) oxoanion is determined, the dissociation constant may be found by plotting, for a series of such systems, the bound *sulphur dioxide* as a function of the ratio of bound to free *sulphur dioxide*.

In contrast to the relatively straightforward approach in simple aqueous systems, theoretical analysis of the distribution of the additive between the various carbonyl components of dehydrated sulphited foods is impossible. The only conceivable approach is that of solvent extraction followed by a separation technique, the duration of which is short compared with the lifetime of the adducts in question. No such investigations have yet been reported.

There has been some interest in the formation of organic products, derived from the additive, which are stable under conditions of Monier-Williams analysis and which, therefore, lead to irreversible combination of the additive. The remainder of this chapter will be concerned with the analysis of such products.

2.7.2.2 Stable organic products

If it is assumed that the highest level of use of *sulphur dioxide* as a food additive is probably in the region of 5000 ppm on a dry-weight basis in vegetables, for a target in the range 2000–2500 ppm after processing, the maximum amount, on a molar basis, is approximately $0.08\ \text{mol kg}^{-1}$. A 50% conversion to organic products ($0.04\ \text{mol kg}^{-1}$) is possible and if it is assumed that less than 5% of the reacted additive may be present as one particular product, the analytical method should be capable of isolating that product when its abundance is less than $0.002\ \text{mol kg}^{-1}$, which, for a product with a molecular weight of, say, 300, is 0.06% by weight. Furthermore, this value represents a maximum expectation since the amounts of the additive permitted in many foods are an order of magnitude smaller than the limit stated, and two orders of magnitude in some cases. It is very desirable, therefore, that the work-up procedure prior to analysis should involve selective concentration of the components of interest.

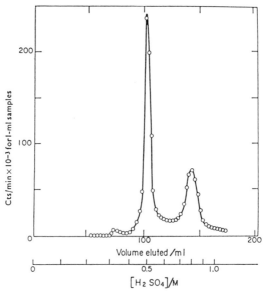

Fig. 2.15. Separation of ^{35}S-sulphite derived products in the reaction between ascorbic acid and sulphur(IV) oxoanion in the presence of glycine using a strongly basic resin in the $H_2PO_4^-$ form and the sulphuric acid gradient shown (Wedzicha and McWeeny, 1974a).

Fortunately all the known products formed by reaction between sulphur(IV) oxoanions and food components have the distinctive characteristic of being salts of acids stronger than any other acid present in the food and are, therefore, amenable to separation by ion exchange methods.

When working with solid foods such as dehydrated vegetables, an initial extraction operation is required, the use of water being, perhaps, the most sensible. Although salts of sulphonic acids are expected to be readily soluble in water, transfer to the aqueous phase should not be assumed to take place automatically, since these products may be associated with macromolecules or cell components which are insoluble in water. In the case of dehydrated cabbage, washing with cold water leads to the extraction of 94 % of free and combined *sulphur dioxide* (Panahi, 1981). In the case of other solid foods, this information may be obtained by pilot plant preparation in the presence of ^{35}S-labelled sulphur(IV) oxospecies as the additive. The product is subsequently analysed for total ^{35}S by oxygen-flask combustion and scintillation counting, extracted with solvent and the activity in the extract determined by scintillation counting.

Organic sulphonic acids are held firmly by strongly basic ion exchangers. Ingles (1960) separated glucose-6-sulphate from glucose and gluconic acid using Dowex-1 anion exchange resin in the carbonate form. Both gluconic acid and the sulphate interact with the column but may be eluted with 0·1 M sulphuric acid, the gluconic acid appearing first. Similarly, 3-sulphopropionic acid and 3,4-dideoxy-4-sulphohexosulose may each be separated from reaction mixtures containing glucose and its degradation products by this procedure (Ingles, 1962). To separate sulphonates into components, gradient elution is required. Good separation of product mixtures, formed by the reaction of sulphur(IV) oxoanions with ascorbic acid in the presence of glycine, may be achieved with a non-linear sulphuric acid gradient as shown in Fig. 2.15 (Wedzicha and McWeeny, 1974a), similar profiles being observable when extracts of dehydrated sulphited vegetables such as cabbage or swede are subjected to the analysis. Milder conditions for the separation of sulphonates are required when a DEAE medium (Pharmacia DEAE Sephacel or DEAE bonded to HPLC medium) is used as the anion exchanger. In the case of the former, linear gradient elution with sulphuric acid to a maximum concentration of 0·1 M is capable of liberating all the free and combined additive from the column, when extracts of sulphited dehydrated cabbage are analysed. A typical elution profile showing perhaps nine components derived from the additive is shown in Fig. 2.16 (Panahi, 1982). Low recoveries of sulphur(IV) oxoanions and reaction products may be found when fractionated column effluent is analysed. When ion exchange resins are employed, a possible reason for non-quantitative desorption of reaction products is their presence in a highly coloured band at the top of the column, presumably consisting of high molecular weight material, which cannot be removed with strong acid or alkali. This band is observed when products of 'sulphite inhibited' non-enzymic browning or extracts of a number of sulphited foods are subjected to this chromatographic analysis. No such irreversible adsorption is evident when DEAE Sephacel is used. A second cause of low recovery is the loss of gaseous sulphur dioxide from the acid column effluent arising either from the presence of free sulphur(IV) oxoanions or as a result of the decomposition of any labile adducts. A simple remedy is to mix column effluent with a dilute solution of hydrogen peroxide and to pass the mixture through a short mixing coil. Whilst such oxidation may lead to changes in the products, the method will still allow quantitative analysis of the distribution of the additive amongst the product fractions when ^{35}S-labelled additive is used.

The use of sulphuric acid as the eluant is particularly convenient since it

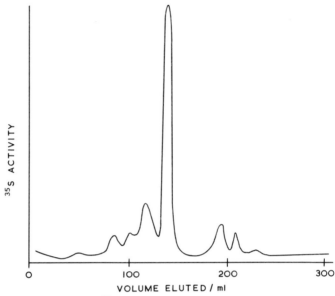

Fig. 2.16. Separation of ^{35}S-sulphite derived products in dehydrated cabbage using DEAE Sephacel and a linear sulphuric acid gradient (Panahi, 1982).

may subsequently be removed from the fractions by treatment with barium carbonate,

$$BaCO_3 + H_2SO_4 \rightarrow BaSO_4 + CO_2 + H_2O$$

the reactant being sparingly soluble in water whilst the product is insoluble. Thus, fractions treated in this way will have an ionic strength largely determined by ionic components eluted from the column, all acids being converted to the barium salts. One disadvantage of the use of barium carbonate is the practical problem connected with an insoluble reactant and product, the conversion of carbonate to sulphate presumably taking place at the surface of particles. Thus, for efficient conversion small particle size is required. A second disadvantage is that once neutralisation is complete an excess of barium carbonate will be present. The dilute solution will be hydrolysed to some extent and may catalyse the hydration of glyoxal-type groups which are known to be present in intermediates of carbohydrate breakdown:

$$\begin{array}{ccc} CHO & & COOH \\ | & & | \\ C{=}O & \rightarrow & CHOH \\ | & & | \\ R & & R \end{array}$$

The normal method of carrying out this conversion is to use limewater, but it is found that appreciable amounts of the acid are formed in the presence of barium carbonate (Wedzicha and McWeeny, 1974a). The presence of weakly alkaline conditions will, however, promote the decomposition of any remaining hydroxysulphonate-type adducts, the liberated sulphur(IV) oxoanion either being precipitated as barium sulphite or oxidised to barium sulphate and precipitated, or remaining as barium sulphite in solution.

The products isolated in this way include 2,5-dihydroxy-4-sulpho-pentanoic acid, from the thermal degradation of xylose in the presence of sulphur(IV) oxoanions at pH 6·5 (Yllner, 1956); glucose-6-sulphate from the reaction of glucose with sodium hydrogen sulphite (Ingles, 1960); 3,4-dideoxy-4-sulphohexosulose from the reaction of glucose with sulphur(IV) oxoanions at pH 6·5 (Ingles, 1962); and 3,4-dideoxy-4-sulphopentosulose, from the degradation of ascorbic acid in the presence of sulphur(IV) oxoanions at pH 3·0 (Wedzicha and McWeeny, 1974a). The products are not crystallisable as simple salts but as salts with organic bases such as brucine; crystallisation, nevertheless, is difficult, but is facilitated by the addition of acetone to concentrated aqueous solutions.

Further separation of products in chromatographic fractions has been reported for components containing the carbonyl grouping. Thus, the 2,4-dinitrophenylhydrazone of 3,4-dideoxy-4-sulphopentosulose runs as a characteristic purple spot by thin layer chromatography on silica gel G at R_f 0·42 when developed with butanol:0·880 ammonia (4:1) solvent (Wedzicha and McWeeny, 1974a), and may be separated from 3-deoxypentosulose, its precursor in ascorbic acid–sulphur(IV) oxoanion reactions, whose derivative runs at R_f 0·86. It is interesting that the hydroxysulphonate of 3-deoxypentosulose is eluted from strongly basic ion exchangers under the same conditions (using sulphuric acid as the eluant) as the 4-sulpho- compound and, therefore, after treatment with barium carbonate during the subsequent work-up, the product containing the sulphonate grouping is contaminated with one containing no sulphur. The hydroxysulphonate-type adduct of 3,4-dideoxy-4-sulphopentosulose is eluted at higher acid concentration than the free aldehyde; subjecting the column effluent containing this adduct to the barium treatment gives rise to a product which, when re-run on the ion exchange column, co-chromatographs with free aldehyde (Wedzicha and Imeson, 1977). The conclusion from further analysis of the two major fractions shown on the chromatogram in Fig. 2.15 is that both contain the 4-sulpho- compound, the splitting into two fractions being a result of reversible combination at the aldehyde grouping. Thus, it is expected that chromatograms of mixtures of such reaction products may well appear more complex than warranted by the number of distinct, non-labile products, a feature which may also be variable according to the experimental conditions of the separation (e.g. acid concentration, time). Apart from providing an indication of the position at which the inorganic components run, no identification of fractions shown on the chromatogram in Fig. 2.16 has been carried out.

The structure of 2,5-dihydroxy-4-sulphopentanoic acid may be demonstrated by oxidative degradation with lead tetraacetate followed by oxidation of the resulting aldehyde with peroxoacetic acid (Yllner, 1956),

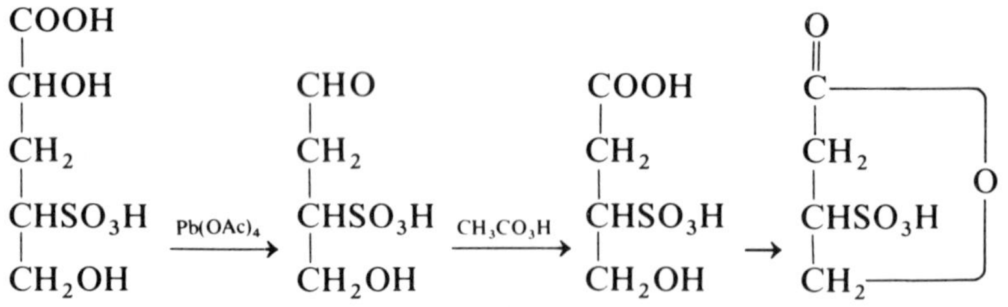

The resulting lactone is identical to that which may be prepared by reaction of the unsaturated compound, γ-hydroxyisocrotonolactone, with sulphur(IV) oxoanions:

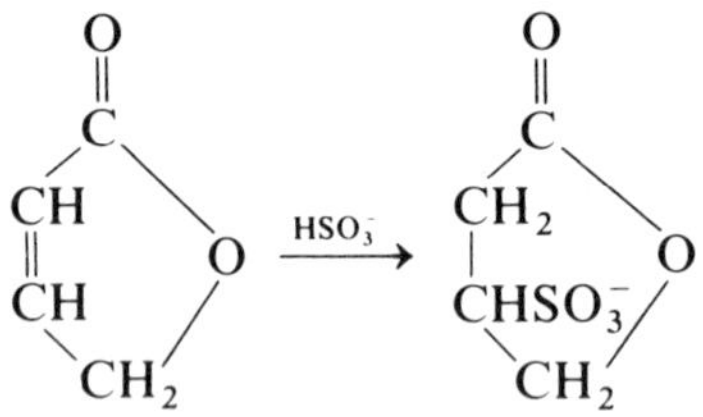

The structure of 3,4-dideoxy-4-sulphohexosulose may also be demonstrated by classical methods. The formation of a 2,4-dinitrophenylhydrazone which is blue in alkaline solution is indicative of the presence of an α-dicarbonyl grouping. Oxidation of the carbonyl compound with hypoiodite shows that the two carbonyl groups are found at C_1 and C_2. Acetylation of the 2,4-dinitrophenylhydrazone gives a diacetate, implying two hydroxy groups, and periodate oxidation leads to the formation of 0·8 mol formaldehyde per mol compound, confirming that the groups are adjacent and that one is terminal, that is, they are located in positions 5 and

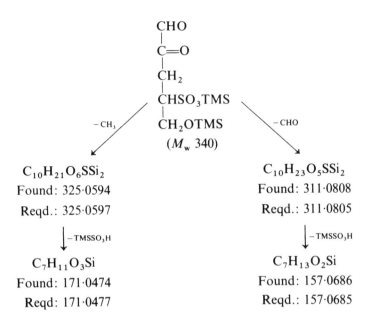

Fig. 2.17. Primary mass spectrometric fragmentation of the trimethylsilyl (TMS) derivative of 3,4-dideoxy-4-sulphopentosulose (Wedzicha and McWeeny, 1974a).

6. Oxidation of the underivatised compound with periodate appears to give anomalous results with the formation of formaldehyde and formic acid (Ingles, 1962). Although the successful isolation of sulphosuccinic acid from the chromic acid oxidation of the compound in question has not been reported, Cordingly (1959) has isolated the product from mixtures of sulphonic acids produced from xylose and arabinose by reaction with sulphite ion. The assignment of the sulphonic acid group to position 4 is on the basis of possible pathways for the formation of sulphonated products.

Both 3,4-dideoxy-4-sulphohexosulose and 3,4-dideoxy-4-sulphopento-sulose may be rendered sufficiently volatile for analysis by mass spectrometry by conversion to trimethylsilyl derivatives, the reaction being conveniently carried out using an excess of trimethylsilyl imidazole. The primary fragmentation pattern of the five-carbon compound is shown in Fig. 2.17 and is very similar to that of the six-carbon compound (Knowles and Eagles, 1971). Trimethylsilyl derivatives rarely give rise to molecular ions, but the presence of an M-15 ion is typical of the group. The loss of TMS-SO_3H (m/e 154) appears to be characteristic of sulphonic acids of this type. High resolution mass spectra at m/e 157, 171, 311 and 325 of

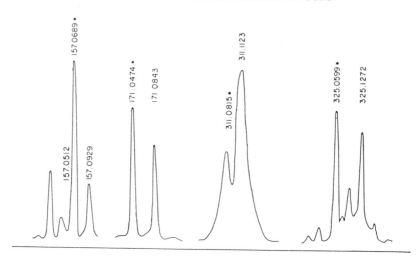

Fig. 2.18.   High resolution mass spectra at m/e 157, 171, 311 and 325 for the trimethylsilyl derivative of extract from stored, sulphited cabbage. Peaks marked * correspond to the derivatives of 3,4-dideoxy-4-sulphopentosulose for which expected masses are 157·0685, 171·0477, 311·0805 and 325·0597 respectively (Wedzicha and McWeeny, 1974b).

trimethylsilyl derivatives of acidic components of an aqueous extract of stored, sulphited, dehydrated cabbage are shown in Fig. 2.18 and confirm, beyond doubt, the presence of 3,4-dideoxy-4-sulphopentosulose in the dehydrated vegetable. The method also demonstrates the presence of 3,4-dideoxy-4-sulphohexosulose and some 2,5-dihydroxy-4-sulphopentanoic acid, the latter being formed during the isolation of acidic components with the help of barium carbonate (Wedzicha and McWeeny, 1974b). Mass spectra of a control consisting of an extract of fresh cabbage subjected to the same work-up procedure show no peaks at any of the masses in question. The low resolution spectrum of the trimethylsilyl derivatives of acidic components of sulphited cabbage shows a not very surprising distribution of fragments to at least $m/e = 1000$. An interesting observation is the presence of high molecular weight fragments to m/e 655 when the first major chromatographic fraction of the separation shown in Fig. 2.15 is converted to the trimethylsilyl derivative and subjected to mass spectrometry. This mixture contains 3,4-dideoxy-4-sulphopentosulose and 3-deoxypentosulose, the latter being liberated as such during the treatment with barium carbonate in the work-up procedure. A possible scheme for the formation of a product whose trimethylsilyl derivative has a molecular weight of 670 (that is, $655 + 15$) and involving both components of the mixture is as follows:

$$
\begin{array}{c}
\text{CHO} \\
| \\
\text{C}=\text{O} \\
| \\
\text{CH}_2 \\
| \\
\text{CHSO}_3\text{H} \\
| \\
\text{CH}_2\text{OH}
\end{array}
\;+\;
\begin{array}{c}
\text{CHO} \\
| \\
\text{C}=\text{O} \\
| \\
\text{H}_2\text{C} \\
| \\
\text{CHOH} \\
| \\
\text{CH}_2\text{OH}
\end{array}
\;\rightarrow\;
\begin{array}{c}
\quad\quad\quad\text{CHO} \\
\quad\quad\quad| \\
\text{CHO}\quad\text{C}=\text{O} \\
\backslash\;\diagup \\
\text{C}=\text{C} \\
\diagup\quad\backslash \\
\text{CH}_2\quad\text{CHOH} \\
| \quad\quad | \\
\text{CHSO}_3\text{H}\;\;\text{CH}_2\text{OH} \\
| \\
\text{CH}_2\text{OH}
\end{array}
$$

$$\Updownarrow$$

$$
\begin{array}{c}
\quad\quad\text{CHO} \\
\quad\quad| \\
\text{CHO}\quad\text{C—OTMS} \\
\backslash\;\diagup \\
\text{C—C} \\
\diagup\quad\backslash \\
\text{CH}\quad\quad\text{CHOTMS} \\
\| \quad\quad | \\
\text{CHSO}_3\text{TMS}\;\;\text{CH}_2\text{OTMS} \\
| \\
\text{CH}_2\text{OTMS}
\end{array}
\;\;\xleftarrow[\text{reagent}]{\text{Silylating}}\;\;
\begin{array}{c}
\quad\quad\text{CHO} \\
\quad\quad| \\
\text{CHO}\quad\text{C—OH} \\
\backslash\;\diagup \\
\text{C—C} \\
\diagup\quad\backslash \\
\text{CH}\quad\quad\text{CHOH} \\
\| \quad\quad | \\
\text{CHSO}_3\text{H}\;\;\text{CH}_2\text{OH} \\
| \\
\text{CH}_2\text{OH}
\end{array}
$$

The accurate mass of the ion at m/e 655 is 655·2066, which compares favourably with a value of 655·2100 calculated for the above product with the mass of CH_3 taken off. The proposed structure is further supported by an accurate mass of 697·4744 for the derivative prepared using the fully deuterated silylating agent, in good agreement with a value of 697·4736 calculated for M-CD_3. Both the normal and deuterated compounds also show ions resulting from loss of CH_2OTMS, presumably by cleavage of a terminal group. If this group was the one adjacent to the sulphonate group of the product, resonance stabilisation of the resulting ion could account for its relatively high abundance (Wedzicha and McWeeny, 1975). The stage at which this product is formed is not known, but it is likely that the high temperature and basic environment during the formation of the trimethylsilyl derivative and subsequent analysis on the mass spectrometer probe could be favourable to aldol condensation followed by dehydration to give the first product shown in the above scheme.

   The S-sulphonates produced through cleavage of simple aliphatic and aromatic disulphides by sulphite ion can be separated on paper, cellulose and silica gel; thus, for example, on paper chromatography a mixture of cysteine-S-sulphonate and glutathione-S-sulphonate, using butanol:acetic acid:water (5:2:3) as solvent, is well resolved and separated from sulphur(IV) oxospecies (Nakamura and Tamura, 1975b). In foods it is expected that any S-sulphonates formed will consist of the functional group attached to peptides or proteins of significant size and, therefore, the methods of separation will be those applicable to the protein fragment rather than to low molecular weight S-sulphonates. A chemical method for the detection of S-sulphonates is based upon nucleophilic displacement of sulphite ion using a thiol and subsequent identification of the liberated sulphur(IV) species. Nakamura and Tamura (1974) describe the use of dithiothreitol, a reagent known to reduce disulphide bonds, for the decomposition of S-sulphonates. The reaction proceeds with a 1:1 stoichiometry, verified for cysteine-S-sulphonate, the products being sulphur(IV) oxoanion, the corresponding thiol and oxidised dithiothreitol:

$$
RSSO_3^- +
\begin{array}{c}
\quad CH_2 \\
CHOH \quad SH \\
CHOH \quad SH \\
\quad CH_2
\end{array}
\rightarrow RSH + HSO_3^- +
\begin{array}{c}
\quad CH_2 \\
CHOH \quad S \\
CHOH \quad S \\
\quad CH_2
\end{array}
$$

No reaction occurs when the reagent is mixed with thiosulphate ion. When components are being detected on chromatography paper or plates, the liberated sulphur(IV) oxoanion may be fixed with mercury(II) chloride solution and visualised by spraying with a mixture of pararosaniline and formaldehyde in acid (Nakamura and Tamura, 1975b). The presence of sulphur(IV) oxospecies causes interference. The procedure is easily adaptable to quantitative analysis of S-sulphonates: thus, after treatment of a sample with dithiothreitol, thiols are precipitated with an excess of mercury(II) chloride and the supernatant assayed with pararosaniline reagent. Alternatively, a procedure in which thiols are masked using p-chloromercuribenzoic acid and sodium arsenite followed by fluorimetric determination of sulphur(IV) oxoanions using N-(p-dimethylamino-phenyl)-1,4-naphthoquinoneimine (see Section 2.3.4) may be adopted (Nakamura and Tamura, 1974). The analysis will, of course, include any sulphur(IV) oxoanion originally present, which must be determined beforehand, and the amount of S-sulphonate is thus found by difference.

There are very few reports of the quantitative analysis of specific compounds derived from the interaction of sulphur(IV) oxoanions with components in foods. The difficulties associated with such analyses may be reduced by the use of ^{35}S-labelled additive in production of the food.

Hence, a useful method for following the production of S-sulphonate during biscuit manufacture involves the use of ^{35}S-sulphite as the additive to flour and analysis based upon the hydrolytic reaction of S-sulphonates to sulphate described in Section 1.11.9. Thewlis and Wade (1974) thus measured the amount of ^{35}S-sulphate present in biscuit samples before and after acid hydrolysis in order to determine, by difference, the amount of S-sulphonate present. In a large number of cases, including the production of dehydrated vegetables where large amounts of the additive may be incorporated into the food, it is not possible to carry out preparations involving radiolabelled additive on a sufficiently large scale for the product to be representative of the commercially produced food, nor is it possible to simulate the variable time/temperature conditions encountered during subsequent storage before sale. In order to be able to state the conversion of sulphur(IV) oxoanion to stable organic product in a commercial sample, it is necessary to devise analytical methods which can be carried out on inactive components. The only such analysis reported is that described by Wedzicha and McWeeny (1975) for 3,4-dideoxy-4-sulphopentosulose and 3,4-dideoxy-4-sulphohexosulose in dehydrated cabbage, swede, white wine and lemon juice. The procedure is an isotope dilution assay based upon:

(a)   the incorporation of ^{35}S-labelled 3,4-dideoxy-4-sulphopentosulose into the food product,

(b)   separation, by means of ion exchange, of the sulphonates from other components,

(c)   conversion of carbonylic sulphonates to their 2,4-dinitrophenyl-hydrazine derivatives,

(d)   separation, by thin layer chromatography, of the derivatives of the compounds of interest from other components,

(e)   comparison of the ^{35}S: colour intensity of the recovered derivative with that of the derivative of the labelled sulphonate as added. (The determination of colour intensity is carried out directly using a thin layer densitometer, the activity in the spot being subsequently found by transferring the component to a scintillation vial and counting).

The success or failure of this procedure is determined by the resolution and specificity of the final separation on the thin layer (silica gel G with butanol:0·880 ammonia (4:1) solvent) and adherence to a calibration curve of the relationship, for unknowns, between amount of derivative present and its integrated optical density on the plate. Unfortunately, the thin layer chromatography is not capable of resolving the two sulphonates of interest, a spot at R_f 0·42 being evident in both cases. However, absorbances for equimolar amounts of derivatives of these sulphonates are found to be almost identical (that is, within 5%) and on the basis of the co-precipitation, co-chromatography and equimolar colour intensity, the single labelled compound can be used as a tracer for both products; in isotope dilution studies it will label the combined pool and the extent of dilution will depend upon the combined concentrations. Up to 4 μg of dinitrophenylhydrazone may be applied to the thin layer plates (250 μm) before deviation from Beer's law becomes apparent, the possibility of high, localised concentration being detectable by the use of a flying spot densitometer. The specific activity of labelled sulphonate required for this analysis is in the region of 1500 cpm/μg and is prepared by the reaction of xylose with labelled sulphur(IV) oxoanion, followed by ion exchange separation and isolation.

A development which would allow facile analysis of such products is that of high resolution chromatography, enabling either the carbonyl compounds or their derivatives to be separated from other components of extracts of foods. This is potentially possible by the use of high performance liquid chromatography, but no such studies have yet been reported.

3

Control of non-enzymic browning

3.1 INTRODUCTION

When the browning of food has taken place, it is generally difficult to ascertain whether the mechanism has been enzymic or non-enzymic unless the conditions under which the food has been treated are such that enzymes are inactivated. In the latter case, only non-enzymic browning is said to occur although, strictly, colour could develop non-enzymically from intermediates formed through enzyme-mediated oxidations which have taken place before the enzymes were inactivated. However, when considering non-enzymic browning, it is generally understood that reference is being made to one of the following reactions:

(a) Maillard browning or the reaction between a reducing sugar and an amino-compound
(b) ascorbic acid browning, that is the spontaneous thermal decomposition of ascorbic acid under both aerobic and anaerobic conditions and either in the presence or absence of amino-compound
(c) caramelisation or the pyrolysis of food carbohydrate
(d) lipid browning, which is probably oxidative deterioration followed by polymerisation.

In many situations the term, non-enzymic browning, is synonymous with Maillard browning (Mauron, 1981), which is the most widely studied of the non-enzymic browning reactions. The list of reactions covers a diversity of chemical types with regard to reactants and involves a variety of reaction conditions. However, a common feature is the participation of carbonyl compounds as the reactive intermediates and also the ability of sulphur(IV) oxoanions to inhibit, at least partially, the formation of coloured products. These products are frequently referred to as melanoidins and probably consist of a mixture of discrete, high molecular weight components (Motai and Inoue, 1974) rather than having the nature of random polymers of indeterminate length. Usually, the yellow colour is due to the tailing of an intense ultraviolet peak into the visible region, and there is no single wavelength at which non-enzymic browning can be quantified. The measurement of the colour is further complicated by the fact that the products mixture is heterogeneous, each component having the possibility of absorbing at a different wavelength. When considering the promotion or inhibition of non-enzymic browning reactions as a result of a particular treatment, it is possible to measure, as a function of the treatment, either the change in subjective colour or change in the amounts of identifiable intermediates which are formed. The two measurements need not bear any simple relationship to one another. In the case of subjective colour measurement, an integrated result incorporating the portion of the spectrum of the product in the visible region and the spectral response of the observer is preferable to a single wavelength measurement at, say, 470 nm. The former measurement is more in keeping with the reason for carrying out such an experiment, that is to assess the effect of a particular treatment on the quality of a food product with respect to perceived colour, but it is somewhat more difficult to carry out than is a single wavelength measurement. It should be evident that absorbance in the visible region of the spectrum cannot be simply correlated to concentration. Also, when a particular treatment gives a considerable reduction in absorbance in the visible region, it does not imply that the treatment has necessarily had a profound effect on the system. Indeed, the loss of colour may be due to the modification of a minor component or the interruption of a minor reaction pathway.

The relevant known features of the various non-enzymic browning reactions will now be reviewed.

3.2 MECHANISMS OF NON-ENZYMIC BROWNING REACTIONS

3.2.1 Maillard browning

This reaction was first described by Louis Maillard (Maillard, 1912) as the formation of coloured products on heating a mixture of reducing sugars

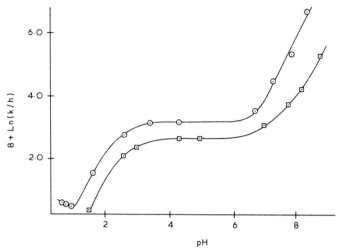

Fig. 3.1. Effect of pH on the rate constant of the xylose–glycine and xylose–alanine reactions at 65 °C when the concentration of all reactants is 0·25 M: glycine, ⊙; alanine, ⊡ (Wolfrom *et al.*, 1953*a*). Reprinted with permission from *J. Amer. Chem. Soc.*, 1953, **75**, page 3473. © 1953 American Chemical Society.

and amino acids. It takes place readily at all pH values found in food systems, Fig. 3.1 showing the pH dependence in the range pH 1–8 when xylose is the aldose and glycine and alanine are the amino-compounds (Wolfrom *et al.*, 1953*a*). The reaction leading to the formation of colour is independent of pH in the range expected for foods (pH 3–7). Although amino acids readily induce the browning reaction, it takes place with most amines, including peptides and proteins, and the reaction involving ammonia is widely used in the manufacture of ammoniated caramels. Reviews on the chemistry of the reaction include those by Hodge (1953), Ellis (1959), Reynolds (1963, 1965) and a series of papers (Mauron, 1981; Feather, 1981; Olsson *et al.*, 1981; Namiki and Hayashi, 1981; Baltes *et al.*, 1981; Kato and Tsuchida, 1981).

The early stages of the Maillard reaction are essentially those of amine assisted sugar dehydration. The sequence of events is initiated by the condensation reaction between the carbonyl group of the aldose and the amine (a result of nucleophilic attack by the amine) to give the corresponding aldosylamine. Such adducts form readily when aldoses are heated with primary amines in water and separate out on cooling (Weygand, 1939). Primary amines should be able to form dialdosylamines; one such product which has been isolated is di-D-glucosylamine in which the amine is ammonia (Ellis and Honeyman, 1955). The aldosylamine next undergoes a facile and irreversible rearrangement, the Amadori rearrangement (Hodge, 1955), to give the 1-amino-1-deoxy-2-ketose or, more simply, ketoseamine. The complete sequence of events from aldose to ketoseamine is, therefore (Mauron, 1981):

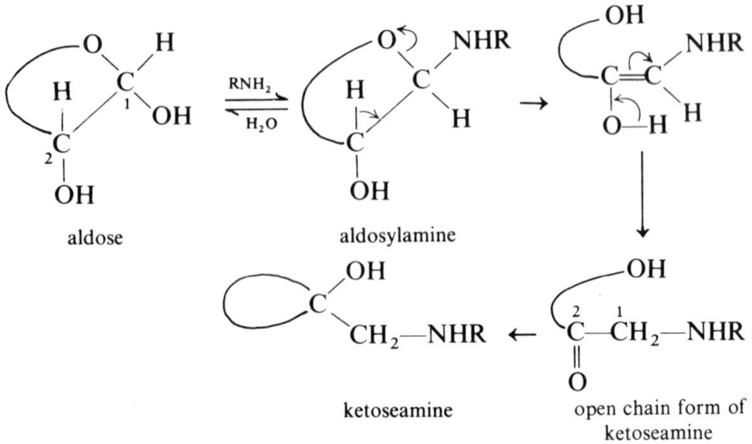

aldose   aldosylamine

ketoseamine   open chain form of ketoseamine

Thus, if the aldose is D-glucose, then the Amadori product will be the 1-amino-1-deoxy-D-fructose. It is evident that the ketoseamine still bears a replaceable hydrogen on its amino group and, since secondary amines react with aldoses to form aldosylamines with a subsequent Amadori rearrangement (Hodge and Rist, 1952), it is not surprising that the Amadori product shown above will react with a second mole of aldose, undergo a second Amadori rearrangement and, thereby, give rise to a diketoseamine. The model of the Maillard reaction most frequently referred to is a mixture of glucose and glycine. The reaction between these two compounds to give the corresponding glycosylamine is rapid and the product rearranges to N-(carboxymethyl)-1-amino-1-deoxy-D-fructose, known usually as mono-fructoseglycine (MFG) (Abrams et al., 1955; Richards, 1956; Anet, 1957; Duborg and Devillers, 1957). Further reaction of MFG with glucose yields the diketoseamine which is commonly referred to as difructoseglycine (DFG) (Anet, 1959a). The mono- and di-ketoseamines are regarded as the key intermediates in Maillard browning, coloured compounds resulting from their degradation and polymerisation. Difructoseglycine appears to be much less stable than the monoketoseamine, decomposing spontaneously with a maximum rate in the region of pH 5·5 to a quantitative yield of the monoketoseamine and nitrogen-free carbonyl compounds (Anet, 1959a). The rate of colour development, as judged from absorbance measurement at 440 nm, is also much greater when difructoseglycine is allowed to decompose in the presence of amino acids compared with when monofructoseglycine or glucose is used (Anet, 1959*b*). The main product of the decomposition of difructoseglycine is 3-deoxy-D-*erythro*-hexosulose in 80–90 % yield (Anet, 1960*a*, *b*). This compound will be referred to simply as 3-deoxyhexosulose or 3DH, and is accompanied by two unsaturated compounds, the *cis*- and *trans*-isomers formed by the removal of water between positions 3 and 4 of 3DH (Anet, 1962*a*) and known collectively as 3,4-dideoxyhexosulos-3-ene. (The *Chemical Abstracts* names for the *cis*- and *trans*-compounds are 3,4-dideoxy-D-*glycero*-hexopyranosulos-*cis*-3-ene and 3,4-dideoxy-D-*glycero*-hexosulos-*trans*-3-ene hydrate, respectively.) The scheme for the decomposition of difructoseglycine is, therefore:

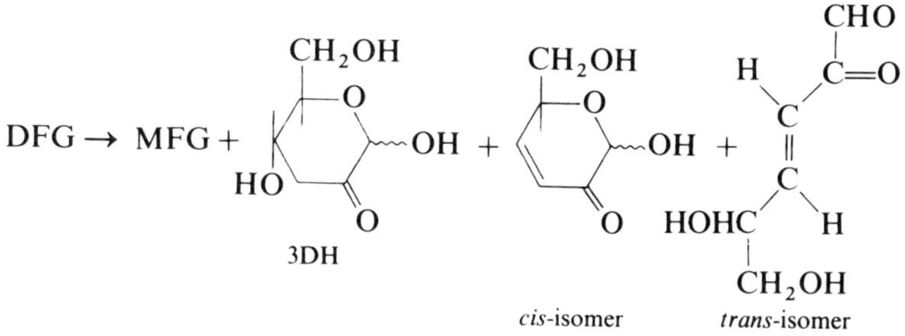

The amount of *cis*-isomer accounts for some 5 % of the product whilst the amount of *trans*-isomer is variable depending on pH, increasing from 1 to 10 % as the initial pH of the difructoseglycine is lowered from 5 to 3·5 (Anet, 1962*a*).

The nature of the product arising from the degradation of the monoketoseamine depends on whether the Amadori compound undergoes 1,2- or 2,3-enolisation, the former leading to the formation of osuloses as follows (Anet, 1960*b*):

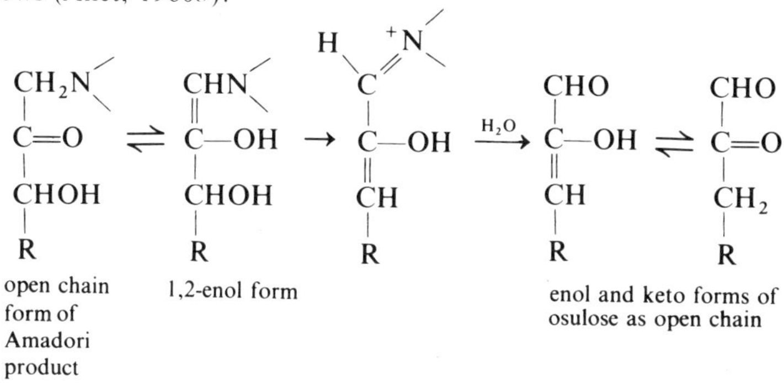

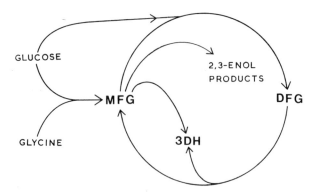

Fig. 3.2. Simplified scheme for the early stages of the reaction between glucose and glycine. Adapted from McWeeny and Burton (1963) with permission of D. J. McWeeny.

In the case when glucose is the aldose, $R = (CHOH)_2CH_2OH$. In contrast, 2,3-enolisation leads to a suspected dehydroreductone intermediate:

$$\begin{array}{cccc}
CH_2N\diagdown & CH_2N\diagdown & CH_2 & CH_3 \\
\mid & \mid & \parallel & \mid \\
C{=}O & C{-}OH & C{-}OH & C{=}O \\
\mid & \parallel & \parallel & \mid \\
CHOH & C{-}OH & C{=}O & C{=}O \\
\mid & \mid & \mid & \mid \\
R & R & R & R \\
 & \text{2,3-enol form} & &
\end{array}$$

Whereas the osuloses have been isolated and identified in non-enzymic browning systems, the existence of the 1-deoxy-2,3-dicarbonyl inter-mediate is proposed on the basis of its possible cyclisation to maltol and *iso*-maltol (Hodge *et al.*, 1963, 1972):

OH
$COCH_3$
iso-maltol

O
OH
CH_3
maltol

1,2-Enolisation is favoured when the amino group of the Amadori compound is almost completely protonated, that is in acid solution, whereas in weakly acidic and neutral media, 2,3-enolisation is favoured (Feather, 1981).

A simplified scheme, based upon one devised by McWeeny and Burton (1963), showing the relationships between products in the early stages of the glucose–glycine reaction is shown in Fig. 3.2. Almost all the known mechanisms whereby sulphur(IV) oxoanions inhibit the consequences of Maillard reactions involve pathways connected with 3-deoxyhexosulose and, therefore, the reactions of this intermediate will now be discussed in detail. 2,3-Enolisation reactions, however, ultimately lead to the formation of low molecular weight sugar fragments such as acetaldehyde, pyruvaldehyde, diacetyl and acetic acid, as well as being able to cyclise to oxygen heterocycles, and are, therefore, capable of producing distinctive flavours, even if the extent to which they occur is relatively small.

The simplest reaction of 3-deoxyosuloses is their hydration to the *meta*-saccharinic acid (Machell and Richards, 1960; Anet, 1960a, 1961):

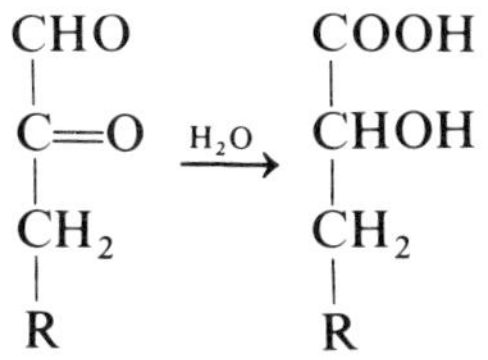

$$\begin{array}{ccc}
\text{CHO} & & \text{COOH} \\
| & & | \\
\text{C}{=}\text{O} & \xrightarrow{\text{H}_2\text{O}} & \text{CHOH} \\
| & & | \\
\text{CH}_2 & & \text{CH}_2 \\
| & & | \\
\text{R} & & \text{R}
\end{array}$$

It takes place readily in limewater, a solution of 0.02 M 3-deoxyhexosulose in 0.48 M calcium hydroxide being some 60% converted to the acid after 80 min at 25 °C and almost quantitatively hydrated after 24 h (Anet, 1961). Perhaps the most well known reaction of 3-deoxyosuloses is their acid catalysed dehydration in which two moles of water are lost leading ultimately to a 2-furaldehyde product:

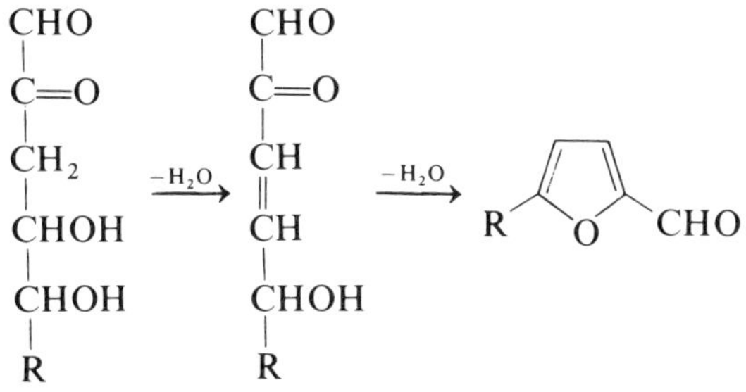

$$\begin{array}{ccccc}
\text{CHO} & & \text{CHO} & & \\
| & & | & & \\
\text{C}{=}\text{O} & & \text{C}{=}\text{O} & & \\
| & \xrightarrow{-\text{H}_2\text{O}} & | & \xrightarrow{-\text{H}_2\text{O}} & \\
\text{CH}_2 & & \text{CH} & & \\
| & & \| & & \\
\text{CHOH} & & \text{CH} & & \\
| & & | & & \\
\text{CHOH} & & \text{CHOH} & & \\
| & & | & & \\
\text{R} & & \text{R} & &
\end{array}$$

When the osulose is derived from glucose, the heterocyclic product is 5-hydroxymethyl-2-furaldehyde ($R = CH_2OH$) or HMF. As shown above, the structure of the osulose is commonly drawn in the keto form at C_2 whether or not a pyranose representation is made. It is, however, likely that the predominant form of the compound in the reaction is the enol form at C_2 with negligible conversion to the keto form on the basis of the following evidence. Aldoses such as glucose undergo acid catalysed dehydration without the participation of amines, albeit under more severe conditions of temperature, and the later stages of the reaction leading to 2-furaldehyde proceed by the same mechanism as proposed for the amine-assisted reaction. When glucose or xylose is dehydrated in acid to HMF or furfural respectively, in deuterated and tritiated water respectively, it is found that no label is incorporated at position 3 in the furan ring (Feather and Harris, 1968, 1970; Feather, 1970; Feather et al., 1972). This position corresponds to position 3 on the osulose which is the activated methylene group of the keto form and is, therefore, labile, capable of exchanging with the solvent. Thus, no keto form appears to be present. On the other hand, when Amadori products are subjected to acid-catalysed decomposition to the furan derivatives, substantial incorporation of label is observed at position 3 on the ring (Feather and Russell, 1969) but is attributable to the sequence of steps involving the amine rather than to the osulose or to its subsequent dehydration (Feather, 1981). In commenting on the work, Anet (1970) points out that primary isotope effects and solvent isotope effects may mask the correct conclusion since deuteron exchange could well proceed several times more slowly than proton exchange. Although these isotope effects are proceeding to some extent, the reaction is too complex to be treated in an exact manner. However, the sensitivity of labelling is of the order of 0·01 atoms of tritium per molecule for experiments carried out in tritiated water (namely, for xylose), and the furfural actually contained 0·5% of the activity of the solvent after acid catalysed dehydration of xylose (Feather et al., 1972), leaving a large margin for kinetic isotope effects.

The 3,4-dideoxyhexosulos-cis-3-ene dehydrates to HMF much more rapidly than does the trans-isomer or 3-deoxyhexosulose, suggesting that the cis-isomer is the intermediate (Anet, 1962b). Also, only the cis-isomer has been extracted in the conversion of 3-deoxyhexosulose to HMF (Anet, 1962a). If account is taken of the fact that 3-deoxyhexosulose probably exists exclusively in a 1,5-pyranose ring form and that it reacts in the enol-form which can only be cis-, then only the cis-form of the unsaturated osulose can be produced if the ring is not to be disrupted:

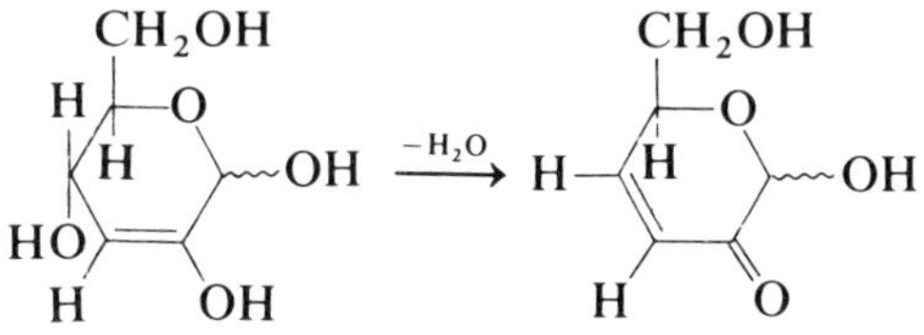

enol form of
3-deoxyhexosulose

The acyclic form of 3-deoxyhexosulose in either the keto- or enol-form (the latter probably being *trans-*) could yield both the *cis-* and *trans-* unsaturated intermediates (Anet, 1962*b*).

It is now pertinent to consider the relationship of the reactive intermediates to the coloured products observable in the later stages of the reaction. Kinetic evidence suggests that the 3-deoxyosulose is the main precursor in pigment formation and that the colour-forming reaction departs from the sequence:

$$\text{3-deoxyosulose} \rightarrow \text{3,4-dideoxyosulos-3-ene} \rightarrow \text{2-furaldehyde}$$

before reaching the furan compound (Reynolds, 1965). The latter conclusion is based upon the relatively low reactivity of hydroxymethyl-furfural towards browning when compared with difructoseglycine or 3-deoxyhexosulose (Anet, 1958, 1959*a, b*; McWeeny and Burton, 1963) and is supported by the observation of Yoshihiro and coworkers (1961) that the rate of the glucose–glycine reaction is not markedly accelerated by the addition of hydroxymethylfurfural. The production of colour also requires the presence of amino compound and, therefore, at the simplest level, the reactions are regarded as taking place between either 3-deoxyosulose or 3,4-dideoxyosulos-3-ene and the amino compound, followed by conden-sation of the products with each other or with nitrogen-free intermediates. There is an extensive amount of evidence confirming that the intermediates in question will, indeed, react with amino acids since N-substituted pyrrole-2-aldehydes have been isolated as one of the most abundant reaction products (Kato, H., 1967; Kato and Fujimaki, 1970; Shigematsu *et al.*, 1971) from Maillard reaction systems, but are assumed to be by-products rather than intermediates in melanoidin formation (Kato and Fujimaki, 1968; Sonobe *et al.*, 1977).

Both pathways for the breakdown of Amadori product lead to the formation of dicarbonyl compounds (diketo or glyoxal groupings) and, therefore, a reaction common to both sets of intermediates is the Strecker degradation reaction (Schönberg and Moubacher, 1952). The mechanism involves formation of a Schiff base adduct between the carbonyl compound and an α-amino acid, enabling facile decarboxylation of the amino acid:

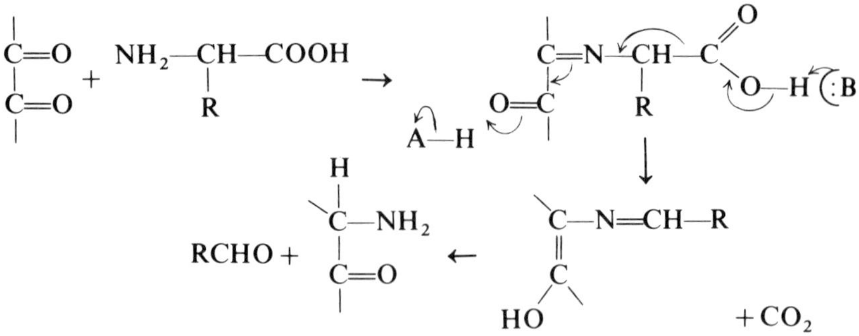

This scheme is after Mauron (1981). In his original report, Maillard (1912) described the formation of carbon dioxide in 36% yield when glucose (1·67 mol) and glycine (1 mol) were heated at 100 °C for 7 h, but the source of the gas as being the carboxyl group of glycine was only demonstrated unequivocally comparatively recently when Stadtman and coworkers (1952) used ^{14}C-labelled reactants. Similar observations were made by Wolfrom and Rooney (1953). The reaction is generally regarded as being favoured at higher temperatures, the aldehyde which is produced contributing directly to the flavour and odour of the mixture. The amine product may well undergo further reaction, such as conversion to pyrazines, again resulting in products with characteristic organoleptic properties (Dawes and Edwards, 1966; Kochler and Odell, 1970).

The majority of aldehydes and ketones encountered in non-enzymic browning reactions have α-hydrogen atoms which are, therefore, activated. Compounds of this type can take part in aldol condensation reactions, the requirement for which is removal of the α-hydrogen by a base (Sykes, 1965):

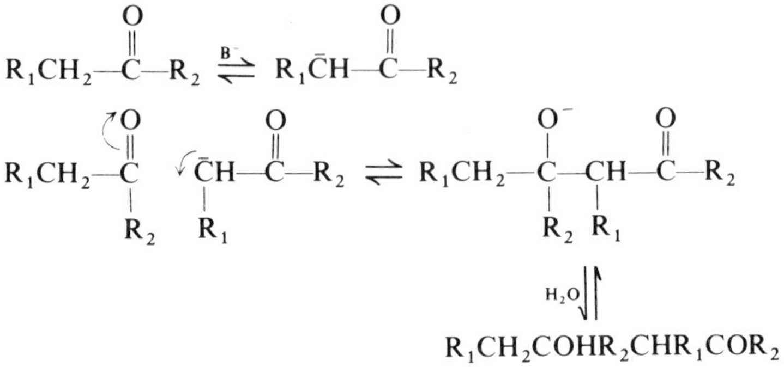

Reactions of this type could well constitute part of the polymerisation reaction leading to high molecular weight products (Olsson *et al.*, 1981).

In contrast to the relatively good understanding of the early stages of Maillard browning, information regarding the structure of the pigments is fragmentary. The products clearly contain a number of different functional groups but the situation is further complicated by the effects of reaction conditions on the composition of the melanoidins. For example, when the molar ratio of glucose to glycine is 1:10, 1·67:1 and 8:1, the ratio of glucose to glycine residues in the melanoidin varies as 1:1, 2:1 and 6:1 respectively. The pH of the medium also appears to have an important influence, for example the melanoidin formed at pH 3·5 is less soluble than that formed at pH 6. In all cases, however, the solubility of the products decreases with increasing reaction time (Reynolds, 1965). The variability of the systems reported in the literature makes it difficult to see an overall picture emerging.

The melanoidins exist as a mixture of discrete products which may be separated from one another using DEAE cellulose. In this way the melanoidins present in ammonia caramel (glucose + ammonia) may be fractionated into six components (Tsuchida and Komoto, 1969) whilst the melanoidin formed from amino acids and aldopentoses, aldohexoses and fructose may be separated into eight components (Motai, 1973). In the case of the reaction between xylose and glycine, the later running components are derived from those running earlier, through oxidation, and evidence of discrete degrees of polymerisation, therefore, exists (Motai and Inoue, 1974). These components span at least an order of magnitude of molecular weights (10^3–10^4) when determined by gel filtration on Sephadex media and using dextrans as calibrants (Motai, 1975).

Data from studies on whole melanoidins have been somewhat crude. Elemental analysis data may be used to calculate the degree of dehydration and loss of carbon dioxide as reactants are converted to melanoidins. If, respectively, a and b moles of aldose and amine combine with liberation of x and y moles of carbon dioxide and water respectively, then the empirical formula for the melanoidin is,

$$C_{la+pb-x}H_{ma+qb-2y}O_{na+rb-2x-y}N_b$$

where l, m, n are, respectively, the numbers of carbon, hydrogen and oxygen atoms in the aldose, and p, q, r are the corresponding numbers of these atoms in the amine. For the glucose–glycine reaction, the extent of dehydration is in the region of three moles of water per mole of glucose, after taking into account one mole of water liberated during the condensation with the amino-group (Maillard, 1916; Enders and Theis, 1938; Wolfrom *et al.*, 1953*b*; Reynolds, 1963; Olsson *et al.*, 1978). A similar result was obtained in the case of the glucose–methylamine reaction (Olsson *et al.*, 1977) but the values were nearer two water molecules per glucose molecule for reactions in which ammonia and *n*-butylamine were the amines, although the difference in dehydration could be attributed to the use of methanolic solution in these cases (Kato and Tsuchida, 1981). The use of infrared and ultraviolet spectroscopy does not provide unambiguous evidence concerning structure; infrared measurements provide the not surprising indication that melanoidins probably contain OH, C=O and C=C moieties (Reynolds, 1965). Perhaps the most interesting spectroscopic study of melanoidins is that mentioned by Olsson and coworkers (1981) where ^{13}C spectra of high molecular weight products from the glucose–glycine reaction and of the Amadori product are compared. This result is shown in Fig. 3.3, the notable features being the relative simplicity of the melanoidin spectrum, its similarity to the spectrum of the Amadori product and the low abundance of olefinic and aromatic carbon atoms.

The most useful information concerning the structure of melanoidins should be obtainable from chemical studies, including the formation of derivatives for functional group analysis, chemical degradation and, perhaps, pyrolysis. Gentle hydrolytic conditions liberate substituted hydroxypyridines (Tsuchida *et al.*, 1973) and pyrazines (Tsuchida *et al.*, 1975) from melanoidins produced in the glucose–ammonia reaction, but these products appear to have only a weak association with the polymer. Degradation by means of permanganate gives oxamic acid (that is, the monoamide of oxalic acid) as the main product from glucose–ammonia melanoidin and *N*-butyloxamic acid in the case when the amine is *n*-butylamine and the aldose is glucose or xylose. No pyrrolecarboxylic acid was found and the same products were indicated when browning reactions were carried out in either methanol or water, with the exception of the glucose–ammonia reaction which was only carried out in water (Kato and Tsuchida, 1981). Pyrolysis gas chromatography is a popular method for the study of polymers (Beroza and Coad, 1967) and has been applied by Tsuchida and coworkers (1976) to the study of melanoidins. Glucose–ammonia melanoidins yield two pyridines and four alkyl-pyrazines, whereas 1-butylaziridine and substituted 1-butylpyrroles were identified after pyrolysis of the reaction products of *n*-butylamine with xylose or glucose (Kato and Tsuchida, 1981). Hydroxyl groups may be determined readily by acetylation with acetic anhydride and pyridine

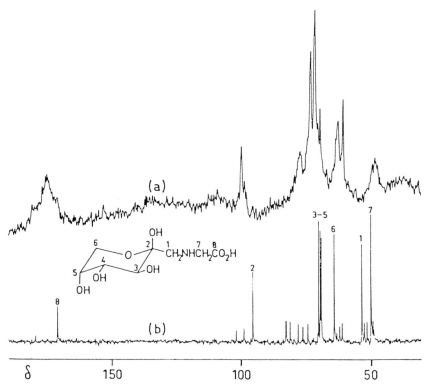

Fig. 3.3. 22·5 MHz ^{13}C-NMR spectra of products from D-glucose and glycine in D$_2$O. (a) High molecular weight products after 4 h at 100 °C and pH 4–5; (b) Amadori compound. Most minor signals are probably due to the β-anomer or the furanose forms (Olsson *et al.*, 1981). Reproduced with permission from *Prog. Food Nutr. Sci.*, 1981, **5**, page 54. © 1981 Pergamon Press Ltd.

followed by determination of combined acetyl groups. If, prior to acetylation, the melanoidin is reduced with sodium borohydride, the extent of acetylation yields the sum of hydroxyl and carbonyl groups and, thus, the number of carbonyl groups may be found by difference. Using this method, Kato and Tsuchida (1981) found that glucose–butylamine melanoidins contained 0·9–1·0 hydroxyl groups and xylose–butylamine melanoidins contained 0·6–0·7 hydroxyl groups per sugar molecule. The number of carbonyl groups was somewhat sparse, perhaps one group per 3–9 sugar molecules. In discussing their results, these authors presented a summary of some possible structures contributing to the melanoidin chain, and their suggestions are reproduced in Fig. 3.4. The furan ring structure

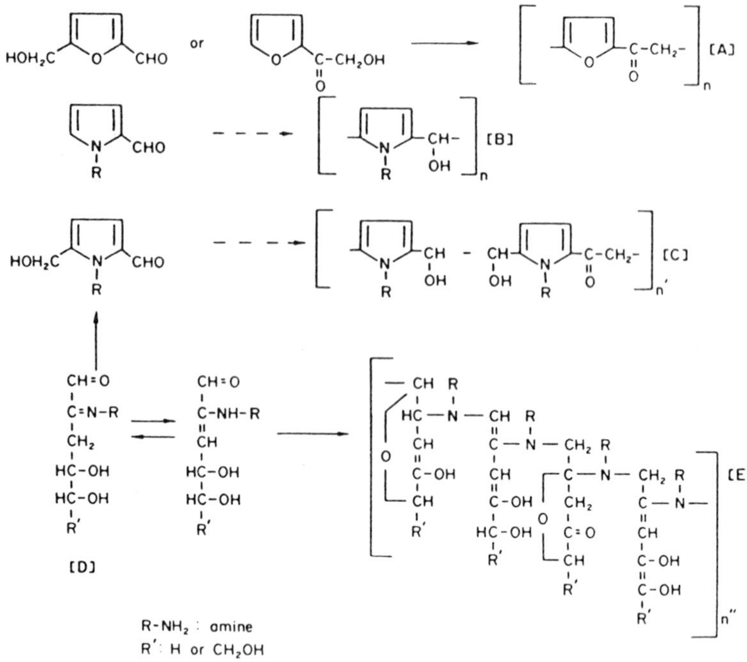

Fig. 3.4. Possible repeating units of polymeric brown products and their precursors (Kato and Tsuchida, 1981). Reproduced with permission from *Prog. Food Nutr. Sci.*, 1981, **5**, page 154. © 1981 Pergamon Press Ltd.

(A) was the postulate of Heyns and Hauber (1970) for the repeating unit of the polymer produced from sorbose in strongly acidic media, the coupled *N*-pyrrole structures (B and C) arise from the fact that the monomers are readily formed in Maillard reactions and structure E is a new proposal to reconcile the new data. In the case of the Maillard reaction, in which browning proceeds under weakly acidic and neutral conditions and the melanoidin contains a high proportion of nitrogen, structure A cannot be regarded as important. Kato and Tsuchida (1981) do not favour the N-substituted pyrroles as repeating units since, although the free N-substituted pyrroles are formed on pyrolysis of melanoidins, permanganate oxidation does not yield pyrrolecarboxylic acids. Also, a dehydration ratio of three moles of water per mole of sugar is required to form the pyrrole compound. Although this may be achieved with some melanoidins, the ratio is often less than this. Stating that a structure is less likely does not, however, preclude the occasional incorporation of the monomer into the polymer. Structure E is postulated as being derived from the Schiff base formed between the 3-deoxyosulose and the amine, and this intermediate is also required for the formation of N-substituted pyrrole-2-aldehydes, regarded as by-products. The structure also helps to explain the increase in reductone content when melanoidins are heated under anaerobic conditions (Gomyo et al., 1972), the formation of oxamic acid on permanganate oxidation and is not inconsistent with the formation of aziridine and pyrroles on pyrolysis. The relationship of these structures to the ^{13}C data referred to above has not yet been discussed.

The accepted mechanisms of the Maillard reaction are heterolytic, although evidence for the participation of free radicals is available (Namiki and Hayashi, 1973, 1974). The radicals in question are 1,4-disubstituted pyrazine cation radicals (Hayashi et al., 1977), believed to originate through dealdolisation of the sugar or Schiff base. Taking the diquaternary salt of 1,4-diethylpyrazine as an example of such a free radical, it was found that at pH 6 the compound was extremely labile, browning rapidly (Namiki and Hayashi, 1981). Melanoidins are also found to contain stable free radicals (Mitsuda et al., 1965).

The Maillard reaction involves intermediates which are derived from ketoses and a reasonable question concerns the reactivity of ketoses, such as fructose, towards browning. This is particularly relevant since fructose often accompanies glucose in foods, a result of the hydrolysis of sucrose. The non-enzymic browning of fructose has also been reviewed by Reynolds (1965). Although it is often thought that the rate of browning of fructose in the presence of amino acids is greater than that of glucose under the same conditions, the relative rates appear to depend upon the extent to which the mixtures are buffered. In unbuffered media, the rate of browning of fructose with amino acids is greater than that of glucose, although eventually the colour in the glucose system exceeds that for fructose. On the other hand, if mixtures are buffered, fructose browns less rapidly with amino acids than does glucose. Therefore, in food systems, which are normally buffered to some extent, the browning of fructose is expected to be less important than that of glucose.

The early products of the addition of amino acid to a ketose are the ketosylamine, the 2-amino-2-deoxyaldose and the ketoseamine. The 2-deoxy-compound is formed as a result of the Heyns rearrangement of the ketosylamine, a reaction which takes place readily with primary alkylamines, ω-amino acids, basic amino acids and α-amino acids with small side chains. When α-amino acids with bulky side chains are present, the rearrangement is less facile and, in such cases, there is a greater tendency towards the formation of ketoseamine (Reynolds, 1965). The ketoseamine is, of course, common to the ketose and aldose browning systems and may be the reactive intermediate in ketose browning.

3.2.2 Ascorbic acid browning

3.2.2.1 Structure of ascorbic acid

The ascorbic acid molecule is usually drawn in the form of the tautomer showing an ene-diol, or reductone grouping as follows:

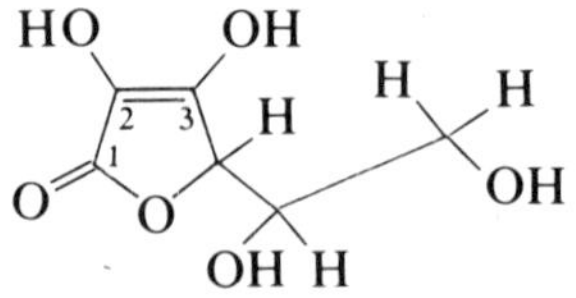

Carbon-13 NMR studies (Berger, 1977) and molecular orbital calculations (Bischof *et al.*, 1981) confirm that, of the four possible tautomers which may be drawn by rearrangement of electrons and hydrogen atoms around carbon atoms 1, 2 and 3 and their associated oxygen atoms, this is, indeed, the most stable form both in solution and in the solid state. Enolic protons are acidic and it is these which render ascorbic acid a weak acid with $pK_a = 4·17$ (Birch and Harris, 1933; Hurd, 1970). The most acidic proton is that on the hydroxyl group attached to carbon 3, confirmed by ^{13}C NMR data (Berger, 1977) using solutions and from X-ray crystallography data on sodium ascorbate (Hvoslef, 1969). Molecular orbital calculations are consistent with this result (Bischoff *et al.*, 1981).

The five membered ring is the result of γ-lactone formation between a carboxyl group at carbon 1 and an hydroxyl group at carbon 4. Lactones are internal esters and are, therefore, formed as a result of attack by the hydroxyl-oxygen on the carboxyl-carbon. Therefore, the ring oxygen is the hydroxyl-oxygen and any subsequent cleavage of the lactone should involve rupture of the bond between carbon 1 and the ring oxygen atom. It is expected that the lactone will undergo base catalysed ring opening in the same way as esters are hydrolysed but, in the case of ascorbic acid, instability of the ring is further promoted by ionisation of the acid group, thus:

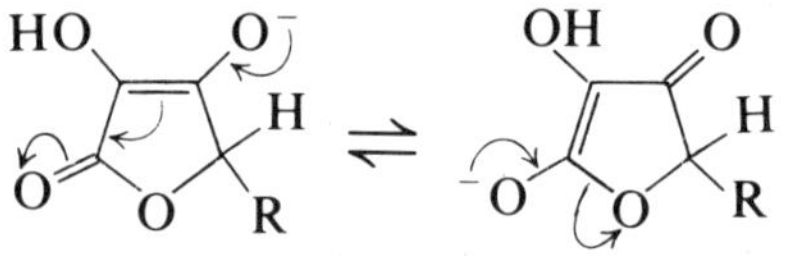

The delocalised form of ascorbate ion was proposed by Euler and Eistert

(1957) and confirmed by X-ray studies (Hvoslef, 1969) in the solid state and using ^{13}C NMR (Berger, 1977) for solutions where it was found that carbons 1 and 3 resonate very close together, at 178 and 176 ppm respectively, compared with their respective chemical shifts of 174 and 157 ppm in ascorbic acid.

The reductone grouping renders ascorbic acid a good reducing agent, complete oxidation involving the transfer of two electrons, thus (from data shown by Clark, 1960):

$$A + 2H^+ + 2e^- \rightleftharpoons AH_2 \qquad E_0 = 0.39 \text{ V}$$

$$A + H^+ + 2e^- \rightleftharpoons AH^- \qquad E_0 = 0.25 \text{ V}$$

where A is the dehydroascorbic acid moiety. The electrode potential for the conversion of ascorbic acid or ascorbate ion to dehydroascorbic acid will be pH dependent, but it is evident that, under normal conditions, it is possible for both species to undergo oxidation by oxygen and a wide range of common oxidising agents. For this reason it is necessary to distinguish between the reactions of ascorbic acid under oxidising and non-oxidising conditions.

The five membered ring in ascorbic acid is planar and the molecule, therefore, has a tendency to become adsorbed at interfaces with consequent modification of stability and redox behaviour. It is, therefore, not strictly possible to predict the reactivity of ascorbic acid in food from model system studies.

3.2.2.2 Anaerobic systems

When held under anaerobic conditions, aqueous solutions of ascorbic acid and ascorbate ion decompose spontaneously producing carbon dioxide, furfural (Herbert et al., 1933) and 2,5-dihydro-2-furoic acid (Coggiola, 1963). The dependence upon pH of the rate of destruction of ascorbic acid and the formation of furfural is shown in Fig. 3·5 for thermal decomposition of buffered solutions at 100 °C, pH control being achieved with mixed phosphate, citrate and phthalate buffers at pH $\geq$ 2·2, whilst phosphoric, sulphuric and hydrochloric acids were used below pH 2·2. Ascorbic acid is relatively unstable in strongly acid media, the rate of decomposition decreasing with increasing pH to a minimum value at around pH 2·3. Above this pH value, stability decreases and passes through a minimum in the region of pH 4 before increasing again. The maximum pH studied was pH 6. On the other hand, the amount of furfural produced, measured in each case at the time of half destruction of ascorbic acid, decreases with increasing pH above pH 1·5, no furfural being

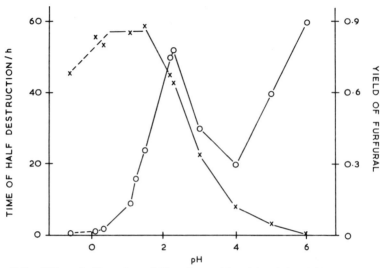

Fig. 3.5.   Effect of pH on the stability of ascorbic acid and the yield of furfural at 100 °C. Ascorbic acid, ◯; furfural, × (Huelin *et al.*, 1971). Reproduced from *J. Sci. Food Agric.*, 1971, **22**, page 540. © The authors.

detectable at pH 6. Similar trends in both the loss of ascorbic acid and formation of furfural are obtained if ascorbic acid solutions are heated to 50 °C instead of 100 °C. The yield of carbon dioxide is quantitative at pH 2·2 and 3·0, and 91 % at pH 6·0, but falls to around 60–70 % in the region pH 4–5. It may, however, be concluded that, over the whole range of pH investigated, decarboxylation is relatively independent of pH and takes place whether or not furfural is a major product. The reduction in amount of furfural with increasing pH cannot be attributed to instability of this product since its half life for destruction is at least an order of magnitude greater than that found for ascorbic acid. It is, however, possible that at the lowest pH investigated (pH < 0·34) the loss may be due to a lowering of furfural stability (Huelin *et al.*, 1971).

The decomposition of ascorbic acid is accompanied by the formation of colour, the rate of colour development being increased in the presence of amino compounds (Mori *et al.*, 1967).

Perhaps the most significant development regarding the understanding of the mechanism of the decomposition of ascorbic acid is the isolation of 3-deoxy-L-*threo*-2-pentosulose during degradation at pH 2·2 (Kurata and Sakurai, 1967*a*). This product will be referred to simply as 3-deoxypentosulose, consistent with the naming of the 3-deoxyhexosulose in Maillard browning (Section 3.2.1). It is suggested that lactone ring opening takes place at an early stage to give an open chain form of ascorbic acid, in fact the enol of 3-ketogulonic acid or the keto-acid itself. The subsequent degradation of this carboxylic acid could take place in two ways:

1.   Dehydration of the 2,3-ene-diol form followed by rearrangement to a 2,3-diketo-acid which, presumably, undergoes facile decarboxylation to 3-deoxypentosulose.

2.   Decarboxylation of the 3-keto-acid to a pentose-1,2-ene-diol which then dehydrates to 3-deoxypentosulose, enol form.

Feather and coworkers (1972) commented on these two possibilities as a result of work on the incorporation of tritium into furfural when ascorbic acid is decomposed in tritiated water. If the aldehydic group originates from carbon 2 of ascorbic acid, then the amount of tritium present at the aldehyde group will reflect the extent to which 2,3-enolisation of ascorbic acid contributes to the mechanism. Conversely, the activity acquired in position 3 on the furan ring indicates exchange in position 4 of ascorbic acid or the activated methylene group of 3-deoxypentosulose. Extensive incorporation of label onto the aldehyde group of furfural is observed, consistent with either mechanism, but the specific activity of tritium in the ring is only 3 % of that of the solvent. This is somewhat larger than the specific activity of tritium in the ring (0·5 % of that of the solvent) when furfural is produced by dehydration of xylose. If the latter result indicates that the 3-position of 3-deoxypentosulose does not exchange rapidly with the solvent, then it could be argued that the results favour a pathway for the degradation of ascorbic acid in which the proton at position 4 is labile. Alternative 1 provides such an opportunity. However, Feather and coworkers consider that the low level of incorporation of label in the ring indicates that exchange is not taking place and prefer mechanism 2 on these grounds, but the result is not conclusive on account of the unknown kinetic isotope effect.

A fundamentally different pathway, not requiring initial opening of the lactone ring, has been proposed by Tokuyama, Goshima and their coworkers (Tokuyama *et al.*, 1971; Goshima *et al.*, 1973). This proposal is based upon the observation that, if ascorbic acid is allowed to decompose in methanolic solution containing boron trifluoride etherate, the following two products could be isolated:

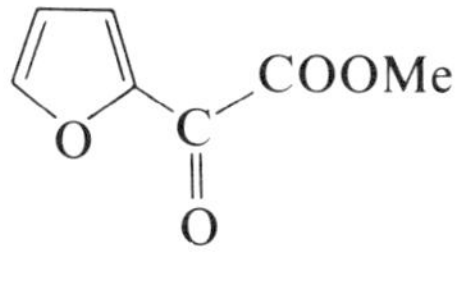

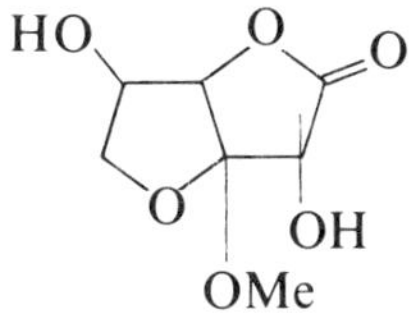

It is suggested that the mechanism involves initial cyclisation of ascorbic acid to form a bicyclic product which loses water before the lactone ring opens to give the substituted furan compound:

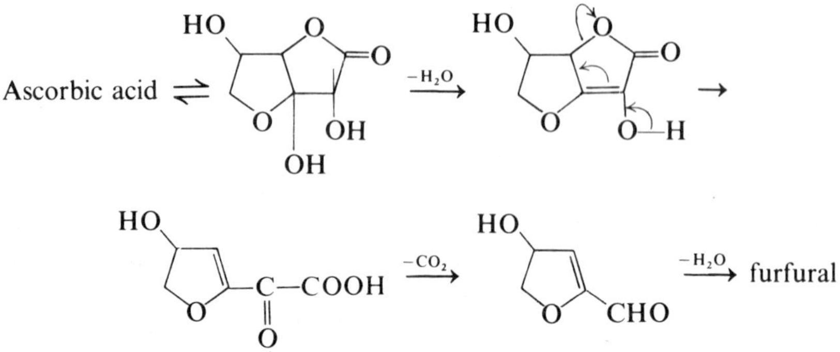

These authors consider that 3-deoxypentosulose is only formed as a by-product of this scheme by hydration at the double bond of the precursor of furfural.

No simple route leading to 2,5-dihydro-2-furoic acid has been proposed.

The mechanism for the formation of coloured products is more speculative than that advanced in Maillard browning. The observation that 3-deoxyhexosulose is more reactive towards browning than hydroxy-methylfurfural could well be extended to the reactivities of 3-deoxypento-sulose and furfural. N-substituted pyrrole-2-aldehydes are formed when either pentoses or hexoses take part in Maillard browning. Therefore, reaction of both the C_5 and C_6 3-deoxyosulose with amine is possible. It is not unreasonable to suggest that, in the presence of amino-compounds, the important intermediate with respect to browning in the degradation of ascorbic acid is 3-deoxypentosulose and that the subsequent reactions resemble those in the Maillard reaction, particularly when pentoses are involved. Since decarboxylation of ascorbic acid is extensive, it may be concluded that little or no six-carbon intermediate is incorporated into the final product. The question of the participation of amino acids in Strecker degradation reactions also arises. There seems to be no firm evidence supporting their occurrence in the ascorbic acid–amino acid system. Lalikainen and coworkers (1958) found that when ascorbic acid and glycine were heated at 50 °C at both pH 3·7 and 7·0, less than 3 % of the carbon dioxide evolved originated from carbon 1 of glycine. The source of coloured product in the absence of amine is not known but could also be attributed to 3-deoxypentosulose.

3.2.2.3 Oxidative browning

The browning of ascorbic acid in the presence of oxygen proceeds by way of dehydroascorbic acid. The latter may be conveniently prepared by the oxidation of ascorbic acid using iodine (Kenyon and Munro, 1948) or oxygen in the presence of catalyst (Dietz, 1970; Ohmori and Takagi, 1978), the preferred solvent being methanol. Under these conditions the product is the dimer (Albers *et al.*, 1963) which dissociates in aqueous solution to the hydrated monomer (Berger, 1977):

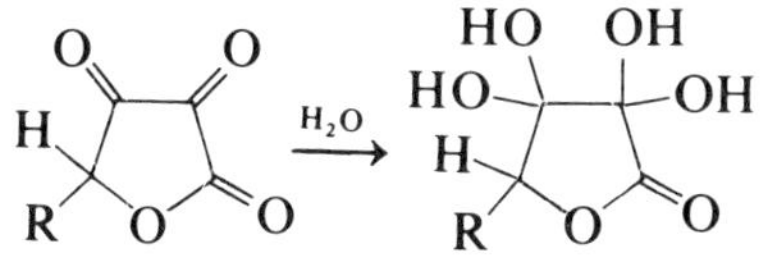

Dehydroreductone form   Hydrated form

$$(R = -CHOH-CH_2OH)$$

Furthermore, Pfeilsticker and coworkers (1975) consider that in aqueous solution dehydroascorbic acid exists as a more complex bicyclic structure, 3,6-anhydro-L-*xylo*-hexulono-1,4-lactone hydrate, thus:

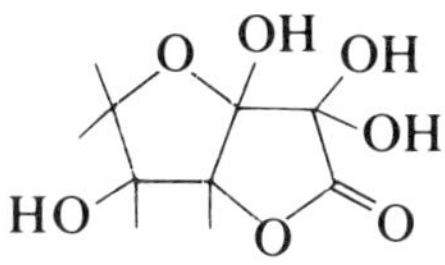

formed by the elimination of water between positions 3 and 6 with ring closure. The similarity between this structure and the bicyclic intermediate proposed in ascorbic acid degradation is striking.

The mechanisms of oxidation of ascorbic acid have been reviewed by Mushran and Agrawal (1977). Autoxidation is strongly catalysed by metal ions in trace amounts and the rate is pH dependent showing a maximum at pH 5. The autoxidation proceeds by way of two consecutive one-electron transfer operations, the free radical intermediate, monodehydroascorbic acid, being formed:

$$AH_2 \rightleftharpoons AH^- + H^+$$
$$AH^- \rightleftharpoons AH^{\cdot} + e^-$$
$$AH^{\cdot} \rightleftharpoons A + e^- + H^+$$

The free radical also disproportionates into ascorbic acid and dehydro-ascorbic acid, the equilibrium constant for the reaction,

$$AH_2 + A \rightleftharpoons 2AH^{\cdot}$$

being, at 25 °C, of the order of 5×10^{-9} at pH 6·4 and 6×10^{-12} at pH 4 (Foerster *et al.*, 1965).

Dehydroascorbic acid browns on its own much more readily than does ascorbic acid, the pathway involving opening of the lactone ring to 2,3-diketogulonic acid followed by degradation. The latter compound is a β-keto-acid which will undergo facile decarboxylation leading to L-*erythro*-pentosulose (referred to usually as L-xylosone). Kurata and Sakurai (1967b) rationalised the formation of ethylglyoxal, 2-furoic acid, 2-keto-3-deoxy-L-pentono-γ-lactone and two reductones, that is 2,3,4-trihydroxy-2-pentenoic acid and its dehydration product, 5-methyl-3,4-dihydroxy-tetrone, in terms of successive dehydrations, proton shifts and de-carboxylation of the pentosulose. The formation of ethylglyoxal and the reductones requires also an allylic rearrangement of the 2-keto-5-hydroxy-3-pentenoic acid intermediate. In a later paper Kurata and Fujimaki (1976) accounted for the formation of a tricarbonylic product, 3-keto-4-deoxypentosulose, by similar operations starting with either diketogulonic acid or L-*erythro*-pentosulose. The main aroma compound formed during the degradation of dehydroascorbic acid is 3-hydroxy-2-pyrone which has been identified in stored orange juice powder (Tatum *et al.*, 1967). It is suggested that the pyrone is formed by lactonisation of 2-keto-5-hydroxy-3-pentenoic acid, a dehydration product of 2-keto-3-deoxy-L-pentonic acid, and an intermediate in the formation of ethylglyoxal and the reductones (Kurata and Sakurai, 1967b). A similar mechanism for the formation of 3-hydroxy-2-pyrones has been put forward by Charon and Szabo (1973) who investigated the syntheses and acid degradation of 3-deoxyketoaldonic acids.

The manner in which these intermediates contribute to colour formation is not known but, presumably, high molecular weight products arise from condensation reactions in the absence or in the presence of amines. However, perhaps the most characteristic observation is the development of a wine red or cherry red coloration when dehydroascorbic acid and amino acid are mixed. This phenomenon was first described by Koppanyi and coworkers (1945) and reported in foods such as dehydrated cabbage (Ranganna and Setty, 1968, 1974a, b). Kurata and coworkers (1973) suggest that the first step in the formation of the coloured compound is a Strecker reaction between dehydroascorbic acid and the amino acid, thus:

$$\text{(diketone lactone structure)} + \text{RCHNH}_2\text{COOH} \rightarrow \text{(HO, NH}_2\text{ enamine lactone structure)}$$

$$R = -\text{CHOHCH}_2\text{OH}$$

followed by oxidation of the product to the imine. The red product is then formed by transalkylidenation of the imine with the unoxidised amine. The dehydroascorbic acid–amino acid system also gives rise to free radicals (Namiki *et al.*, 1974; Kurata and Fujimaki, 1974), identified by Kurata and Fujimaki to be monodehydro-2,2′-iminodi-2(2′)-deoxy-L-ascorbic acid formed by reduction of the red compound described above. A similar structure was proposed by Yano and coworkers (1974). The group at the Department of Food Science and Technology, Nagoya (Yano *et al.*, 1976*a*, *b*, 1978; Hayashi *et al.*, 1981) further demonstrated the presence of two radical species, one associated with a blue and the other with a red component of the mixtures. The blue compound is derived from tris-(2-deoxy-2-L-ascorbyl)amine by oxidation, further oxidation leading to the red pigment with the elimination of one molecule of ascorbic acid. The actual mechanism of formation of the red compound from dehydro-ascorbic acid is not yet fully resolved but it does not involve the opening of the lactone ring. The product is, however, labile and ultimately leads to the formation of a brown pigment.

3.2.3 Caramelisation

The degradation of carbohydrate in the absence of amines proceeds by way of the Lobry de Bruyn–Alberda van Ekenstein transformation and has been reviewed in detail by Feather and Harris (1973). D-Glucose is the most stable of the aldoses, maximum stability occurring on the acid side of neutrality in the region pH 3–4. To demonstrate degradation, aldoses and ketoses are usually boiled in dilute acid when the formation of acyclic ene-diols is the initial reaction leading to dehydration, thus:

$$
\begin{array}{cccc}
\text{CHO} & \text{HC}-\text{OH} & \text{CHO} & \text{CHO}\\
|&\;\;\;\|&|&|\\
\text{CHOH} & \text{C}-\text{OH} & \text{C}-\text{OH} & \text{C}=\text{O}\\
|\quad\rightleftharpoons&|\quad\xrightarrow{-\text{H}_2\text{O}}&\|\quad\rightleftharpoons&|\\
\text{CHOH} & \text{CH}-\text{OH} & \text{CH} & \text{CH}_2\\
|&|&|&|\\
\text{R} & \text{R} & \text{R} & \text{R}\\
& \text{1,2-enol form} & &
\end{array}
$$

The product of dehydration is the already familar 3-deoxyosulose which will dehydrate further, losing one molecule of water, to form the 3,4-dideoxyosulos-3-ene, and a second molecule of water to give a 5-substituted furan-2-aldehyde, the substituent being —CH_2OH in the case of glucose or fructose. There is, therefore, much in common between this reaction and the Maillard reaction, the difference being the requirement for much more vigorous conditions in the absence of amine and the formation of nitrogen-free coloured products. Thermal decomposition of carbohydrate material leads to the production of a wide range of volatile components. For example, when cellulose, cellobiose or glucose is heated at 250 °C, the products mixture contains acetaldehyde, furan, propionaldehyde, acrolein, acetone, diacetyl, furfural and 5-methylfurfural, all of which have characteristic odours (Kato, K., 1967).

In a food it is not possible to state the amount of browning taking place as a result of caramelisation unless the conditions are such that amines are rigorously excluded. The higher energy requirement of the reaction in the absence of amines renders it less favourable than Maillard browning and, therefore, the latter will predominate if composition allows. In any event, should the reactive intermediates be formed without the intermediacy of the amine but in its presence, then colour formation will still take place as in Maillard browning.

3.2.4 Lipid browning

It is generally accepted that lipids brown as a result of oxidation of unsaturated glyceride components, and the browning reaction is accelerated by the presence of ammonia, amines or proteins. Since low molecular weight aldehydes are the products of fat autoxidation, the immediate conclusion is that amino–carbonyl reactions are important, but the function of the amine is by no means clear. The products of reaction between components arising from fat degradation and amine do not necessarily need to be complex or polymeric to absorb light close to the blue end of the visible spectrum. For example, malonaldehyde, the major constituent of oxidised methyl linolenate, reacts with amino acids such as glycine, valine or leucine to form a fluorescent product which absorbs light at 435 nm. This reaction involves one mole of the aldehyde and two moles of amine as follows:

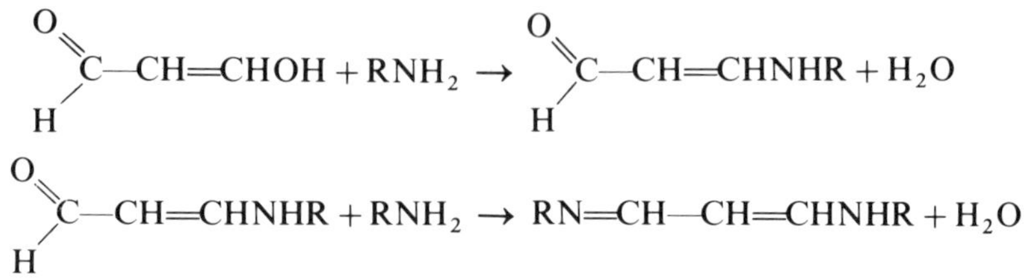

$$\underset{H}{\overset{O}{\diagdown}}C-CH{=}CHOH + RNH_2 \rightarrow \underset{H}{\overset{O}{\diagdown}}C-CH{=}CHNHR + H_2O$$

$$\underset{H}{\overset{O}{\diagdown}}C-CH{=}CHNHR + RNH_2 \rightarrow RN{=}CH-CH{=}CHNHR + H_2O$$

the spectroscopic characteristics of the product depending upon whether it is *cis* or *trans* (Chio and Tappel, 1969). Shimasaki and coworkers (1977) demonstrated the production of fluorescent products through the reaction of methyl linolenate hydroperoxide with glycine. They found a linear correlation between decrease in diene conjugation and increase in fluorescence, demonstrating a direct relationship between peroxide decomposition and formation of fluorescent substances. The alternative mechanisms for this reaction have not been resolved. It has been found that lysine availability in protein is reduced in the presence of low molecular weight aldehyde, but whether the amine is irreversibly bound or not in the coloured product and, indeed, whether or not its presence in such combination is important is not established (Reynolds, 1965).

3.3   INHIBITION BY SULPHUR(IV) OXOANIONS

3.3.1   General trends of behaviour

When sulphur(IV) oxoanions are added to a mixture of aldose and amino acid, the time required for browning to become apparent is extended when compared with browning in the system containing no sulphur(IV) oxoanion. During the period of inhibition of browning, a proportion of the additive becomes irreversibly bound.* The relationship between the amounts of free, reversibly bound and total (that is, free + reversibly bound) sulphur(IV) oxoanion and the development of colour is illustrated in Fig. 3.6 (McWeeny *et al.*, 1969). These concentration data were obtained by iodine titration, as described in Section 2.6.4.1, and the development of colour determined as absorbance at 490 nm for reaction mixtures containing D-glucose (1 M), glycine (0·5 M) and hydrogen sulphite ion (0·039 M) at pH 5·5 and 55 °C. In the early stages of the reaction the rate of irreversible loss of sulphur(IV) oxoanion follows the amount of free additive present in solution, the amount of reversibly bound additive remaining relatively constant. If it is assumed that the reversible binding is a result of the formation of hydroxysulphonate type adducts between carbonyl groups and sulphur(IV) oxoanions, then the stability of these adducts increases with time. This could be interpreted as the

---

* The words 'irreversible' and 'reversible' are used here in the same sense as in Section 2.6.3. Thus, irreversibly bound sulphur(IV) oxoanion cannot be recovered by any of the recognised analytical methods, whilst reversibly bound additive is released either on raising the pH of the medium or on prolonged treatment with boiling acid.

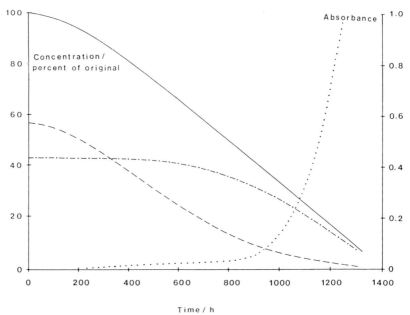

Fig. 3.6. Concentration–time behaviour of total, reversibly bound and irreversibly bound sulphur(IV) oxoanion during the inhibited reaction of glucose and glycine. Absorbance data superimposed. Reaction conditions: [glucose] = 1 M; [glycine] = 0·5 M; [S(IV)] = 0·039 M; pH 5·5; 55 °C. Graphs plotted from the data published by McWeeny and coworkers (1969) to reproduce figure shown by these authors. Total S(IV), ———; free S(IV), - - - -; reversibly bound S(IV), —·—·; absorbance at 490 nm, ·······. Reprinted with permission from *Journal of Food Science*, 1969, **34**, pages 641–3. © 1969 Institute of Food Technologists.

formation of carbonyl compounds which bind the additive more strongly than do the reactants. Consequently, the amount of free sulphur(IV) oxoanion present in the mixture in the later stages of reaction is small, and it is apparent also that the rates of loss of reversibly bound additive and of total recoverable additive become similar. It is reasonable to assume that irreversible combination of sulphur(IV) oxoanions is a result of reactions between the free additive and intermediates in the browning reaction. If the initial period of reaction is neglected as a period during which the Maillard sequence is being established, the rate of loss of additive is constant over a large part of the reaction. The zero order kinetic behaviour suggests that formation of intermediate in the Maillard reaction is rate determining as far as loss of sulphur(IV) oxoanion is concerned and, therefore, reaction between sulphur(IV) oxoanion and the intermediate to form compounds

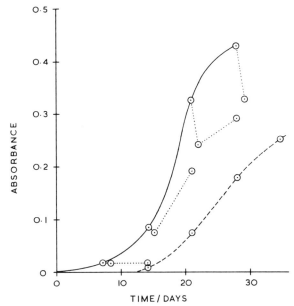

Fig. 3.7. Development of colour in the glucose–glycine model systems as measured by absorbance at 470 nm. Reaction conditions: [glucose] = 2·5 M; [glycine] = 1 M; 50 °C. No S(IV), ——; reaction mixture containing 0·031 M S(IV), – – – –; effect of addition of S(IV) after browning had been allowed to proceed, · · · · · ·. [S(IV)] = 0·031 M. Plotted from data of Burton and coworkers (1962) with permission of D. J. McWeeny.

which are not dissociated under the conditions of analysis is relatively fast. The commencement of colour development coincides with that of the decomposition of labile adducts at a reaction time of around 800 h. This dissociation of the adducts is presumably a simple consequence of the depletion of free sulphur(IV) oxoanion and the maintenance of a composition consistent with the equilibrium constant for adduct formation. The release of reactive carbonyl compounds then allows browning to proceed. It is evident also from these data and the results illustrated in Fig. 3.7 (Burton *et al.*, 1962) that the effect of sulphur(IV) oxoanion is only to delay colour formation. Once browning has begun, it proceeds at similar rates in both the sulphited and non-sulphited systems. When mixtures of glucose and glycine are allowed to undergo browning and are then treated subsequently with sulphur(IV) oxoanion, an almost instantaneous reduction in colour is noted, although this reduction is by only a fraction of the difference in absorbances of the non-sulphited

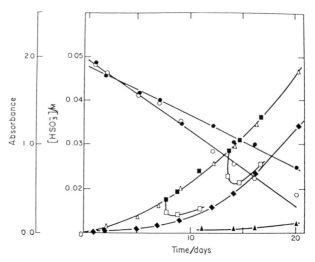

Fig. 3.8. Kinetic measurements on ascorbic acid–glycine–hydrogen sulphite mixtures: △, absorbance curve for 1.00 M ascorbic acid $+ 0.50$ M glycine; ◆, absorbance curve for 1.00 M ascorbic acid $+ 0.50$ M glycine $+ 0.05$ M HSO_3^-; ▲, absorbance curve for 1.00 M ascorbic acid $+ 0.05$ M HSO_3^-; ■, absorbance curve for 2.00 M ascorbic acid $+ 1.00$ M glycine, browned and diluted with an equal volume of water; □, absorbance curve for 2.00 M ascorbic acid $+ 1.00$ M glycine, browned and diluted with an equal volume of 0.10 M HSO_3^-; ○, $[HSO_3^-]$ for 1.00 M ascorbic acid $+ 0.50$ M glycine $+ 0.05$ M HSO_3^-; ●, $[HSO_3^-]$ for 1.00 M ascorbic acid $+ 0.05$ M HSO_3^-.

mixture and that which contained the additive from the beginning of the experiment. The colour of melanoidins is, therefore, partially bleached by sulphur(IV) oxoanions. However, subsequent development of colour, in the partially bleached sample, occurs at the same rate as in the sample to which sulphur(IV) oxoanion had been added at zero time. It is not known if reaction between the additive and melanoidin leads to reversible or irreversible combination of the additive but it is evident that the bleaching reaction is not the only mechanism by which Maillard browning is inhibited.

Kinetic data for the anaerobic browning reaction of ascorbic acid in the absence and presence of sulphur(IV) oxoanions are shown in Fig. 3.8. The reaction mixtures contained ascorbic acid (1 M), glycine (0.5 M) and, where sulphur(IV) oxoanions were added, the concentration was 0.05 M. The reaction was followed at pH 3.5 and 40 °C. Colour development was monitored by single wavelength measurement at 470 nm and the amount of sulphur(IV) oxoanion determined spectrophotometrically using 5,5'-dithiobis(2-nitrobenzoic acid) as described in Section 2.3.3.2. This analysis is carried out at pH 7, under which conditions hydroxysulphonate adducts are labile (Section 1.11.1), and the method is, therefore, expected to give the total free and reversibly bound sulphur(IV) oxoanion concentration in the mixture. Any losses are, therefore, a result of 'irreversible' combination with components of the browning system. It may be seen immediately that this system behaves in a similar manner to the inhibited Maillard reaction in a number of respects. The effect of the additive is again that of delaying colour development, colour formation in the later stages of reaction being independent of sulphur(IV) oxoanions. The linearity of concentration/time data is consistent with the rate determining step involving ascorbic acid with or without glycine but not sulphur(IV) oxoanion. The slight increase in rate of loss of additive in the presence of glycine is negligible when compared with the increase in concentration of the latter from zero to 0.5 M and, hence, it may be said that the reaction leading to the loss of sulphur(IV) oxoanion is independent of this component. At this preliminary stage it is, therefore, possible to state that, in the anaerobic sulphite inhibited browning reaction of ascorbic acid, the reaction leading to loss of the additive involves a spontaneous rate determining reaction of ascorbic acid followed by a reaction of intermediates, so formed, with the additive. No data concerning the distribution of sulphur(IV) oxoanions between free and reversibly bound forms are available. The effect of partial bleaching of melanoidin is particularly striking in the case of the ascorbic acid system, the rapid loss of colour amounting to at least 75% of the difference in absorbance between the non-sulphited system and that in which the additive was present from zero time. It is, therefore, possible that, in the ascorbic acid system, the melanoidin could be the reactive intermediate towards sulphur(IV) oxoanion reactivity and it could be speculated that the pathway:

$$\text{Ascorbic acid} \rightarrow \text{intermediate} \rightarrow \text{melanoidin} \xrightarrow[\text{fast}]{\text{S(IV)}} \text{products}$$

constitutes the route for the inhibition of browning. This is, however, disfavoured for two reasons. First, the rate of loss of sulphur(IV) oxoanion from reaction mixtures is finite at zero time when no melanoidin is formed and is the same in the later stages when melanoidin has accumulated. Since the rate of loss of additive is not determined by its concentration but by the concentration of the relevant intermediate, it is unlikely that this intermediate is the melanoidin. Secondly, glycine has little effect on the rate of loss of sulphur(IV) oxoanion but a profound effect on colour development, virtually no browning being detectable in mixtures of ascorbic acid and sulphur(IV) oxoanions alone.

There is no published investigation of the dehydroascorbic acid/amino acid/sulphur(IV) oxoanion system.

The general trends in behaviour of the non-enzymic browning systems in the presence of sulphur(IV) oxoanions will now be explained.

3.3.2 Reaction with carbonyl components

The reactivity of intermediates in non-enzymic browning reactions is dependent upon the presence of the carbonyl group, the most reactive components being those with a dicarbonyl function. Therefore, reaction of the carbonyl compounds with sulphur(IV) oxoanions to form the hydroxysulphonate adduct is expected to inhibit effectively the browning reaction, the degree of inhibition being a function of the amount of free carbonyl compound present in equilibrium with hydroxysulphonate. It is suggested also that melanoidin structures contain carbonyl chromophores and their effective removal through hydroxysulphonate formation could also lead to a bleaching effect on the pigment. The chemistry of hydroxysulphonate formation has been considered in Chapter 1 (Section 1.11.1) but the situation for many of the intermediates in non-enzymic browning is more complex, since the carbonyl group is often in equilibrium with species which do not react directly with sulphur(IV) oxoanions, that is the carbonyl group is involved in forming the hemiacetal ring structure of carbohydrates. In such instances, therefore, the activity of the carbonyl compound is reduced and equilibrium constants are likely to be small when they are calculated using the analytical concentration of the carbonyl compound.

The most elementary reaction which could lead to the control of Maillard browning is that between glucose and sulphur(IV) oxoanion to form glucose hydroxysulphonate. (The systematic name of the sodium salt is D-*glycero*-D-*ido*-1,2,3,4,5-hexahydroxyhexylsodiumsulphonate.) The equilibrium constant for the reaction is close to unity over the pH range 2–6 (Vas, 1949), a value which may be compared with equilibrium constants of the order of 10^5–$10^6 \, \text{M}^{-1}$ for adduct formation with simple aldehydes and ketones. A steric effect is also evident in the reaction with aldoses. When hydroxysulphonate adducts of D-glucose, D-galactose, D-mannose and L-rhamnose are hydrolysed with sodium hydroxide, the α-isomers of the aldoses are formed. Assuming that inversion of configuration takes place on hydrolysis, Ingles (1959b) suggested that the configuration at carbon 1 of the adduct is β, that is:

$$\text{HO} \underset{\mid}{\overset{\text{SO}_3\text{Na}}{\overset{\mid}{\rule{2.5cm}{0.4pt}}}} \text{H}$$

In contrast, hydrolysis of L-arabinose hydroxysulphonate leads to the formation of the β-isomer, suggesting that in this case the configuration of hydroxysulphonate is α.

The only available mechanism for the formation of aldose hydroxysulphonate involves ring opening followed by addition to the carbonyl group and, therefore, the early stages of the reaction are similar to those taking place during the mutarotation of aldose. The relationship between the rate of mutarotation and rate of hydroxysulphonate formation has apparently not been studied but it is evident that sulphur(IV) oxoanions could act as acids or bases capable of catalysing the interconversion of α- and β-forms of glucose (Ivanov and Lavrent'ev, 1967).

With regard to the inhibition of Maillard browning through reaction between glucose and hydrogen sulphite ion, the effect may be examined by considering the extent of adduct formation in mixtures of the two components when the concentration of glucose to hydrogen sulphite ion is in the ratio of, say, 1:10 at pH 4. For a glucose concentration of 0·01 M, only some 7% is present in the form of hydroxysulphonate, whereas approximately 40% conversion takes place at a concentration of 0·1 M. A similar result may be obtained at a glucose concentration of 0·01 M but with a hundredfold excess of the sulphur(IV) oxoanion. In the browning system considered by McWeeny and coworkers (1969) the concentration of glucose exceeded that of sulphur(IV) oxoanion by a factor of approximately 20 and, in this case, the observed inhibition of browning cannot be attributed to formation of hydroxysulphonate adduct. The only effect of combination of sulphur(IV) oxoanion with glucose is, therefore, the reduction of the amount of the former available for other purposes. In conclusion, therefore, the aldose–hydrogen sulphite reaction will not inhibit the browning reaction of glucose since it is not expected that the concentration of sulphur(IV) oxoanion will be sufficiently high to lead to a significant reduction in glucose content. Other monocarbonyl compounds known for browning activity are, for example, pyruvic acid and furfural. The former is known to give a more stable hydroxysulphonate adduct with a formation constant of $4·5 \times 10^3$ M^{-1} (Burroughs and Sparks, 1973a) at pH 4 and 20 °C.

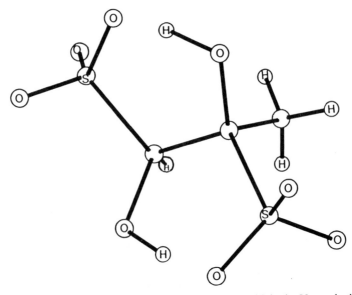

Fig. 3.9.   Model of the dihydroxysulphonate of pyruvaldehyde. Unmarked atoms
are carbon.

The most significant intermediates in Maillard and ascorbic acid
browning have the dicarbonyl function and the reaction of glyoxal-type
and diketo-compounds with sulphur(IV) oxoanions is, therefore, of
interest. The simplest dicarbonyl compound is glyoxal itself and
equilibrium measurements at pH 7·3 show that the dissociation of the
diadduct takes place much more readily than that of the monoadduct. At
this pH the equilibrium constant for formation of dihydroxysulphonate
from the monoadduct is in the region of $3·7 \times 10^3 \, M^{-1}$ at 20 °C and at an
ionic strength estimated to be in the range 0·5–1·0 M (Salomaa, 1956). This
value represents a reasonably stable diadduct, but it has been suggested
previously (Knowles, 1971) that dihydroxysulphonates of this type should
be particularly stable as a result of hydrogen bonding between hydroxyl
and sulphonate groups on adjacent carbon atoms. A molecular model of
the dihydroxysulphonate of pyruvaldehyde is shown in Fig. 3.9 to
illustrate this suggestion. Of the four possible isomers of this compound,
only two are capable of giving the desired arrangement of atoms. The
largest S—O---H angle (approximately 110°) is obtained with all atoms in
the six-membered rings coplanar, but the resulting O—H---O distance is
only some 0·2 nm, that is just twice the sum of the covalent radii of an

oxygen atom and a hydrogen atom. This distance of approach is, therefore, too small but may be increased by rotation about the bonds, resulting also in loss of colinearity of O—H---O and reduction in angle between the line joining the hydroxyl oxygen to the sulphonate oxygen, and the sulphur–oxygen bond. At an oxygen–oxygen separation of 0·27 nm, this angle is approximately 90°, that is the following arrangement exists:

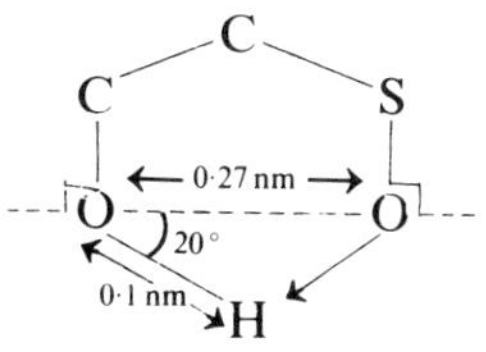

A similar situation exists in the case of 2-hydroxybenzaldehyde and 2-hydroxybenzoic acid where intramolecular hydrogen bonding takes place, and it is, therefore, reasonable to expect that the same will be true of the dihydroxysulphonates of compounds with a glyoxal grouping. These considerations lead also to the expectation that the diadduct will be formed in preference to the monoadduct.

Four reasons why dihydroxysulphonate formation may be less favourable than that of monohydroxysulphonate may be advanced. First, the addition of a hydrogen sulphite ion to the monohydroxysulphonate needs to be stereospecific and is, therefore, less probable. Secondly, the monohydroxysulphonate may be already stabilised through hydrogen bonding between the free carbonyl group and the adjacent hydroxyl group, thus:

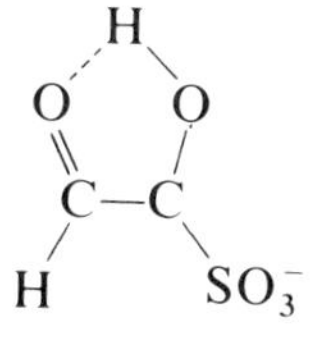

This type of stabilisation is expected to exist in the hydrate of glyoxal. Thirdly, the hydrogen bonds in the dihydroxysulphonate will be formed between a negatively charged sulphonate group and a hydroxyl group. Since most hydrogen bonds are basically electrostatic, the situation in the diadduct will lead to a larger negative charge being induced on the hydroxyl oxygen than would be the case if the hydrogen bond had been to, say, an uncharged carbonyl group. The development of negative charge on the oxygen of the hydroxyl group of hydroxysulphonates is the first step in the alkaline hydrolysis of these compounds. It may be significant that, although the diadduct of glyoxal is less stable than the monoadduct, it is formed and decomposed more rapidly than the latter (Salomaa, 1956). The hydroxysulphonate type adducts of monocarbonyl compounds become similarly labile as pH is increased. Fourthly, the polarity of the diadduct will be reduced through hydrogen bonding and, therefore, formation of the diadduct could be more favourable in less polar solvents. It may be relevant that diadducts of glyoxal and of hydroxypyruvaldehyde may be prepared in the crystalline state (Ronzio and Waugh, 1944; Ingles, 1961a).

It is striking that most of the dicarbonyl compounds which are encountered in the non-enzymic browning pathways, and which are capable of being studied, form only monoadducts with hydrogen sulphite ion. Thus, monohydroxysulphonates are known for L-*erythro*-pentosulose (equilibrium constant for formation = $710 \, \text{M}^{-1}$ at pH 4 and $20\,^\circ\text{C}$; Burroughs and Sparks, 1973a) and dehydroascorbic acid (Ingles, 1961a; equilibrium constant for formation = $1750 \, \text{M}^{-1}$ at pH 4 and $20\,^\circ\text{C}$; Wisser *et al.*, 1970). The reason why the first compound appears to react as a monocarbonyl compound is explained by the participation of the carbonyl group, at carbon 1, in hemiacetal formation. This effect may also be illustrated with other diketo-compounds such as D-*threo*-2,5-hexodiulose and 2,5-diketogluconic acid, both of which apparently form only monoadducts with hydrogen sulphite ion. In both cases the carbonyl group reactive towards the additive is in position 5, since that in position 2 is involved in the same pyranose ring structure as occurs with fructose, which does not form a hydroxysulphonate (Burroughs and Sparks, 1973a). The result may be extended to 3-deoxyhexosulose, the structure of which is shown in Section 3.2.1, and to 3-deoxypentosulose which should, similarly, use the carbonyl group at carbon 1 for hemiacetal formation. No equilibrium constant data are available for these two important compounds but, on the basis of crude experiments carried out by Ingles (1961a), it appears that the hydroxysulphonate of 3-deoxyhexosulose is somewhat more dissociated than that of dehydroascorbic acid. The additional factor which needs to be taken into account when considering the addition of hydrogen sulphite ion to the carbonyl groups of 3-deoxyhexosulose and 3-deoxypentosulose is that both compounds are likely to exist with the carbonyl group at carbon 2 in the enol form (Section 3.2.1), which will lead to a further reduction in their reactivity towards the additive. The effectiveness of removal of these important intermediates in browning by sulphur(IV) oxoanions cannot be fully predicted without equilibrium constant data, since the products of reaction are clearly neither very stable nor as unstable as the product of reaction between, say, glucose and hydrogen sulphite ion.

In the case of the addition of hydrogen sulphite ion to dehydroascorbic acid, the reason for the formation of monoadduct only may be appreciated with the aid of a model of the diadduct. Unlike the situation found in pyruvaldehyde, rotation about the bond between carbon atoms 2 and 3 of dehydroascorbic acid dihydroxysulphonate is restricted by the lactone ring, and steric hindrance between the hydroxyl group and the sulphonate group on adjacent carbon atoms can be relieved only by rotation around C—S and C—O bonds. A hydrogen bonding situation does not develop satisfactorily since suitable interoxygen distances lead to S—O---H angles of $90°$ or less.

When dicarbonyl compounds and amino acids are mixed with sulphur(IV) oxoanions, products with both the amino moiety and sulphonate group attached to each of the carbon atoms may be formed. Ingles (1961a) prepared such a disulphonate from glyoxal, glycine and hydrogen sulphite ion; in the case of diacetyl, only the monosulphonate was produced, but with glycine attached to both of the carbon atoms in question. It may be speculated that in these cases there is contribution from hydrogen bonding between the NH-group of the bound amine and adjacent sulphonate groups, but equilibrium constant data are not available to enable evaluation of the importance of such a contribution or, indeed, of the contribution of these reactions to the inhibition of non-enzymic browning by sulphur(IV) oxoanions.

Sulphur(IV) oxoanions react rapidly with melanoidins formed in both the Maillard and ascorbic acid browning systems and reaction is accompanied by a reduction of absorbance in the visible region (Section 3.3.1). It is not known if this reaction represents reversible or irreversible addition of the oxoanion but, since melanoidin structure is shown consistently with carbonyl groups (or their enol forms) comprising part of the chromophores as, for example, in Fig. 3.4, it is reasonable to expect that bleaching of the pigment may, at least in part, be a consequence of the carbonyl–hydrogen sulphite ion reaction.

Although very different from hydroxysulphonate formation, the reduction of sulphur(IV) oxoanion by the reducing group of aldoses and ketoses falls under the heading of carbonyl group–sulphur(IV) oxoanion reactions. The possibility of this reaction was first demonstrated by Hägglund and coworkers (1930) and later considered by Ingles (1959c, 1960) in relation to food systems. It is found that when equimolar amounts of glucose and hydrogen sulphite ion are heated at $100°C$ for 8 h at high concentration (25 g water/100 g solids), the glucose is converted to gluconic acid (0·08 mol) and glucose-6-sulphate (0·19 mol), whilst the inorganic products are sulphur (0·25 gram atom) and sulphate (0·39 mol), where the yields are expressed per mole of reactant. The reaction is given also by galactose, xylose, arabinose, rhamnose, lactose, sorbose, fructose, ascorbic acid and dihydroxyacetone, but not by mannitol, sorbitol, α-methyl-D-glucoside, mannose or glyoxal. The reason why neither mannose nor glycoxal participate is attributed to the tendency of these compounds to form somewhat stable hydroxysulphonates, the free carbonyl compound being that which shows the desired reactivity. Indeed, fructose, which does not form a hydroxysulphonate, reacts more rapidly than do aldoses, as is also the case with sorbose. The reduction of sulphur(IV) oxoanions leads to dithionite ion which can disproportionate yielding sulphur (Section 1.10.7). Dithionite will also react with glucose and mannitol to form sugar sulphates (Ingles, 1960) and the suggested mechanism is, therefore, the reduction of sulphur(IV) oxoanion to this intermediate by the aldose or ketose, followed by sulphation. Since mannitol does not possess a reducing group, it will not react with hydrogen sulphite ion. Although most of these studies have been carried out at 100 °C, some data on the formation of sulphur and sulphate ion at lower temperatures are available (Ingles, 1960) for a 4:1 molar ratio of hydrogen sulphite ion to reducing compound in concentrated solution (25 g water/100 g solids). The most reactive of the three compounds, glucose, fructose and ascorbic acid, was ascorbic acid, yielding 0·09 gram atom sulphur after 24 weeks at 25 °C, and as much as 0·75 gram atom sulphur when mixtures were heated at 37 °C for 16 weeks. The corresponding amounts of sulphate ion produced were 0·54 and 1·4 mol, all yields being expressed per mole of reducing compound. Fructose was more reactive than glucose, giving 0·06 gram atom sulphur and 0·21 mol sulphate at 37 °C after 16 weeks, glucose giving no sulphur but a small quantity of sulphate ion (0·1 mol) under these conditions. Neither glucose nor fructose produced sulphur at 25 °C after 24 weeks but a small amount of sulphate ion (0·1 mol) was measured. It is interesting also to note that monofructoseglycine (Section 3.2.1) will reduce hydrogen sulphite ion, its reactivity being comparable to that of fructose.

    The oxidation products of the aldoses, that is the aldonic acids, are less reactive towards browning than the aldoses but it is found that glucose-6-sulphate is much more reactive towards Maillard browning than is glucose (McWeeny and Burton, 1963). Elemental sulphur has not been isolated from sulphite inhibited non-enzymic browning reactions and, therefore, there is no evidence to suggest that reduction of the additive takes place under model browning conditions. The reaction described by Ingles (1960) could, however, become significant in concentrated systems.

3.3.3 Irreversible formation of sulphonates

The only products which are known to account for the irreversible combination of sulphur(IV) oxoanions during their inhibition of Maillard and ascorbic acid browning are organic sulphonates. Methods of separating these compounds from reaction mixtures and their subsequent identification have been considered in Chapter 2 (Section 2.7.3.2). The mechanism of the formation of the most important of these sulphonates is effectively the replacement of the hydroxyl group by sulphite at position 4 of the 3-deoxyosuloses, which are known intermediates in the browning reactions. Thus, the reaction between glucose and sulphur(IV) oxoanion at elevated temperature (100 °C) and pH 6·5 leads to the formation of 3,4-dideoxy-4-sulpho-D-*erythro*-hexosulose (or the corresponding *threo*-isomer), shown as structure I below, with R = CHOH—CH$_2$OH (Ingles, 1962; Lindberg *et al.*, 1964), although the products may include the corresponding 2-hydroxyhexanoic acid, shown as structure II, with R = CHOH—CH$_2$OH, and formed as a result of the benzilic acid rearrangement of I, and also the 2-keto-acid, structure III, with R = CHOH—CH$_2$OH, a result of the oxidation of I:

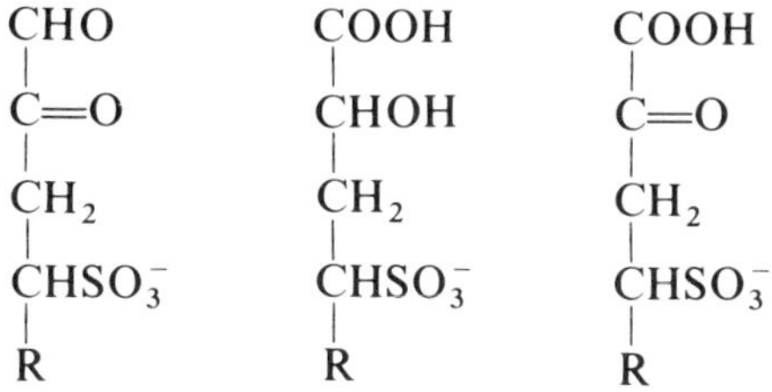

$$
\begin{array}{ccc}
\text{CHO} & \text{COOH} & \text{COOH} \\
| & | & | \\
\text{C}{=}\text{O} & \text{CHOH} & \text{C}{=}\text{O} \\
| & | & | \\
\text{CH}_2 & \text{CH}_2 & \text{CH}_2 \\
| & | & | \\
\text{CHSO}_3^- & \text{CHSO}_3^- & \text{CHSO}_3^- \\
| & | & | \\
\text{R} & \text{R} & \text{R} \\
\text{(I)} & \text{(II)} & \text{(III)}
\end{array}
$$

Similarly, the reaction of xylose with sulphur(IV) oxoanions at 135 °C and pH 6·5 leads to the formation of 2,5-dihydroxy-4-sulphopentanoic acid and 5-hydroxy-2-oxo-4-sulphopentanoic acid (Yllner, 1956; Hardell and Theander, 1971) which have structures II and III respectively, with R = CH$_2$OH. It is anticipated that both compounds are the result of the initial formation of a product with structure I (R = CH$_2$OH). The identity of the sulphonates was established in systems to which amino acid had not been added and, hence, the need for somewhat more vigorous conditions to promote reaction than those required in Maillard browning. The formation of compound **I**, with $R = CHOH-CH_2OH$, in glucose/glycine mixtures containing sulphur(IV) oxoanions after heating at $60\,^{\circ}C$ and pH 5·2 was confirmed by Knowles (1971). When ascorbic acid is allowed to decompose in the presence of sulphur(IV) oxoanions at $40\,^{\circ}C$ and pH 3·0, the sulphur-containing product is compound **I**, with $R = CH_2OH$ (Wedzicha and McWeeny, 1974a).

Ingles (1962) suggested two possible schemes for the formation of sulphonated product, as follows:

Scheme 1

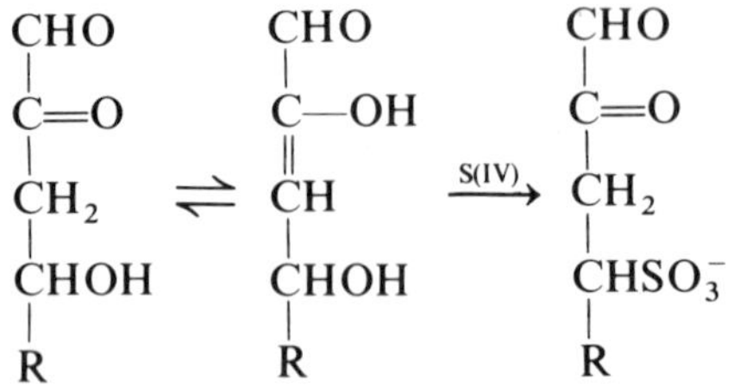

$$
\begin{array}{ccc}
\text{CHO} & \text{CHO} & \text{CHO} \\
| & | & | \\
\text{C}{=}\text{O} & \text{C}{-}\text{OH} & \text{C}{=}\text{O} \\
| & \| & | \\
\text{CH}_2 & \text{CH} & \text{CH}_2 \\
| & | & | \\
\text{CHOH} & \text{CHOH} & \text{CHSO}_3^- \\
| & | & | \\
\text{R} & \text{R} & \text{R}
\end{array}
$$

(with $\rightleftharpoons$ between first two structures, and $\xrightarrow{\text{S(IV)}}$ to the third)

Scheme 2

$$
\begin{array}{ccc}
\text{CHO} & \text{CHO} & \text{CHO} \\
| & | & | \\
\text{C}{=}\text{O} & \text{C}{=}\text{O} & \text{C}{=}\text{O} \\
| & | & | \\
\text{CH}_2 & \text{CH} & \text{CH}_2 \\
| & \| & | \\
\text{CHOH} & \text{CH} & \text{CHSO}_3^- \\
| & | & | \\
\text{R} & \text{R} & \text{R}
\end{array}
$$

(with $\xrightarrow{-H_2O}$ between first two structures, and $\xrightarrow{\text{S(IV)}}$ to the third)

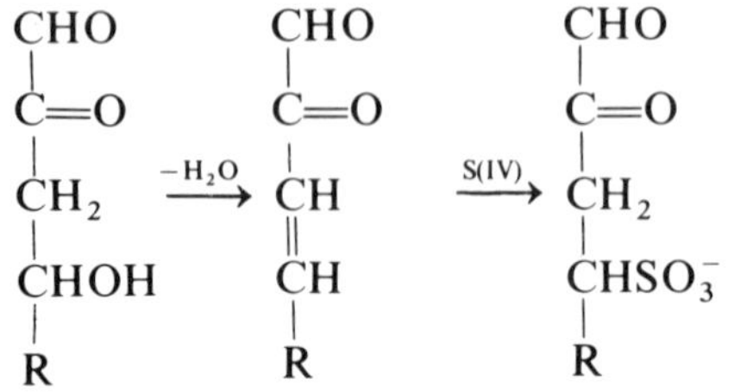

(IV)

Evidence for the probability of an enol structure at carbon 2 of the five and six carbon 3-deoxyosuloses has already been given in this chapter (Section 3.2.1). Hydroxyl groups which are β to an enol or ene-diol grouping are very labile and Scheme 1, therefore, represents the replacement of such a labile group by sulphite ion. This could be by way of an S_N2 displacement and could well lead to inversion of configuration at carbon 4. Alternatively, dehydration of the 3-deoxyosulose to the 3,4-dideoxy-compound **(IV)** is well established as is the tendency of HSO_3^- to undergo directed addition to double bonds. The mechanism of addition of sulphur(IV) oxoanions to α,β-unsaturated carbonyl compounds was considered in Chapter 1 (Section 1.11.2), and a combination of dehydration and addition steps constitute Scheme 2 above. The involvement of 3-deoxyosulose in the formation of sulphonate may be demonstrated by the fact that, when 3-deoxy-D-*erythro*-hexosulose is allowed to react with sulphur(IV) oxoanions at 100 °C and pH 6·5, the yield of sulphonate is greater than that obtained from D-glucose under the same conditions (48 % and 13 % respectively, Lindberg *et al.*, 1964). This result does not, of course, distinguish the mechanistic possibilities. Burton and coworkers (1963) favour the mechanism involving addition to the α,β-unsaturated intermediate, a suggestion supported by Anet and Ingles (1964) who found that, if the dehydrated intermediate in question was prepared by a route which did not involve the 3-deoxyosulose precursor, then the desired product would be obtained on treatment with sulphur(IV) oxoanion. The preparation may be accomplished through the allylic rearrangement of 3-deoxy-2,4-di-*O*-methyl-D-*erythro*-hexopyranos-2-ene (**V**), as follows (Anet, 1963):

$$\text{CH}_2\text{OH} \quad \rightarrow \quad \text{CH}_2\text{OH}$$

MeO — OH — OMe (**V**) → OH — O

and the complete reaction was effected by allowing a 10 % solution of **V** in water saturated with sulphur dioxide to stand at 20 °C for 24 h (Anet and Ingles, 1964). The six carbon sulphonate retains at least one asymmetric carbon atom whose configuration is the same as that of glucose and it is, therefore, not surprising that it is optically active (Lindberg *et al.*, 1964). In the case of the five carbon compound (structure **I**, with R = CH$_2$OH), optical activity would result from the stereospecific addition of sulphite ion at carbon 4 and the contributions from α- and β-forms of possible pyranoside or, less likely, furanoside structures. If this compound is converted to the corresponding 2-hydroxypentanoic acid (structure **II**, with R = CH$_2$OH), then any optical activity in the product will be a result of either retention or inversion of the configuration at carbon 4 during the sulphonation reaction. Yllner (1956) found that the product of the reaction between xylose and sulphur(IV) oxoanion was optically inactive and, therefore, more consistent with a mechanism which involves dehydration of the 3-deoxypentosulose to the α,β-unsaturated intermediate followed by non-stereospecific addition than with direct replacement of the hydroxyl group by sulphite ion, in which case an optically active product is expected. Recently, in unpublished work, Wedzicha and Galsworthy (1982) found that 3,4-dideoxy-4-sulphopentosulose, obtained from the decomposition of ascorbic acid in the presence of sulphur(IV) oxoanions, had a specific rotation of $0 \pm 0·04°\,\text{ml}\,\text{g}^{-1}\,\text{dm}^{-1}$, the error representing the sensitivity of the measurement. Since the 3-deoxypentosulose formed from ascorbic acid will retain the original configuration, this result appears to support the idea that the racemic compound is formed when sulphur(IV) oxoanions add to this intermediate. It is not known, however, if the hydroxyl group in position 4 is sufficiently labile to allow racemisation at this carbon atom under the conditions used for the sulphonation reactions.

The structures of sulphonated 3-deoxyosuloses have normally been shown as straight chain compounds and it is not known if there is any appreciable contribution from cyclic forms. Anet and Ingles (1964) recorded proton NMR spectra of 3,4-dideoxy-4-sulphohexosulose in deuterium oxide and noted a complex band assigned to two protons attached to carbon 3. This result indicates that it is probably correct to regard this sulphonated product as one with a well defined 2-keto group rather than as one which has an enol-OH in this position. Hardell and Theander (1971) report the presence of a methylene group in position 3 of 5-hydroxy-2-oxo-4-sulphopentanoic acid, also by means of NMR.

Sulphonated fragments of aldoses have also been isolated, particularly when reactions between aldose and sulphite ion have taken place at high pH. Ingles (1961b) identified 3-sulphopropanoic acid after glucose or fructose had been heated with sulphite ion at pH 9·6 and 100 °C for $4\frac{1}{2}$ h. When the reaction was carried out with glucose labelled (^{14}C) in positions 1, 2 or 6, the product was derived from various parts of the glucose molecule in the following yields:

Carbon atoms	Yield
1, 2, 3	32·1%
2, 3, 4	23·2%
4, 5, 6	34·5%
3, 4, 5	10·2%

These results indicate that dealdolisation of the sugar and Cannizzaro-type rearrangements occurred in the alkaline environment but no distinction can be made regarding the addition of sulphur(IV) before or after the fragmentation process. 3-Sulphopropanoic acid has also been isolated after the heating of D-erythrose with sulphur(IV) oxoanions at pH 6·5 and 135 °C for $1\frac{1}{2}$ h (Hardell and Theander, 1971).

The implications of the formation of stable sulphonates, by reaction between 3-deoxyosuloses and sulphur(IV) oxoanions, with regard to non-enzymic browning reactions of the Maillard and ascorbic acid types are clear. The pathway to the formation of colour in both types of reaction proceeds by way of the 3-deoxyosulose, through the 3,4-unsaturated dehydration product which shows the greatest reactivity towards colour formation. Reaction of the latter compounds with sulphur(IV) oxoanions at position 4 is irreversible and leads to the formation of saturated products which are not able to participate in the reactions of their precursors; the sulphonated 3-deoxyosuloses brown much less readily than the un-sulphonated compounds. The rate of irreversible loss of sulphur(IV) oxoanion from non-enzymic browning systems involving glucose and glycine, or ascorbic acid is of zero kinetic order and, therefore, reaction with intermediates is rapid compared with the rate of their formation. The function of the additive is, therefore, one of 'mopping up' the reactive intermediates of browning. Even at concentrations of 0·002 M free sulphur(IV) oxoanion compared with an initial glucose concentration of 1 M, Fig. 3.6 shows that there is still sufficient additive present to preclude a limiting of the rate of its irreversible combination when the system undergoes the Maillard reaction. Although rate constant data are not available it may be stated that, on the basis of there being a reasonable yield of stable sulphonate (for example, at least 66% of reacted sulphur(IV) oxoanion is converted to such products in the ascorbic acid system), the sulphonation reaction is much faster than the non-enzymic browning of the reactive intermediates. Nothing is known of the contribution of irreversible addition of sulphur(IV) oxoanions to unsaturated parts of melanoidin molecules to the loss of the additive or to its effect upon the colour of these compounds.

3.3.4 Sulphur(IV) oxoanion inhibited Maillard browning

The mechanisms of the inhibition of non-enzymic browning reactions by sulphur(IV) oxoanions have been reviewed by McWeeny and coworkers (1974) and McWeeny (1981). The interactions discussed in Sections 3.3.2 and 3.3.3 have been summarised as follows:

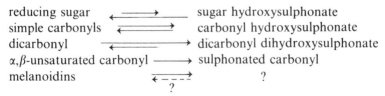

and there is the following additional step:

$$\text{dicarbonyl} \rightleftharpoons \text{dicarbonyl monohydroxysulphonate}$$

It is necessary now to consider how these possible reactions lead to the observed general trends described in Section 3.3.1. The rate of loss of additive to form products in which it is reversibly or irreversibly bound is zero at zero time, therefore both types of reaction require for their progress intermediates which are slow to be formed. In the early stages the intermediates and products which are potentially capable of reversibly binding sulphur(IV) oxoanion more strongly than glucose are 3-deoxyhexosulose, 3,4-dideoxyhexosulos-3-ene and 3,4-dideoxy-4-sulpho-hexosulose. The most reactive intermediates towards colour formation are the unsulphonated osuloses, and the inhibition of browning takes place by the effective removal of these compounds. If this inactivation of intermediate is through hydroxysulphonate formation, then the same intermediate will be released once the hydroxysulphonates start to decompose in the later stages of inhibition (Fig. 3.6). Therefore, as the intermediate in question will be formed during the normal course of reaction and additional intermediate will be present as a result of this release, its concentration at the end of the period of inhibition of browning by sulphur(IV) oxoanions will be greater than that present during the early stages of the glucose/glycine reaction to which no sulphur(IV) oxoanion had been added. It would be expected, therefore, that the rate of browning observed once sulphur(IV) oxoanion has been exhausted should be greater than that in the reaction containing glucose and glycine alone. This is found not to be the case. Burton and coworkers (1962) found that the rate of colour development after the end of inhibition is either equal to or slower than that in the uninhibited process and the implication of this result is that the compound or compounds which are released towards the end of the period of inhibition of browning have a lower reactivity towards colour formation than do the normal intermediates in the Maillard reaction. This observation is supported by Wedzicha (1984) who proposed a simple kinetic model to account quantitatively for the data of McWeeny and coworkers (1969) plotted in Fig. 3.6. The essential features of this model are the two-step formation of product, incorporating sulphur(IV) oxoanion in irreversibly bound form, which passes through a rate controlling intermediate, and the reversible binding of the additive resulting from combination with both glucose and the final stable sulphonated product. The latter is presumably 3,4-dideoxy-4-sulphohexosulose.

The kinetic model is of the form:

glucose + glycine $\xrightarrow{k_1}$ intermediate 1 $\xrightarrow{k_2}$ intermediate 2 $\xrightarrow[S(IV)]{fast}$ stable

$K_1 \updownarrow \Big| S(IV)$ products

hydroxysulphonate $K_2 \updownarrow \Big| S(IV)$

                                                                                        hydroxysulphonate

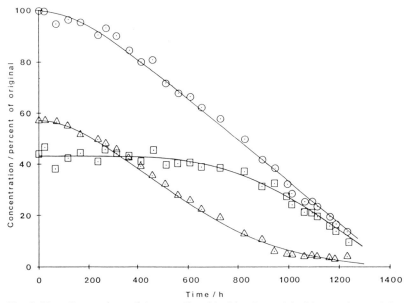

Fig. 3.10. Comparison of data predicted by kinetic model with experimental data of McWeeny and coworkers (1969). Lines denote calculated values. Reaction conditions as in Fig. 3.6. Total S(IV), ⊙; free S(IV), △; reversibly bound S(IV), ⊡.

where intermediate 1 is formed at constant rate during the whole of the period of observation, and formation of intermediate 2 is of first order with respect to intermediate 1. The rate constants, k_1 and k_2, are $3 \cdot 19 \times 10^{-5}\,\text{M}\,\text{h}^{-1}$ and $5 \cdot 5 \times 10^{-3}\,\text{h}^{-1}$ respectively at pH 5·5 and 55 °C. The hydroxysulphonate formed with 3,4-dideoxy-4-sulphohexosulose is assumed to be the monoadduct for the reasons discussed in Section 3.3.2 and the value of the equilibrium constant for its formation, consistent with the observed kinetic data, is $200\,\text{M}^{-1}$, representing a moderately stable adduct compared with the formation constants of simple hydroxy-sulphonates (Section 1.11.1). The closeness with which this model fits the experimental data of McWeeny and coworkers (1969) is illustrated in Fig. 3.10 and the lower browning potential of 3,4-dideoxy-4-sulphohexo-sulose, released towards the end of the period of inhibition of browning, is consistent with the observed rates of colour formation in the inhibited and uninhibited systems. Of course, once all the sulphur(IV) oxoanion has been lost from the system, the intermediates in the normal Maillard reaction, which are more reactive towards browning than the sulphonated product, will presumably form melanoidins, by-passing the sulphonate. In critically appraising this model is it necessary to draw attention to a possible shortcoming. In his studies on the model system glucose–glycine–sulphur(IV) oxoanion, Knowles (1971) reports the finding of the hydroxysulphonate adduct of 3-deoxyhexosulose mixed with the stable sulphonate. His conclusion is based upon the observation that the 3-deoxyhexosulose is apparently adsorbed onto strongly basic ion exchange resins as an acidic product. A similar observation has been made by Wedzicha and Imeson (1977) during work on the sulphur(IV) oxoanion-inhibited anaerobic browning of ascorbic acid which proceeds through 3-deoxypentosulose to the sulphonated product, 3,4-dideoxy-4-sulpho-pentosulose. There are no quantitative data concerning the amounts of these products which are observed, nor are there any published equilibrium constant data for hydroxysulphonate formation. Detailed comment on this apparent difficulty is, therefore, not possible, but the apparent in-significance of such adduct formation in the kinetic model could be a function of the low stationary concentrations of these intermediates and, perhaps, a somewhat lower than expected stability of the adduct. The latter is suggested on the grounds of a possible 1,5-pyranose structure for 3-deoxyhexosulose and also the reduced carbonyl group reactivity with respect to sulphur(IV) oxoanions as a result of a large contribution from an enolic OH-group in position 2, as discussed in Section 3.2.1.

No conclusions may be drawn concerning the interaction of sulphur(IV) oxoanions with melanoidins, since this area has not yet been explored in sufficient detail.

3.3.5 Sulphur(IV) oxoanion inhibited ascorbic acid browning
3.3.5.1 Anaerobic systems
Despite the differences between the structures of glucose and ascorbic acid there are many similarities in the methods by which browning is inhibited in the two systems, the main reason for these similarities being the participation of 3-deoxyosuloses in both the browning reactions. In both cases sulphur(IV) oxoanions are expected to have little or no effect upon the primary reaction, since they do not associate sufficiently strongly with the reactants, and in model system studies the concentrations of glucose and of ascorbic acid are generally in large excess over that of sulphur(IV) oxoanion. In the case of the ascorbic acid/sulphur(IV) oxoanion reaction, zero order kinetic behaviour with respect to the additive is observed from zero time and, therefore, the kinetic model in this case is expected to be somewhat simpler than that for the glucose/glycine/sulphur(IV) oxoanion reaction. Thus, a rate determining reaction of ascorbic acid, which

proceeds at constant rate during the period of observation of loss of sulphur(IV) oxoanion, is followed by rapid addition of sulphur(IV) oxoanion to form a product in which the additive is irreversibly bound. Thus, the scheme is:

$$\text{ascorbic acid} \xrightarrow{\text{slow}} \text{intermediate} \xrightarrow[\text{S(IV)}]{\text{fast}} \text{product}$$

where the product is 3,4-dideoxy-4-sulphopentosulose. Considering the mechanistic possibilities for the degradation of ascorbic acid discussed in Section 3.2.2.2, the only plausible route to 3,4-dideoxy-4-sulphopentosulose is that of decarboxylation and dehydration of ascorbic acid followed by addition of sulphur(IV) oxoanion to 3,4-dideoxypentosulos-3-ene. The precursor of this intermediate may also be formed from the cyclic precursor of furfural, as suggested by Tokuyama, Goshima and their coworkers (Tokuyama *et al.*, 1971), but it is difficult to see how any of their proposed intermediates could react directly with sulphur(IV) oxoanions to form the known or any new product. The important difference between inhibited ascorbic acid browning and inhibited glucose/glycine browning is that, whereas the glycine is essential for the occurrence of irreversible loss of sulphur(IV) oxoanion in the Maillard system, the presence of amino acid has only a small effect upon the rate of combination of the additive with ascorbic acid degradation products, although glycine accelerates the rate of browning of ascorbic acid in the absence of the additive. This result is reconciled easily with the initial stages of the two browning reactions, the formation of reactive intermediate from ascorbic acid being a spontaneous process.

There are no data on the concentrations of free and reversibly bound sulphur(IV) oxoanion during the ascorbic acid/sulphur(IV) oxoanion reaction but, in view of the similarity of the important stages of inhibition to those in the inhibited Maillard reaction, the conclusions obtained in the previous section could well apply.

3.3.5.2 Oxidative systems

The effect of sulphur(IV) oxoanion upon oxidative browning of ascorbic acid is two-fold. First, the additive inhibits autoxidation of ascorbic acid and secondly, it prevents subsequent browning of dehydroascorbic acid. The mechanism of the autoxidation effect is not understood but could well involve interactions with free radical intermediates. Kalus and Filby (1977) note that, at pH 9·5, the addition of sodium sulphite to ascorbic acid solutions undergoing oxidation increases the amount of ascorbate radical. The maximum effect is observed at pH 9·5, possibly coinciding with maximum rate of formation of intermediate, but its accumulation could indicate interruption of the final step in the pathway to dehydroascorbic acid. This reaction was not studied in the pH range of food systems.

If dehydroascorbic acid is formed, sulphur(IV) oxoanions will effectively reduce its carbonyl reactivity through hydroxysulphonate formation. Only the monoadduct is known, the rigid ring structure of dehydroascorbic acid sterically hindering approach of nucleophiles to the free carbonyl group of the hydroxysulphonate and, therefore, the adduct should show little carbonyl activity. Thus, such hydroxysulphonate formation will reduce the tendency for reaction with amines and subsequent development of the characteristic wine red or cherry red coloration (Section 3.2.2.3). There is no evidence to suggest that dehydroascorbic acid or compounds derived therefrom react with sulphur(IV) oxoanions, binding them irreversibly. In the absence of such information the only conclusion reached regarding the mechanism of inhibition of dehydroascorbic acid browning by sulphur(IV) oxoanions is that it takes place as a result of hydroxysulphonate formation with mono- and dicarbonylic intermediates in colour formation. Examples of such intermediates have been given in Section 3.2.2.3.

3.3.6 Sulphur(IV) oxoanions and lipid browning

The simple description of lipid browning as the interaction of carbonylic oxidation products of unsaturated fats with amines (Section 3.2.4) leads to the expectation that, in common with other forms of browning reaction, inhibition of lipid browning by sulphur(IV) oxoanions will take place through hydroxysulphonate formation. If, on the other hand, it is the earlier products of oxidation, such as hydroperoxides, which interact with amines to form coloured products (Shimasaki *et al.*, 1977), then reduction of these hydroperoxides by sulphur(IV) oxoanions (Section 1.11.3) could constitute a pathway to the reduction of colour formation. It may also be argued that, since free radical sulphonation is known (Section 1.11.2), then addition of the sulphonate group to unsaturated fatty acids would lead to products with reduced browning potential. Such sulphonated oils are well known in the leather industry but their production leads also to peroxidation of the lipid (Küntzel and Schwörzer, 1957) and this process is, perhaps, the first example of the sulphur(IV) oxoanion-induced oxidation of unsaturated fats. The ability of sulphur(IV) oxoanions to promote the browning of linolenic acid has been demonstrated by Lamikanra (1982). Mixtures of linolenic acid and sulphur(IV) oxoanion in aqueous ethanol containing glycine and a catalytic amount of manganese(II) ion at an apparent pH of 5·7 take up oxygen and show more rapid colour development than is the case in control experiments containing no sulphur(IV) oxoanion, after a period of apparent inhibition. The ability of sulphur(IV) oxoanion to induce peroxidation in corn oil emulsions (Kaplan *et al.*, 1975) and in lung lipid extracts, mitochondria or microsomes (Inouye *et al.*, 1980) is indicated by the formation of products which give a positive reaction with thiobarbituric acid. The most detailed investigation of this effect has been that of Lizada and Yang (1981) who describe the effects of sulphur(IV) oxoanion and oxygen upon Tween-stabilised emulsions of linoleic and linolenic acids. Sulphite-induced oxidation of linoleic acid proceeds by way of hydroperoxide, the amount of hydroperoxide formed being equal to the amount of conjugated diene formed, suggesting that the intermediates are conjugated diene hydroperoxides, both 9- and 13-hydroperoxides being observed. Also, the yield of conjugated diene is high, some 0·64 mol per mole of sulphur(IV) oxoanion oxidised. This reaction takes place between pH 4 and pH 7 and leads also to the formation of products containing bound sulphur. When linolenic acid is used, the reaction proceeds only below pH 6 and the reaction products give a positive test with thiobarbituric acid. In both cases the oxidation of sulphur(IV) oxoanions and hydroperoxide formation are inhibited by potent radical scavengers such as hydroquinone, butylated hydroxytoluene and α-tocopherol, but the addition of hydroxyl-radical scavengers and superoxide dismutase does not affect significantly the rate of oxidation. An interesting observation is that conjugated diene formation is inhibited by 4-thiouridine. Sulphite ion is known to react with 4-thiouracil derivatives in an oxygen-mediated radical reaction which involves the addition of the SO_3^- radical (Hayatsu and Inoue, 1971) and the inference is, therefore, that the oxidation of the fatty acid also proceeds by way of this species, which presumably leads also to sulphonation.

There are insufficient data available for any conclusion to be drawn concerning the effect of sulphur(IV) oxoanion on lipid oxidation in more complicated systems, since an important effect is likely to be that of other components on the oxidation of sulphur(IV) oxoanion. It has been shown already that such autoxidation is very sensitive, in the catalytic and inhibitory senses, to the presence of impurities (Section 1.10.1) but the results reported here indicate that, under appropriate conditions, not far removed from those found in food systems, sulphur(IV) oxoanion-mediated autoxidation of lipids is possible.

4

Enzymology

4.1 INTRODUCTION

The interactions between sulphur(IV) oxoanions and enzyme systems may be divided broadly into those effects which modify the reactivity of the enzyme towards its substrate and those which result in specific changes in the oxidation state of sulphur. The majority of reported cases of the former effect are of enzyme inhibition, and examples may be drawn to cover a wide range of enzyme function, as indicated in the compilation of representative literature shown in the Appendix to this chapter. The absence of an effect of sulphur(IV) oxoanion is regarded usually as a negative result and, hence, little information has been published regarding enzymes which are resistant to sulphur(IV) oxoanion. It is, however, important to note that the inhibitive effect is by no means universal; Moutonnet (1967), for example, reports that the addition of sulphur(IV) oxoanion at levels of up to 250 mg potassium disulphite per gram of substrate has no effect upon the rates of hydrolysis by pepsin and trypsin of skimmed milk powder, gelatin or gluten. A further example in which no effect is observed also illustrates some potential pitfalls encountered in making generalisations concerning the inactivation of enzyme types. Perhaps the most well known example of enzyme inhibition by sulphur(IV) oxoanion is the inhibition of polyphenol oxidase browning but it is evident that within a single plant the isoenzymes of polyphenol oxidase show differential activity towards the additive, to the extent that some components are resistant to inhibition of browning (Wong *et al.*, 1971; Montgomery and Sgarbieri, 1975). Also, the activity of a small number of enzymes may be increased by sulphur(IV) oxoanion, for example the activation of phosphofructokinase by sulphite and other divalent anions (Foe and Trujillo, 1980) and the presence of a sulphite dependent adenosine triphosphatase in some bacteria and in extracts of animal tissues (Lambeth and Lardy, 1971; Tomigana, 1978). Sulphur(IV) oxoanion dependent enzymes are widely present in the sulphur utilising bacteria of genus, *Thiobacillus*, an example which involves sulphur(IV) oxoanion as an electron source being the sulphite dependent nitrate reductase of *Thiobacillus denitrificans* (Adams *et al.*, 1971).

Two classes of enzyme, the purpose of which is to change the oxidation state of sulphur(IV) oxoanion, are the sulphite oxidases and reductases. The oxidation of sulphur(IV) oxoanion through the action of sulphite oxidase is distinguishable from that arising from the action of other oxidases, such as peroxidase (in its oxidative capacity) or xanthine oxidase, which promote oxidation by the generation of the superoxide ion intermediate (Section 1.10.1). With these enzymes oxidation proceeds by way of a free radical mechanism in contrast to sulphite oxidase catalysed oxidation which does not involve free radical intermediates. Therefore, whereas the free radical pathway is inhibited by a large number of organic compounds and is not efficient at low sulphur(IV) oxoanion concentrations, the sulphite oxidase reaction is found to take place under such conditions in the presence of radical inhibitors in a wide variety of animal and plant tissues and in bacteria. It is clear that the role of sulphite oxidase is of considerable physiological importance and is responsible for the detoxification of ingested sulphur(IV) oxoanion. The sulphite reductases are found in plants, bacteria and fungi and are linked with sulphur metabolism in both assimilatory (biosynthetic) and dissimilatory (respiratory) capacities. The product of dissimilatory action is hydrogen sulphide at low sulphur(IV) oxoanion concentration and polythionate at high concentration, whilst assimilatory action leads ultimately to the incorporation of sulphur into organic molecules, the reaction proceeding through sulphide.

4.2   ENZYME INHIBITION

It is reasonable to expect that enzyme inhibition by sulphur(IV) oxoanion takes place as a result of one or more of the following reactions:

(a)   reaction with disulphide groups on proteins leading to modification

of tertiary structure which is important in protein function. Reaction with thiols under oxidising conditions;
(b) reaction with coenzymes;
(c) inactivation of prosthetic groups and cofactors;
(d) competition for active site of enzyme;
(e) removal of reactants.

A sixth reaction which may lead to apparent enzyme inhibition, but which does not prevent enzyme action, is that of sulphur(IV) oxoanion with the products of enzyme action to give new compounds which do not show the characteristics of normal reaction products. The relative importance of these reactions to the modification of enzyme activity in any particular case is unique to that enzyme or isoenzyme and, therefore, consideration of enzyme inhibition under the headings of enzyme systems is an investigation of a large number of special cases. For this reason, the subject of enzyme inhibition will be discussed here under the headings of the possible reactions.

4.2.1 Reaction with disulphide and thiol groups on proteins

The reactions between sulphite ion and simple disulphide compounds and in systems in which the disulphide bond is part of a protein have been discussed in Section 1.11.9. It is evident that the availability of disulphide bonds and the ease with which they are cleaved by sulphite ion is dependent to a large extent on protein structure and on the environment in which the protein is placed. Therefore, the presence of disulphide groups in an enzyme does not indicate certain cleavage of such groups by sulphite ion and the presence of cleavage may not be assumed automatically to lead to enzyme inactivation. The interpretation of the kinetic effects of disulphide bond cleavage by sulphite ion depends upon the reversibility of the process. The reaction of disulphides with sulphite ion is reversible and may, therefore, be described by an equilibrium constant, but it is unlikely that the reaction between protein disulphides and sulphite ion is generally so, since the ability of the disulphide bond to reform depends upon the re-establishing by the molecule of its original conformation. It is, therefore, also a question of the degree to which the structure has been changed by disulphide cleavage.

There is no apparent experimental evidence to link unequivocally disulphide cleavage by sulphite with reduced enzyme activity, although numerous possiblities of such inhibition have been discussed. Cooperstein (1963) reports the reversible inactivation of cytochrome oxidase by reagents, including sulphur(IV) oxoanion, which react with disulphide bonds, the reversal being accomplished by the addition of either oxidised glutathione or cystine. Yoshikawa and Orii (1972a), on the other hand, suggest that the inhibiting action is a result of reaction with the haem moiety of cytochrome a, this reaction competing also with those of cyanide and azide which are regarded as haem specific ligands (Yoshikawa and Orii, 1972b). The position with regard to this system is, therefore, unclear. The state of aggregation of plant malate dehydrogenase is altered by the addition of sulphur(IV) oxoanion, possibly as a result of disulphide cleavage, but the modification in structure is not held to be responsible for the strong inhibitory effect of sulphur(IV) oxoanion (Ziegler, 1974a). Indeed, it is possible that disaggregation at low sulphur(IV) oxoanion concentration leads to a slight increase in enzyme activity. A change in the electrophoretic pattern of aggregates of glutamate dehydrogenase has been observed when pea seedlings are treated with gaseous sulphur dioxide (Pahlich, 1972), an effect which arises possibly from the specific action of the additive on disulphide bonds of the aggregates (Ziegler, 1974a), but in this case the relationship between the state of aggregation and enzyme activity is not known. Glutamate dehydrogenase is also inhibited by sulphur(IV) oxoanion when present in assay media (Pierre, 1977), but the activity of the enzyme extracted from leaves fumigated with sulphur dioxide is increased (Pahlich et al., 1972). Another enzyme system which is sensitive to disulphide cleaving reagents, including sulphur(IV) oxoanion, is microbial nitrate reductase (from Chlorella fusca), the inactivation by these reagents being partly reversible through oxidation with ferricyanide ion. Although such behaviour is characteristic of disulphide cleavage, doubt is cast on the conclusion that disulphide modification is an important contributor to enzyme inactivation by the presence of other moieties in the protein which may interact with the additive. Thus, the reductase in question is a flavomolybdoprotein and may contain a cytochrome (Gomez-Moreno and Palacian, 1974). Phelps and Hatefi (1981) discussed the possibilities for the inhibition of D(−)-β-hydroxybutyrate dehydrogenase by modifiers of disulphides, thiols and vicinal dithiols. These studies implied that the enzyme contains a thiol group essential for catalysis in both forward and reverse directions and that this thiol is one of a pair in a vicinal dithiol grouping. Applying the treatment of kinetic data normally used to describe reversible inhibition, these authors found the sulphite inhibition to be competitive or possibly mixed non-competitive with respect to β-hydroxybutyrate, an observation which implies that a group in the enzyme active site, required for the binding of this substrate, might be modified by the additive. Since sulphide is found also to inhibit the enzyme competitively with respect to β-hydroxybutyrate, it is possible that the two inhibitors have the same site of action. An interesting possibility, alternative to that of disulphide bond cleavage by sulphite, was also considered. This is the addition of sulphite ion to the thiol group under oxidising conditions, as follows:

$$E{-}SH + SO_3^{2-} + \tfrac{1}{2}O_2 \rightarrow E{-}SSO_3^- + OH^-$$

Such a mechanism constitutes part of the reaction used for the preparation, in good yield, of S-sulphonates from disulphides and sulphite ion (Section 1.11.9). Since sulphite ion may be displaced from S-sulphonate through nucleophilic attack by thiolate ion, it is possible that the thiol group vicinal to the S-sulphonate may displace the sulphite ion to form a disulphide bond, thus:

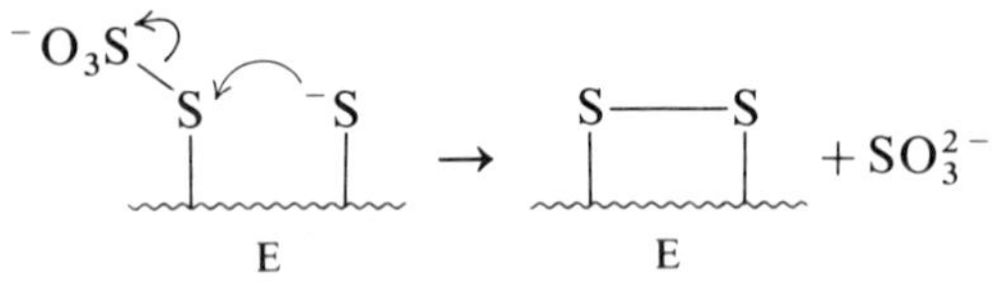

This possibility was not supported by experimental evidence in the case of β-hydroxybutyrate dehydrogenase, for two reasons: first, sulphite inhibition is reversible by dilution of the enzyme–sulphite mixture and secondly, the extent of inhibition is unaffected by the presence of oxygen. This type of mechanism could, however, participate in other cases of protein modification by sulphur(IV) oxoanion.

In the absence of unequivocal evidence, the inhibition of enzyme action through disulphide cleavage by sulphur(IV) oxoanion has been relegated to the level at which it is used to account tentatively for otherwise unexplained inhibition.

4.2.2 Reaction with coenzymes

The most important coenzymes which react with sulphur(IV) oxoanion are the pyridine coenzymes, nicotinamide adenine dinucleotide (NAD$^+$), nicotinamide adenine dinucleotide phosphate (NADP$^+$) and the reduced forms of these compounds, NADH and NADPH, respectively, which are important hydrogen carriers associated with the dehydrogenases. The involvement of these coenzymes in a straightforward enzyme reaction may be summarised in the following simplified scheme (Whitaker, 1972):

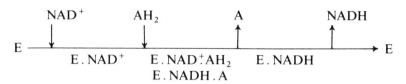

in which E denotes the enzyme, A is the oxidised form of substrate, and the complexes involving the enzyme are shown below the line. The coenzyme is liberated in a reduced form and may be oxidised enzymically to replenish the supply of NAD^+.

The reaction of 1-substituted nicotinamides with sulphur(IV) oxoanion has been considered in Section 1.11.7, the pH independent equilibrium constant for the reaction,

$$NAD^+ + SO_3^{2-} \rightleftharpoons NADSO_3^-$$

being in the region of $40\text{--}70\,M^{-1}$ at $25\,°C$. Therefore, at typical concentrations of NAD^+ and sulphite ion in the region of 1 mM, very little of the coenzyme is present in the form of sulphonate (approximately 5 %, when $[NAD^+] = [SO_3^{2-}] = 1$ mM) and clearly the formation of such an adduct will have little kinetic effect upon the enzyme reaction, unless the concentration of sulphite ion is high. The possible additional reactions which may take place when mixtures of enzyme and coenzyme are mixed with sulphur(IV) oxoanion are as follows:

$$E + NADSO_3^- \rightleftharpoons E \cdot NADSO_3^-$$

$$E \cdot NAD^+ + SO_3^{2-} \rightleftharpoons E \cdot NADSO_3^-$$

$$\left(\begin{array}{c} E + SO_3^{2-} \rightleftharpoons E \cdot SO_3^{2-} \\ E \cdot SO_3^{2-} + NAD^+ \rightleftharpoons E \cdot NADSO_3^- \end{array} \right)$$

If $NADSO_3^-$ competes effectively with NAD^+ for the coenzyme binding site on the enzyme, then it is conceivable that a low concentration of the sulphonate may prove to be strongly inhibiting. Similarly, a high affinity of $E \cdot NAD^+$ for sulphite ion would result in inhibition of reaction of the enzyme–coenzyme complex. In their studies of the addition of sulphite ion to 1-substituted nicotinamides, including poly(1-(4-vinylbenzyl)nicotin-amide chloride), Shinkai and Kunitake (1977) report an increase in equilibrium constant for the formation of sulphonate by a factor of approximately 500 when the polymeric substrate is compared with the monomeric substrate, in aqueous solution at an ionic strength of $0\cdot1$ M. This result may well be the effect of the sulphonate substituent reducing the energy of interaction between the hydrophobic polymer and its aqueous environment, but a large enhancement of binding of sulphite ion has also been found when NAD^+ is bound to lactate dehydrogenase or to malate dehydrogenase (Parker et al., 1978). Under these conditions the extent of activation of the nicotinamide ring towards attack by sulphite ion was in the region of $1-2 \times 10^5$ in equilibrium constant, the suggested reason for the activation being the presence of a protonated histidine residue in close proximity to the bound NAD^+. The observation that the apparent stability of $E . NADSO_3^-$ passes through a maximum at around pH 6·6 is consistent with this reason, since the observed maximum will be the result of two opposing effects: the increase in the amount of sulphite ion present in equilibrium with other sulphur(IV) oxospecies and the reduction in the amount of protonated histidine, as the pH is raised. An estimate of the extent of conversion of enzyme–coenzyme complex to inactive sulphonate, assuming an equilibrium constant of $1 \times 10^7 M^{-1}$ for the formation of sulphonate from complex and sulphite ion, shows a very high conversion when, for example, the sulphite concentration is $100 \mu M$ and the concentration of complex is $1 \mu M$ or less. Such combination of sulphite ion with enzyme reaction intermediate offers a very effective method of enzyme inhibition. Parker and coworkers (1978) were unable to distinguish the formation of $E . NADSO_3^-$ by reaction of $E . SO_3^{2-}$ with NAD^+ from the reaction between $E . NAD^+$ and SO_3^{2-}. The former reaction requires an interaction between the coenzyme binding site of the enzyme and sulphite ion. Although the apoenzyme of lactate dehydrogenase will interact with sulphite ion, it is not known if this takes place at the site in question or if it is a general interaction with positively charged groups. The increase in equilibrium constant for formation of adduct on going from NAD^+ to enzyme bound NAD^+ results from an increase, by three to four orders of magnitude, in rate constant for the addition reaction, such an increase being noted also by Shinkai and Kunitake (1977) for addition to polymeric nicotinamide, in which case the enhancement was by a factor of approximately 2×10^4. The possible structure of the lactate dehydrogenase–NAD^+–SO_3^{2-} complex has been discussed also by Trommer and Glöggler (1979).

If enzyme inhibition takes place as a result of reaction between $E . NAD^+$ and SO_3^{2-}, then the inhibition will be uncompetitive with respect to NAD^+ and, since the sulphite ion and substrate, AH_2, react with the same form of enzyme, $E . NAD^+$, then the inhibition will be competitive with respect to AH_2. An important consequence of uncompetitive inhibition with respect to NAD^+ is that the addition of excess coenzyme will not reverse the inhibitory effect of sulphite ion. Pfleiderer first suggested that the inhibition of lactate and malate dehydrogenases was a result of ternary complex formation involving the enzyme, coenzyme and sulphite ion (Pfleiderer and Hohnholz, 1959) and an example of a kinetic investigation which gives a result consistent with the prediction is that by Egorov and coworkers (1980) on formate dehydrogenase. The effect of the possible reaction between the reduced form of coenzyme (NADH) and sulphite ion (Tuazon and Johnson, 1977) on the kinetics of the enzyme reaction is not known, neither is there evidence for reaction between enzyme bound NADH and sulphite ion.

A rather unusual case of enzyme inhibition by binding to NAD^+ is observed in the enzyme, urocanase, which contains a tightly bound NAD^+ moiety. The unusual feature is that the inactive enzyme–coenzyme–sulphite complex may be photoactivated, the mechanism of activation being the photodissociation of the $NADSO_3^-$ adduct. Evidence for this process comes from the presence of the spectrum of $NADSO_3^-$ in the difference spectrum of active and inactive enzyme. Imidazole propionate, which acts as a competitive inhibitor with respect to substrate, also protects against inhibition by sulphite (Hug *et al.*, 1978).

4.2.3 Reaction with prosthetic groups and cofactors
Prosthetic groups and cofactors are distinguished from coenzymes by the fact that the complete enzyme reaction leads to regeneration of the prosthetic group or cofactor without it becoming detached from the enzyme. On the other hand, the coenzymes, NAD^+ or $NADP^+$, are released, after enzyme reaction, in a reduced form and a second enzyme reaction is required to regenerate the coenzyme. The prosthetic groups or cofactors of significance with respect to reactions of sulphur(IV) oxoanions are flavin, folate, haem, pyridoxal and thiamine. These will be considered in turn.

4.2.3.1 Flavin cofactors
Flavoproteins contain a prosthetic group which incorporates an isoalloxazine moiety with the following structure:

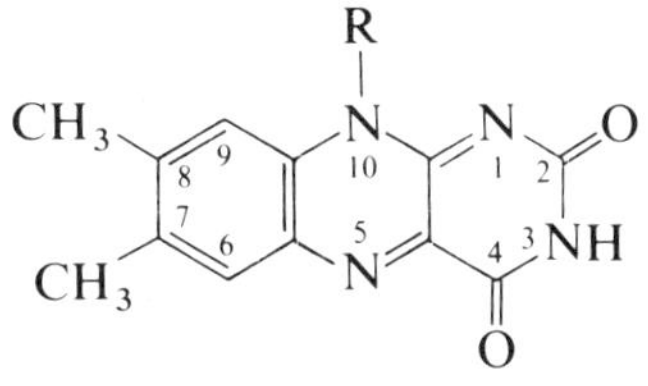

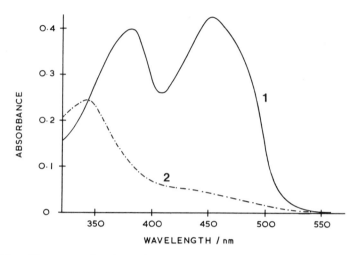

Fig. 4.1.   Changes to the absorption spectrum of glucose oxidase caused by the addition of a large molar excess of sulphur(IV) oxoanion at pH 6·3 and 25 °C. Curve 1, 3×10^{-5} M glucose oxidase; curve 2, 3×10^{-5} M glucose oxidase + 0·09 M sulphur(IV) oxoanion (Swoboda and Massey, 1966). After *J. Biol. Chem.*, 1966, **241**, page 3410.

in which the substituent, R, is ribitol-5-phosphate in the case of flavin mononucleotide (FMN) or ribitol-diphosphate-D-ribose-adenine in the case of flavin adenine dinucleotide (FAD), the molecule with only D-ribitol as the substituent being riboflavin. Flavoenzymes include oxidases, reductases, dehydrogenases and hydroxylases and may include another prosthetic group such as haem and metal ions.

The flavin component colours flavoproteins yellow and the immediate consequence of the addition of sulphur(IV) oxoanion is reversible bleaching of the pigment. The loss of the visible spectrum is illustrated in Fig. 4.1 for the glucose oxidase–sulphur(IV) oxoanion reaction, which is the first reported case of a flavoprotein–sulphur(IV) oxoanion interaction (Swoboda and Massey, 1966). A similar result was found subsequently for the oxidases: D-amino acid, L-amino acid, glycollate and lactate (decarboxylating) and also for oxynitrilase (Massey *et al.*, 1969) and, as shown in Fig. 4.1, the reactions are all characterised by an isosbestic point around 340 nm. There appears to be a wide range of stability constants for these sulphite adducts, which may be grouped into those reflecting moderate stability, with a dissociation constant of the order of 10^{-3}–10^{-4} M, namely glucose, D-amino and L-amino oxidases, oxynitrilase (Massey *et al.*, 1969) and thiamine dehydrogenase (Gomez-Moreno *et al.*,

1979), and those which bind sulphite much more tightly, with adduct dissociation constants less than 10^{-5} M, such as lactate (decarboxylating) and glycollate oxidases (Massey *et al.*, 1969) and cytochrome b_2 (Lederer, 1978).

However, it is clear that the presence of a flavin moiety in an enzyme does not imply that the prosthetic group will react necessarily with sulphur(IV) oxoanion. Indeed, Massey and coworkers list eleven flavoenzymes which do not appear to take part in such a reaction. When compared with a dissociation constant in the region of 2 M for free FAD and FMN (Muller and Massey, 1969), it is evident that the reactivity of the flavin towards sulphur(IV) oxoanion is greatly enhanced when the flavin is bound to protein. As part of work on the influence of microheterogeneous environments such as micelles and water soluble polymers on the reactivity of coenzymes and prosthetic groups, referred to in Section 4.2.2 in connection with nicotinamides, Shinkai (1978) reports an enhancement of association constant for the reaction between 3,9-disubstituted flavins and sulphite ion by one order of magnitude when the substituent in the 3-position is part of a polystyrene molecule. The microheterogeneous environment of protein bound flavin may contribute, therefore, to the enhanced binding of sulphite ion over free flavin.

The structure of the flavin moiety shown above is in its oxidised form, reduction leading to 1,5-addition of two hydrogen atoms. The oxidised form is electron deficient and is capable, therefore, of reaction with sulphite ion, whilst the reduced form does not react in this way. Theoretical calculations referred to by Hevesi and Bruice (1973) suggest that the nitrogen atom in position 5 is the most electrophilic, supporting the original suggestion made by Muller and Massey (1969) that addition of sulphite ion takes place to this position. This suggestion was made on the basis of the close correspondence between the ultraviolet spectrum of the adduct formed between sulphite ion and 3,10-disubstituted flavin and that of the $N(5)$-acyl derivative of 3,10-disubstituted flavin, the respective structures being as follows:

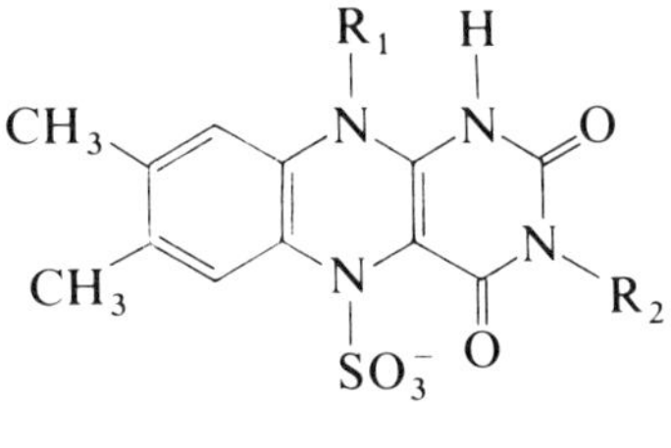

S(IV) adduct 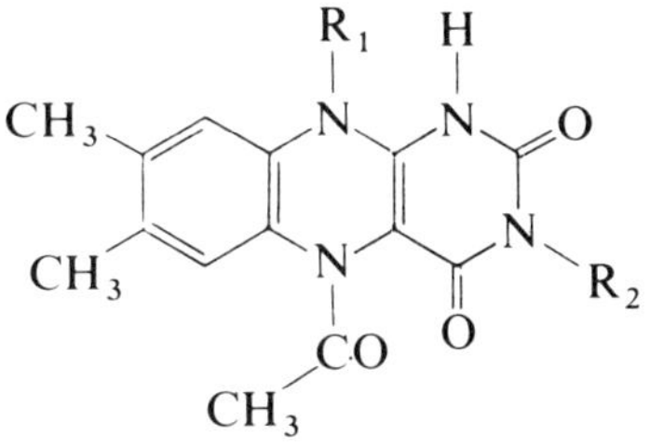   *N*-acyl derivative

The structure of the $N(5)$-acyl derivative of 3,10-dimethyl flavin was shown subsequently by Kierkegaard and coworkers (1971) to be that given above. The ultraviolet spectrum of the sulphite adduct is also similar, but not identical to the 1,5-dihydro form of the flavin. It is interesting to note that there is a good linear relationship between the logarithm of the dissociation constants of a range of flavin derivatives and their redox potentials (Muller and Massey, 1969), suggesting that the processes of sulphite addition and reduction are analogous, that is they both proceed through addition at position 5. Similar linear free energy relationships for other, simpler heterocyclic compounds are described in Section 1.11.8.

The flavin moiety is known to be bound covalently through the methyl group in position 8 to histidine or to cysteine residues on the protein (Singer *et al.*, 1976). However, the electronic effects of such bond formation are unlikely to have a profound effect upon sulphite adduct stability, but the presence of positively charged groups in the vicinity of the flavin could affect greatly the addition of sulphite in a manner similar to that described for enzyme bound NAD^+ in the previous section. The pH dependence of the dissociation constant of the D-amino acid oxidase–sulphite adduct has been reported (Massey *et al.*, 1969) and indicates that the stability of the adduct does, indeed, pass through a maximum at pH 7·6, indicating the presence of a neighbouring charged group with a pK_a of 8·2. It is interesting to note that there is a good deal of evidence for the presence of a positively charged amino acid residue close to the site at which strong binding of benzoate and other carboxylic acids takes place. Benzoate ion is found to displace sulphite ion from the adduct. The stability of the sulphite adduct of glucose oxidase, which has a dissociation constant slightly lower than that of the sulphite adduct of D-amino acid oxidase, does not pass through a maximum in the range pH 4–9. The stability of the complex falls with increasing pH, an inflexion being evident from a graph of log(dissociation constant) against pH around pH 5·5.

Massey and coworkers (1969) used the sulphite binding power of the flavoenzymes which they investigated to make a further prediction. Their observation was that only oxidases which formed also a red, anionic semiquinone underwent reaction with sulphite ion; those enzymes which did not react were dehydrogenases and formed blue, neutral semiquinones. This observation was explained by the presence of a positively charged group which facilitated sulphite binding and stabilised the anionic semiquinone in the oxidases, whereas the dehydrogenases had a negative charge in the vicinity of the binding site, reducing the tendency for attack by sulphite ion and not inhibiting the formation of neutral semiquinones. A

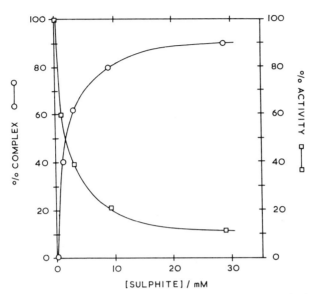

Fig. 4.2. Relationship between formation of complex between the flavin moiety of oxynitrilase and sulphur(IV) oxoanion and reduction of enzyme activity (Jorns, 1980). Reproduced with permission from K. Yagi and Y. Yamano (Eds), *Flavins and Flavoproteins*, page 163. © 1980 Japan Scientific Societies Press.

correlation between the reactivity of the reduced form of enzyme with oxygen and the ability of the oxidised form to react with sulphite ion was also found and was suggested to be connected, perhaps, with the charge in the vicinity of the prosthetic group. Since these suggestions were published there has been only one exception reported in the literature, namely the flavoenzyme baker's yeast L-lactate dehydrogenase (cytochrome b$_2$). This enzyme forms a red semiquinone and reacts with sulphite ion, consistent with the idea of charged group participation, but does not react with oxygen.

The reaction between the flavin moiety of flavoenzymes and sulphite ion inhibits effectively the enzyme reaction. Reassuring evidence for such a statement has been published by Jorns (1980) for the inhibition of oxynitrilase by sulphur(IV) oxoanion and is shown graphically in Fig. 4.2. It is evident that an excellent linear relationship exists between the extent of inhibition of enzyme activity and the amount of flavin–sulphite complex formed. Since both sulphite ion and the substrate are likely to react with the same form of the enzyme, the inhibition is expected to be competitive with respect to substrate. Sulphite ion reacts competitively with benzoate towards oxynitrilase. Since benzoate is a competitive inhibitor of this enzyme, then sulphite ion should act in a similar manner. Lederer (1978) reports competitive inhibition by sulphite ion, with respect to lactate, of cytochrome b_2.

4.2.3.2 Folate
Folic acid is a substituted pteridine with the following structure:

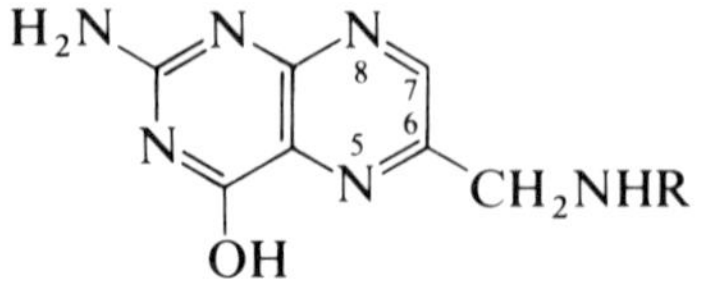

in which R is p-aminobenzoylglutamate with as many as seven residues of glutamic acid. In its reduced form, that is as dihydrofolate, this cofactor has functions which include the transfer of methyl, hydroxymethyl, formyl and formimino groups. Reduction of folate involves two steps. The first is addition of hydrogens to positions 7 and 8 to give dihydrofolate. A further addition of two hydrogens to positions 5 and 6 yields tetrahydrofolate.

There is little information on reactions between this cofactor and sulphur(IV) oxoanions. It is evident that both folate and dihydrofolate react reversibly with hydrogen sulphite ion in the following ways:

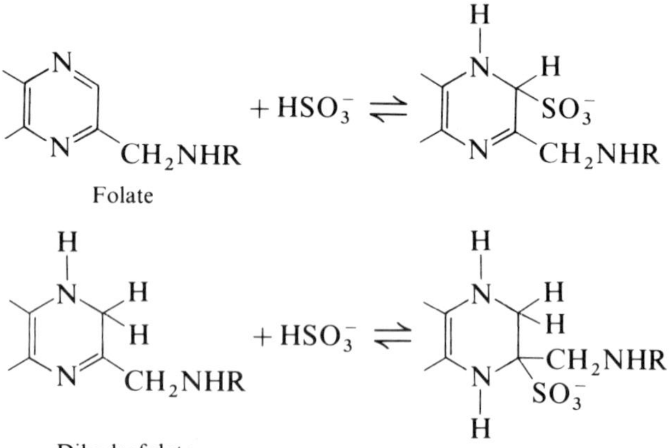

The adducts are relatively unstable, the equilibrium constant for formation of the complex between dihydrofolate and hydrogen sulphite ion being $83 \, \text{M}^{-1}$ at 31 °C and pH 6·5; the adduct of folate is some 30 times less stable (Vonderschmitt et al., 1967).

4.2.3.3 Haem prosthetic groups

Iron porphyrins are prosthetic groups in several enzyme systems including the cytochromes and peroxidases, the differences between the haem moieties in these enzymes arising from the presence of different substituents around the following common iron(III) porphyrin complex:

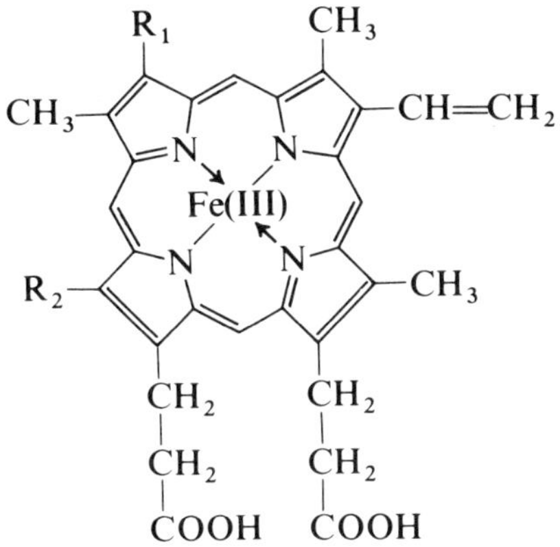

The haems which will be considered in connection with the inhibition of iron porphyrin containing enzymes are haem a, the prosthetic group of cytochrome oxidase ($R_1 = \alpha$-hydroxyfarnesylethyl and R_2 = aldehyde) and ferriprotoporphyrin (R_1 = vinyl and R_2 = methyl), the prosthetic group of the majority of peroxidases.

Haem a reacts with sulphur(IV) oxoanions with an accompanying hypsochromic shift of the porphyrin spectrum attributed to hydroxysulphonate formation involving the aldehyde group, with consequent loss of conjugation between the latter and the ring system. This reaction is advocated, in fact, as one of the available tests for the presence of carbonyl substituents conjugated with a tetrapyrrole nucleus in the characterisation of unknown haems. The apparent stability constant of the hydroxysulphonate adduct of haem a is in the region of unity (pH 6·0, 25 % aqueous pyridine, 20 °C) representing only a weak association (Filippi et al., 1979), presumably a result of the energy requirement for disruption of the porphyrin–carbonyl conjugation. The haemoprotein, cytochrome a, forms a 1:1 adduct with sulphur(IV) oxoanion with a dissociation constant, determined kinetically, of 1×10^{-2} M at pH 6·0, and it is suggested by Yoshikawa and Orii (1972a) that formation of this adduct is responsible for the inhibition of cytochrome oxidase activity by sulphur(IV) oxoanion. According to Cooperstein (1963) it is possible to inhibit cytochrome oxidase by 50 % as a result of pre-incubation with sulphur(IV) oxoanion at a concentration of 0·5 mM at pH 7, whereas Yoshikawa and Orii (1972a) find this concentration to be of the order of 10 mM. In the absence of equilibrium constant data for the interaction between sulphur(IV) oxoanion and the enzyme preparation used by Cooperstein (1963) at pH 7, it is not possible to comment further, but it is evident that this equilibrium constant would need to be at least an order of magnitude smaller, with respect to dissociation of the adduct, if the enzyme inhibition is of the type reported by Yoshikawa and Orii (1972a).

The original suggestion that sulphur(IV) oxoanions react with the prosthetic group of peroxidase was made by Embs and Markakis (1969). More extensive investigations, relating to the oxidised forms of the enzyme, were reported subsequently by Roman and Dunford (1973) and Araiso and coworkers (1976). In its peroxidative capacity, peroxidase reacts with hydrogen peroxide to yield a product (compound I) in which the porphyrin iron is oxidised from Fe(III) to Fe(V). This intermediate, which is coloured green, then undergoes two one-electron transfers to the reducing agent, a pale red compound II resulting from the first electron transfer and peroxidase being regenerated after the second electron has been transferred. The scheme may be summarised as follows:

$$\text{Per–Fe(III)} \xrightarrow{\text{H}_2\text{O}_2} \text{Per–Fe(V)}$$
$$\text{(compound I)}$$
$$\text{Per–Fe(V)} + \text{AH} \longrightarrow \text{Per–Fe(IV)} + \text{A}^{\cdot}$$
$$\text{(compound II)}$$
$$\text{Per–Fe(IV)} + \text{AH} \longrightarrow \text{Per–Fe(III)} + \text{A}^{\cdot}$$

In the cases in which the substrate requires two electrons for oxidation the reactions of peroxidase may still proceed by way of two single-electron transfers, followed by disproportionation of the resulting radical intermediates, thus:

$$2\text{AH}^{\cdot} \rightarrow \text{A} + \text{AH}_2$$

When peroxidase compound I reacts with sulphite ion, it appears to do so by a combination of one- and two-electron transfer mechanisms, the two-electron transfer being favoured at low pH (Araiso et al., 1976). Under two-electron transfer conditions, the scheme is:

$$\text{compound I} + \text{SO}_3^{2-} \rightleftharpoons \text{compound I}\text{-}\text{-}\text{SO}_3^{2-}$$
$$\text{compound I}\text{-}\text{-}\text{SO}_3^{2-} \rightarrow \text{peroxidase} + \text{SO}_4^{2-}$$

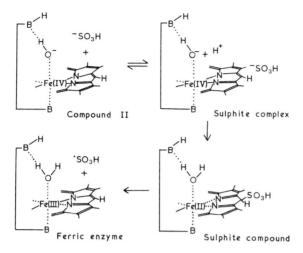

Fig. 4.3. Mechanism proposed to account for the reaction of peroxidase compound II with sulphite (Araiso *et al.*, 1976). Reprinted with permission from *Biochem.*, 1976, **15**, page 3062. © 1976 American Chemical Society.

When the enzyme is horseradish peroxidase, the transition from two-electron to one-electron mechanism, as pH is increased, depends upon the isoenzyme considered, one-electron transfer predominating above pH 4·5 for a mixture of isoenzymes A_1 and A_2 whilst the pH at which this effect is observed for isoenzyme C is pH 7·7. It is evident that the reaction between peroxidase and sulphur(IV) oxoanion will involve the participation of at least two intermediates, with the possibility of there being two different compound I---SO_3^{2-} complexes according to whether the reaction is to be a one- or two-electron transfer. Although no firm suggestion concerning the nature of the adduct between compound I and sulphur(IV) oxoanion has been made, Araiso and coworkers (1976) put forward a scheme, shown in Fig. 4.3, for the reduction by sulphite of peroxidase compound II. The hydroxyl ion stabilises the iron(IV) state and charge balance is achieved by addition of H^+ to give water, which is one of the ligands in peroxidase. Electron transfer from sulphur(IV) oxoanion takes place through the porphyrin ring system, the most appropriate points of attachment being the methine bridges.

The earlier work of Embs and Markakis (1969) provides evidence of a different nature. These workers observed that the effect of sulphur(IV) oxoanion at low pH (pH 3–5) was that of stabilisation of horseradish peroxidase with respect to acid catalysed hydrolysis of the prosthetic group and led, therefore, to an apparent increase in activity, as measured with guaiacol, compared with a blank containing no added sulphur(IV) oxoanion. They assumed that a peroxidase–sulphur(IV) oxoanion complex is formed, which is rapidly oxidised by the addition of hydrogen peroxide, and the apparent activity of the adduct is equal, therefore, to that of the peroxidase itself. The dissociation constant of the adduct was found to be $0.02 \, \text{M}$. A similar protecting effect was observed when cyanide, azide and fluoride ion were used in place of sulphur(IV) oxoanion but, in these three cases, the added ion was removed by dialysis prior to enzyme assay, since the peroxidase–cyanide, peroxidase–azide and peroxidase–fluoride adducts are not enzymically active.

4.2.3.4 Pyridoxal

Pyridoxal phosphate has the following structure:

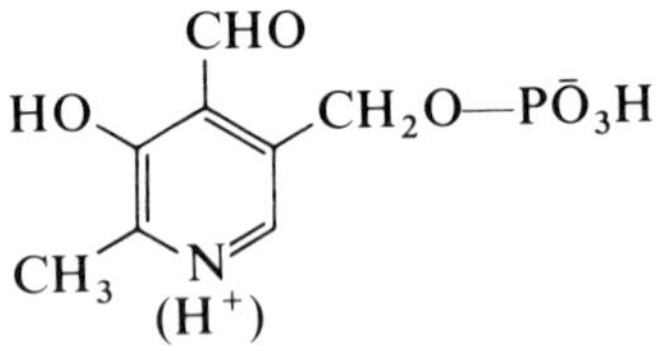

$$
\begin{array}{c}
\text{CHO} \\
\text{HO} \diagdown \diagup \text{CH}_2\text{O}\text{—}\bar{\text{P}}\text{O}_3\text{H} \\
\text{CH}_3 \quad \text{N} \\
(\text{H}^+)
\end{array}
$$

and is the cofactor of a large number of enzymes, most of which catalyse transformations of amino acids. The cofactor is bound to the enzyme as a result of Schiff base formation between the carbonyl group of the cofactor and an ε-lysyl group of the enzyme. The first step in the reaction of an enzyme with this cofactor is the replacing of the internal Schiff base with one involving the amino acid. Subsequently, the bound amino acid undergoes the appropriate transformation (Whitaker, 1972).

The only known reaction between pyridoxal phosphate and sulphur(IV) oxoanion is addition of the latter to the carbonyl group of the cofactor. This carbonyl group is activated by the ring nitrogen atom towards nucleophilic attack by sulphite ion, in the same way as benzaldehyde is activated towards hydroxysulphonate formation when it contains an electron-withdrawing substituent in the *para* position (Section 1.11.1). The equilibrium constant for adduct formation with pyridoxal phosphate is $4 \times 10^4 \, \text{M}^{-1}$ at pH 4.4 and 22 °C (Kozlov *et al.*, 1979), which is of a similar order of magnitude to that for the formation of the hydroxysulphonate of benzaldehyde. Possible reduction of carbonyl group reactivity of pyridoxal through interaction with the adjacent hydroxyl group, or some effect of the phosphate could account for the observed low activation.

The reaction of sulphur(IV) oxoanion with enzyme bound pyridoxal phosphate was reported by Adams (1969) to be much less effective than that for the free cofactor in the cases of glutamic–aspartic transaminase and glutamic–alanine transaminase. No evidence of adduct formation was observed at pH 7·4, whilst a slow reaction took place at pH 5·8. Similarly, α-glucan phosphorylase, which contains this cofactor, shows no change in either absorbance or fluorescence due to pyridoxal phosphate upon addition of sulphur(IV) oxoanion at a concentration of 50 mM and at pH 6·0 (Kamogawa and Fukui, 1973). This enzyme is inhibited by sulphur(IV) oxoanion under these conditions.

An interesting observation is that the hydroxysulphonate adduct of pyridoxal phosphate is hydrolysed at the phosphate ester linkage more slowly, by at least an order of magnitude, at pH 4·4, than is the case in the absence of sulphur(IV) oxoanion (Kozlov et al., 1979), a finding consistent with the participation of the carbonyl group in pyridoxal phosphate hydrolysis.

4.2.3.5 Thiamine

Thiamine, in the form of its pyrophosphate, is a cofactor in the decarboxylation of α-keto-acids, such as the conversion of pyruvate to acetaldehyde and carbon dioxide. The reactivity of thiamine pyrophosphate depends upon the ability of the keto-acid to add to position 2 on the thiazolium ring. Thiamine is, however, cleaved readily by sulphite ion, the simplest representation of this reaction being that of nucleophilic attack on the methylene carbon, activated by the positive charge on nitrogen, thus:

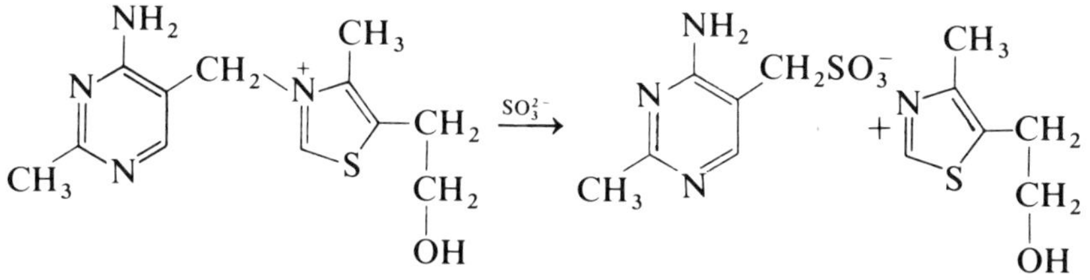

The pyrimidine sulphonate is sparingly soluble in water and insoluble in other solvents, whereas the thiazole-containing product is soluble in water but extractable into chloroform from alkaline solution (Williams *et al.*, 1935). A detailed kinetic study of this reaction was carried out by Leichter and Joslyn (1969) who found that the rate of cleavage was of first order with respect to both thiamine and sulphur(IV) oxoanion, a second order rate constant of $50 \, M^{-1} \, h^{-1}$ being reported at pH 5·5–6·0 and 25 °C. This rate

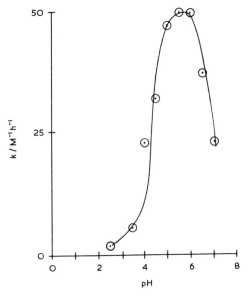

Fig. 4.4.   The effect of pH on the second order rate constant for the decomposition of thiamine (Leichter and Joslyn, 1969). Reprinted with permission from *Biochem. J.*, 1969, **113**, page 613. © 1969 The Biochemical Society.

constant is sensitive to pH, a bell-shaped profile being obtained with a maximum around pH 6, as shown in Fig. 4.4. This profile has been obtained by other workers (Zoltewicz and Kauffman, 1977; Doerge and Ingraham, 1980; Jaroensanti and Panijpan, 1981) and is also found for phosphorylated thiamines, the main difference between the results for these and that for thiamine being the magnitude of the maximum rate constant, which decreases in the following order:

thiamine monophosphate > thiamine diphosphate > thiamine triphosphate

(Jaroensanti and Panijpan, 1981). The activation energy for reaction of thiamine or its phosphates with sulphur(IV) oxoanion is in the region of $60 \, kJ \, mol^{-1}$, the different rates of reaction at any given temperature being a result of different pre-exponential terms when the Arrhenius equation is applied to these systems.

The bell-shaped plots are explained readily on the basis of the reactant being sulphite ion and the substrate the conjugate acid of thiamine. Zoltewicz and Kauffman (1977) draw attention to a number of conflicting observations in the literature which imply that the simple displacement mechanism referred to above may, indeed, be an over-simplification. The first of these concerns the relative importance of the pyrimidine and thiazole moieties to the cleavage reaction, the displacement mechanism attaching greater weight to the thiazole. It is found, however, that thiamine with a reduced thiazole may still be cleaved (Torrence and Tieckelmann, 1968), but a molecule in which the pyrimidine ring is replaced by *ortho-*, *meta-* or *para-*nitrobenzene does not undergo cleavage. The pyrimidine ring is, therefore, of major importance. A second striking observation is that other nucleophiles, such as thiazole, pyridine, anilines or azide ion, will react with thiamine and related compounds, but *only* in the presence of sulphite ion. Sulphite ion has, therefore, a catalytic role in these reactions. On the basis of a study of the competing reactions between sulphite and azide ion, Zoltewicz and Kauffman (1977) suggest that the sulphitolysis of thiamine proceeds by way of a two step mechanism, both steps involving sulphite ion. The first is formation of an intermediate as a result of addition of sulphite ion to the pyrimidine ring, the structure of this intermediate being as follows:

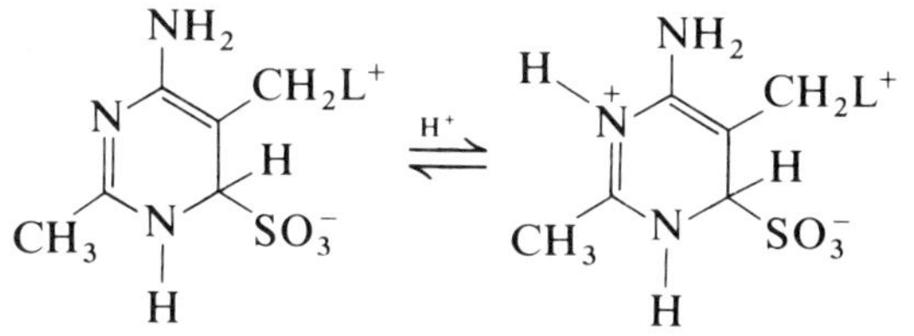

in which L^+ denotes the leaving group. The addition of sulphite to simple pyrimidine compounds is described in Section 1.11.8. The driving force for this adduct formation must be the better electron-donating power of the adduct compared with that of the pyrimidine group, and it is suggested that elimination of the leaving group takes place by way of an S_N1 type reaction to give the following second intermediate:

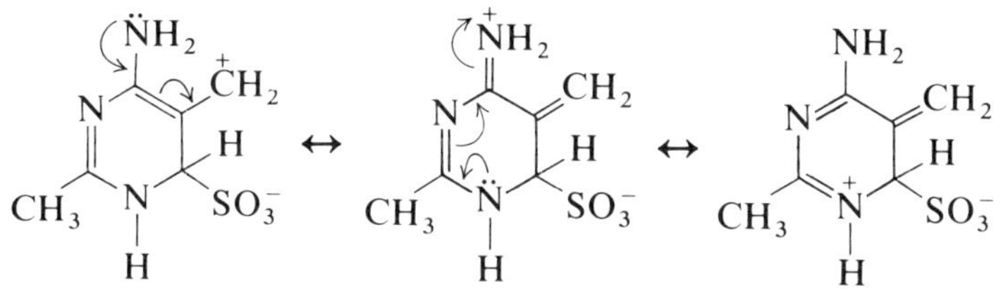

The reaction is completed by rapid combination with a second sulphite ion and expulsion of the sulphonate group in the pyrimidine ring to give the established product. Although the formation of the first intermediate was supported by Doerge and Ingraham (1980), these workers disagreed with the proposed mechanism for the conversion of intermediate to final product, noting that, at low sulphite ion concentration, this final step became apparently rate limited by sulphite ion. However, Zoltewicz and Uray (1981a) explained this result as the successful competition for the intermediate between some other nucleophile and sulphite ion, this nucleophile being thiamine itself, although the effect was actually demonstrated with 4-thiopyridine (Zoltewicz et al., 1980). The explanation was given partly on the basis of the successful isolation of the following product in low yield:

$$\text{NH}_2 \quad N \quad -\text{CH}_2-\overset{+}{\text{N}} \quad \text{CH}_2\text{SO}_3^- \quad \text{CH}_3 \overset{+}{\underset{H}{N}} \quad \text{CH}_3 \quad N \quad \text{NH}_2$$

when thiamine at high concentration was cleaved by sulphite ion at low concentration. It is expected that this product is the result of attack by the second intermediate on a thiamine molecule, followed by sulphite catalysed cleavage of the thiazole grouping. No bispyrimidine compound with a thiazole group could be isolated, which is unfortunate since this would have provided even stronger evidence for the mechanism of Zoltewicz and his coworkers, but their experience has shown that, when the pyrimidine rings of molecules analogous to thiamine bear a substituent, such as methyl, then the molecule is far more susceptible to cleavage. Thus, even at low concentrations of sulphite ion, the thiamine–pyrimidine adduct could be cleaved rapidly. Further work by this group (Zoltewicz and Uray, 1981b) has indicated that hydroxide ion, which is also known to cleave thiamine, probably acts through a similar mechanism, the apparent success of sulphite ion in this reaction being merely its ability to be present at relatively high concentrations at sufficiently low pH, rather than some unique characteristic being attributable to it.

Some interesting speculations have been made by Jaroensanti and Panijpan (1981) concerning the reduced reactivity towards sulphite cleavage of the phosphorylated thiamines compared with thiamine itself. One reason for the differences in rates is the interaction between the phosphate 'tails' of different lengths and the pyrimidine ring, introducing steric hindrance or reducing the effective charge with respect to the initial attack by sulphite ion. Such folded structures have been proposed by several workers in systems with and without divalent cations.

There are no reported studies of the inhibition of thiamine dependent enzymes in the purified state, but the effects of thiamine deficiency in biological systems, whether as a result of an inadequate diet or through sulphite cleavage, are well known and will be considered in Chapter 7.

4.2.4 Competition for active site of enzyme

Perhaps the most widely reported case of competitive inhibition of enzyme reactions by sulphur(IV) oxoanion is that towards hydrogen carbonate ion in carbon dioxide metabolising enzymes, such as malic enzyme (Ziegler, 1974b), phosphoenolpyruvate carboxylase (Ziegler, 1973; Mukerji, 1977) and ribulose-1,5-diphosphate carboxylase (Ziegler, 1972). This effect is observed also in the case of α-glucan phosphorylase, which is affected slightly by hydrogen carbonate ion. Kamogawa and Fukui (1973) consider that hydrogen carbonate and hydrogen sulphite ion compete with phosphate ion at the binding site of the phosphorylase. When the anion carries a third hydroxyl group, such as phosphate (and arsenate), then it can act as a substrate, accepting the glucosyl residue of the polysaccharide. The anions with only two possible hydroxyl groups (hydrogen sulphite and hydrogen carbonate ion) can bind to the active site but do not act as substrates.

4.2.5 Removal of reactants

It is tempting to suggest that all enzyme reactions which involve carbonylic substrates should be inhibited by sulphur(IV) oxoanion as a result of hydroxysulphonate formation, and those reactions which lead to the formation of carbonylic products will be driven, similarly, in the forward direction by the additive. On such a premise, Meyerhof and coworkers (1938) added sodium sulphite to their assay of lactate dehydrogenase in order to remove the pyruvate as hydroxysulphonate. The fact that the additive inhibited enzyme action by combining with enzyme–coenzyme complex was unexpected at the time. Whilst hydroxysulphonate formation between carbonylic substrate and sulphur(IV) oxoanion will undoubtedly occur, it is possible that the kinetic effect is overshadowed by much larger effects upon the enzyme or cofactor. It is for this reason, perhaps, that no conclusive evidence for such inhibition is available.

The oxidising agent for peroxidase activity, hydrogen peroxide, reacts with sulphur(IV) oxoanion yielding sulphate ion and water (Section 1.10.3). The removal of hydrogen peroxide in this way would prevent the peroxidative action of this enzyme (Chmielnicka, 1963). It is possible to make a quantitative estimate of the importance of this reaction by considering the second order rate constants for the following competing reactions when peroxidase is present in a mixture of sulphur(IV) oxoanion and hydrogen peroxide:

$$\text{Peroxidase} + H_2O_2 \rightarrow \text{compound I}$$

$$S(IV) + H_2O_2 \rightarrow S(VI)$$

Whitaker (1972) suggests that the rate constant for the peroxidase–hydrogen peroxide reaction is of the order of $10^7 \, M^{-1} s^{-1}$ at pH 4·6, and it is estimated, from the data of Hoffmann and Edwards (1975), that the rate constant for the sulphur(IV) oxoanion–hydrogen peroxide reaction is of the order of $10^4 \, M^{-1} s^{-1}$. Assuming that all three components are in homogeneous solution, the relative rates of reaction at any hydrogen peroxide concentration will be given by:

$$\frac{\text{Rate of peroxidase reaction}}{\text{Rate of S(IV) reaction}} \approx \frac{10^3 [\text{Peroxidase}]}{[\text{S(IV)}]}$$

Assuming a sulphur(IV) oxoanion concentration in the millimolar range, say 1–10 mM, and a peroxidase concentration of the order of 1 μM, then it is evident that there is a good possibility that removal of hydrogen peroxide will predominate.

4.2.6 Reaction with products of enzyme action

The only important interaction between the products of enzyme action and sulphur(IV) oxoanion is the inhibition of browning as a result of the action of polyphenol oxidase on *ortho*-diphenolic substrates. Embs and Markakis (1965) found that the rate of oxygen uptake by the polyphenol oxidase–diphenol system is unaffected by sulphur(IV) oxoanion at pH 6·5, although the development of colour, as measured by absorbance at 420 nm, is inhibited. The inhibition of browning is accompanied by increase in absorbance at 290 nm and the formation of organic sulphonates which were suggested as quinone–sulphite adducts; the formation of these adducts is considered in Section 1.11.10. Polyphenol oxidase activity in the presence of sulphur(IV) oxoanion is marked by a gradual reduction in the concentration of the additive resulting from adduct formation. The enzyme is also slowly inactivated with respect to its reactivity towards mono- and diphenols, an effect which may be demonstrated by pre-incubation of the enzyme with the inhibitor.

4.3 SULPHUR(IV) UTILISING ENZYMES

4.3.1 Sulphite reductases

Enzymes which catalyse the conversion of sulphite to hydrogen sulphide are present in a wide range of organisms and have been isolated in a purified form from bacteria, fungi, algae and higher plants. Both assimilatory and dissimilatory uses of sulphur(IV) oxoanion involve reduction to sulphide, but the enzymes involved in the two types of process are quite distinct. All sulphite reductases, however, owe their ability to reduce the oxoanion to the presence of a haem moiety which consists of a partially reduced porphyrin ring of the isobacteriochlorin type with methyl, acetic acid and propionic acid side chains (Murphy *et al.*, 1973). It has been demonstrated that, during the reaction catalysed by assimilatory sulphite reductase from *Escherichia coli*, sulphate ion forms a 1:1 complex with this haem moiety, the formation of which is inhibited by cyanide ion and carbon monoxide (Reuger, 1977). The reduction proceeds without the formation of products of intermediate oxidation state in the free form, and the physiological electron donor in assimilatory enzymes appears to be NADPH in fungi and aerobic bacteria and probably ferredoxin in anaerobic bacteria and photosynthetic organisms, the sulphite reductase of higher plants being found in chloroplasts (Siegel, 1975).

Assimilatory sulphite reductases which can use NADPH as electron donor contain other prosthetic groups. The enzymes of yeast, *E. coli* and *Salmonella typhimurium*, contain FMN and FAD prosthetic groups as well as non-haem iron in association with labile sulphide. Thus, the reductase from *E. coli* has a molecular weight of 670 000 and consists of 4 haemoprotein β-subunits, each binding one Fe_4S_4 moiety and one haem group, and 8 α-subunits which together bind 4 moles of each of FMN and FAD. The flavin prosthetic groups are responsible for electron transfer between NADPH and the haem; sulphite reductases, which cannot use NADPH but can use the artificial reducing agent, reduced methyl viologen (MVH), do not possess the flavin moiety. A sulphite reductase which can utilise MVH but not NADPH has also been prepared from *E. coli* NADPH–sulphite reductase by dissociation in 5 M urea. The fragment showing reductase activity contained no flavin moiety. All sulphite reductases are able to use MVH as reducing agent, and a simplified scheme for the electron transfer to sulphite is, therefore (Siegel, 1975):

$$NADPH \rightarrow FAD \rightarrow FMN \dashrightarrow haem \rightarrow SO_3^{2-}$$
$$\nwarrow MVH \nearrow$$

Siegel notes also that a sulphite reductase, such as that from *E. coli*, is a tetramer of the basic electron transfer unit, 1 FAD, 1 FMN, 4 FeS and 1 haem, which is capable of carrying at one time all the six electrons necessary to effect reduction from sulphite to sulphide. This is not possible in other cases, such as in spinach sulphite reductase, which contains not more than one iron atom per enzyme molecule and in which the reduction must proceed through a series of intermediates. The assimilatory sulphite reductases are capable of reducing hydroxylamine to ammonia and, with the exception of those from photosynthetic organisms, are able to reduce nitrite ion to ammonia.

All dissimilatory sulphite reductases contain 16 non-haem iron atoms with associated labile sulphides and 2 haem moieties per molecule and are unable to use NADPH as an electron source, the most likely ultimate source being simple organic compounds, such as lactate, pyruvate or hydrogen. Thiosulphate and trithionate are formed in addition to hydrogen sulphide and a scheme for the reduction, proposed by Kobayashi and coworkers (1972, 1974) and modified by Siegel (1975), suggests that these additional products are formed as a result of side reactions with intermediates in the main reaction as follows:

$S(IV) \rightarrow S(-II)$ *reduction*

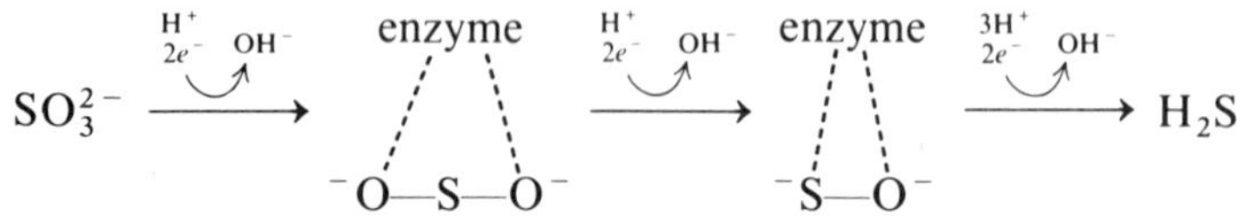

Formation of trithionate

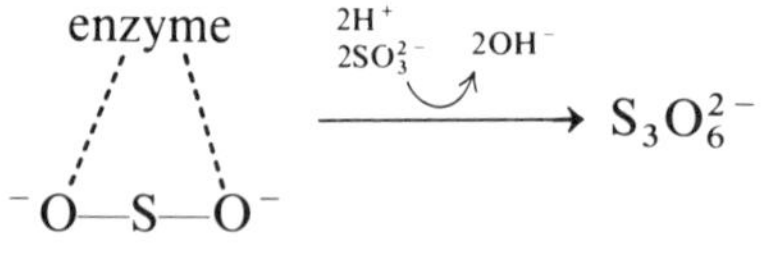

Formation of thiosulphate

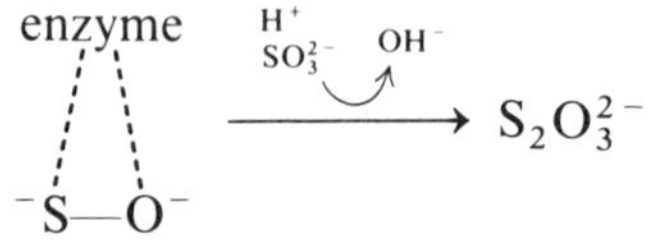

It is suggested also that the action of all sulphite reductases, assimilatory and dissimilatory, proceeds, perhaps, by this mechanism but the turnover rate of assimilatory enzyme–sulphite complexes is much faster than that of dissimilatory enzymes, rendering the intermediates shorter lived and less likely to interact with sulphite to form the thionate products. No thiosulphate or trithionate is formed by assimilatory enzymes.

4.3.2 Sulphite oxidases
The overall reaction for the action of sulphite oxidase is given as:

$$SO_3^{2-} + A + H_2O \rightarrow SO_4^{2-} + AH_2$$

where A represents an electron acceptor. A detailed study of the properties of sulphite oxidase extracted from bovine liver was published by Cohen and Fridovich (1971a, b, c). The enzyme has the ability to transfer two electrons from sulphite ion to either two-electron acceptors, such as oxygen, or to one-electron acceptors, such as cytochrome c or ferricyanide ion. The product of the transfer to oxygen is hydrogen peroxide and there is no evidence for the involvement of superoxide ion. A small proportion of the reaction is inhibited by mannitol, indicating that this part may be free radical in nature; this free radical component is not affected by superoxide dismutase.

Sulphite oxidase contains two types of prosthetic group. The first is haem, there being two functionally different haem moieties in each enzyme molecule, and the second, molybdenum. That this element is a functional part of the enzyme was deduced from:

1.   The fact that the molybdenum of sulphite oxidase was reduced by sulphite ion but not by xanthine, whilst the molybdenum of xanthine oxidase was reduced by xanthine and not by sulphite ion.

2.   Arsenite, which inhibits xanthine oxidase and aldehyde oxidase by acting upon the molybdenum, also inhibits sulphite oxidase and reduces, simultaneously, the amount of molybdenum(V).

3.   Sulphite oxidase is inhibited by cyanide ion with simultaneous reduction in the amount of molybdenum(V). The $+5$ oxidation state of this element is generated when the enzyme is reduced with sulphur(IV) oxoanion and is observable by ESR.

It is suggested that the molybdenum and haem both take part in electron transfer, according to the following scheme:

$$\text{Sulphite} \rightarrow \text{Mo} \rightarrow \text{haem} \rightarrow O_2$$
$$\downarrow$$
$$\text{ferricyanide}$$
$$\text{or}$$
$$\text{cytochrome c}$$

the cytochrome c being the most likely acceptor of electrons from sulphite ions in mitochondria, in which the enzyme is located. Coupled phosphorylation with an ATP/sulphite ratio of 0·8 is observed and the sulphite to oxygen ratio is 2:1, suggesting that the product of reduction of oxygen is water (Cohen *et al.*, 1972). Shibuya and Horie (1980) added further weight to the idea of the involvement of cytochrome c in the action of sulphite oxidase *in vivo* by observing that a sulphite oxidase system of high oxygen affinity can only be reconstituted from mitochondrial fractions if both cytochrome c and cytochrome oxidase are present. Such systems also had catalase activity, which was found to be necessary. Assuming the following K_m values:

sulphite oxidase–ferricytochrome c	$2\,\mu\text{M}$
cytochrome oxidase–ferrocytochrome c	$8\cdot5–10\,\mu\text{M}$
sulphite oxidase–oxygen	$580\,\mu\text{M}$
cytochrome oxidase–oxygen	$<2\,\mu\text{M}$

it seems probable that the electron path *in vivo* is as follows:

$$SO_3^{2-} \rightarrow \text{sulphite oxidase} \rightarrow \text{cytochrome c} \rightarrow \text{cytochrome oxidase} \rightarrow O_2$$

$$ADP \quad ATP$$

with direct reaction between sulphite oxidase and oxygen occurring only in the absence of cytochrome c, as is the case in purified sulphite oxidase preparations.

Cohen and Fridovich (1971c) suggested that sulphite oxidase is the least dispensible of the three known mammalian molybdenum containing enzymes. Whereas the absence of xanthine oxidase in humans is not seriously disabling, and an inhibitor of xanthine oxidase is used for treating gout, and aldehyde oxidase which is present in rabbit liver appears to be absent in extracts of livers of many other mammals, the absence of sulphite oxidase in humans leads to death (Chapter 7).

An alternative pathway for the enzymic oxidation of sulphite involves adenosine monophosphate (AMP), as follows:

$$SO_3^{2-} + AMP^{2-} \xrightarrow{\text{APS reductase}} APS^{2-} + 2e^-$$

$$APS^{2-} + P^{2-} \xrightarrow{\text{ADP sulphurylase}} SO_4^{2-} + ADP^{3-} + H^+$$

$$2ADP^{3-} \xrightarrow{\text{adenylate kinase}} AMP^{2-} + ATP^{4-}$$

where APS is adenosine 5′-phosphosulphate and ADP and ATP are adenosine diphosphate and adenosine triphosphate respectively (Peck, 1960).

APPENDIX: REFERENCES TO LITERATURE

Enzyme inhibited	*Source of information*
Amylase	Chen *et al.*, 1972.
Aryl sulphatase	James, 1979.
Ascorbate oxidase	Bergner, 1947; Ponting and Joslyn, 1948.
Cytochrome oxidase	Cooperstein, 1963; Yoshikawa and Orii, 1972*a*, *b*.
Elastase	Procaccini *et al.*, 1974.
Formate dehydrogenase	Egorov *et al.*, 1980.
α-Glucan phosphorylase	Kamogawa and Fukui, 1973.
Glucose oxidase	Swoboda and Massey, 1966.
Glucose-6-phosphate dehydrogenase	Pierre, 1977.
Glutamate dehydrogenase	Pierre, 1977.
Glyceraldehyde-3-phosphate dehydrogenase	Ziegler *et al.*, 1976.
Glycollate oxidase	Meyers and Jorns, 1981.
β-Hydroxybutyrate dehydrogenase	Phelps and Hatefi, 1981.
Iodothyronine-5-deiodinase	Visser, 1980.
Isocitrate dehydrogenase	Pierre, 1977.
Lactate dehydrogenase	Meyerhof *et al.*, 1938; Parker *et al.*, 1978; Lederer, 1978; Trommer and Glöggler, 1979.
Lipoxygenase	Haisman, 1974.
Malate dehydrogenase	Pfleiderer *et al.*, 1956; Osmond and Avadhani, 1970; Ziegler, 1973; Parker *et al.*, 1978.
Nitrate reductase	Gomez-Moreno and Palacian, 1974.
Oxynitrilase	Jorns, 1980.
Peroxidase (horseradish)	Monikowski and Chmielnicka, 1967; Embs and Markakis, 1969; Haisman, 1974; Araiso *et al.*, 1976.

Enzyme inhibited	*Source of information*
Phosphoenolpyruvate carboxylase	Osmond and Avadhani, 1970; Ziegler, 1973; Mukerji, 1977.
Polyphenol oxidase	Embs and Markakis, 1965; Markakis and Embs, 1966; Wong *et al.*, 1971; Singh and Ahlawat, 1974; Muneta and Wang, 1977.
Pyruvate decarboxylase	Haisman, 1974.
Rhodanase	Volini and Wang, 1973.
Ribulose-1,5-diphosphate carboxylase	Ziegler, 1972.
Thiosulphate reductase	Uhteg and Westley, 1979.
Urocanase	Hug *et al.*, 1978.
Xanthine oxidase	Guilbault *et al.*, 1968.

5

Microbiology

5.1 INTRODUCTION

The reactions between sulphur(IV) oxoanions and components of microbial systems, taken in isolation, have been considered in the foregoing chapters. It may be concluded that there is no part of the microbial cell which will not be affected by the presence of sulphur(IV) oxoanion and, in any particular case, it is not possible to identify a single reason why the activity of a microbe has been modified by the additive. The important difference between systems which have been dealt with in connection with the mechanism of sulphur(IV) oxoanion reactions and the microbial cell is that, whereas in the former transport of reactants is generally not limited by their diffusion, a necessary prerequisite for reaction between the additive and components of a microbial cell is transfer of the oxospecies from the medium surrounding the cell to the intracellular components. For this reason the discussion in this chapter begins with a consideration of the factors affecting the transport of sulphur(IV) oxospecies across cell membranes.

5.2 TRANSPORT

The structure of cell membranes is that of a phospholipid bilayer, in which the hydrophobic fatty acid chains are located in the interior and the polar groups project into the aqueous phase. In most micro-organisms the

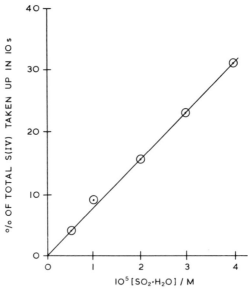

Fig. 5.1. Relationship between initial rate of transport of sulphur(IV) oxoanion into *Saccharomyces cerevisiae* var *ellipsoideus* and the concentration of molecular sulphur dioxide in solution for a total sulphur (IV) oxospecies concentration of 1 mm (Macris and Markakis, 1974). Reproduced with permission of P. Markakis from *J. Sci. Food Agric.*, 1974, **25**, page 26. © The authors.

exterior surface of the membrane has additional layers of protein and/or carbohydrate, and in the case of Gram-negative bacteria an additional highly specialised outer membrane. The transport of solutes across such membranes shows the same characteristics as transfer across model 'oil membranes' and the permeability of such membranes is controlled by the nature of the oil–water interface rather than by the material within the membrane. Since a molecule being transferred across a cell membrane must necessarily leave an aqueous environment and pass into a non-polar environment, the permeability of cell membranes to particular solutes is related to their polarity and, in particular, permeability is satisfactorily related to the number of hydrogen bonds which must be broken during the transfer (Yost, 1972). The driving force for such transport is the concentration gradient across the membrane unless an active transport system is in operation to provide the energy required to oppose such a gradient.

The immeasurably small interaction between dissolved sulphur dioxide and water (Section 1.7) suggests that this sulphur(IV) oxospecies requires less energy for transfer from aqueous to non-aqueous phase than do the ionic forms. Sulphur dioxide is soluble also to some extent in non-polar solvents and it is expected, therefore, that the ease with which sulphur dioxide will pass through cell membranes will be much greater than that with which ionic species will be transported. Much of the reported work in support of this prediction arises from studies of the toxicity of sulphur(IV) oxoanion towards bacteria as a function of pH, toxicity apparently increasing with decrease in pH. Toxicity is a function not only of transport of the toxicant but also of the sensitivity to damage of the cell components. The latter may be modified by the pH of the environment. Therefore, in the first instance, it is advisable to treat the transport phenomenon separately from the effect on the micro-organism as a whole.

A demonstration of the transport of molecular sulphur dioxide across the membrane of the yeast, *Saccharomyces cerevisiae* var. *ellipsoideus*, has been provided by Macris and Markakis (1974). When cells are suspended in a solution containing *sulphur dioxide* (1 mM) at pH 3·8 and 20 °C, uptake of sulphur(IV) oxospecies is rapid, equilibrium being reached after some 2 min. The reversibility of this process has not been demonstrated. If the concentration of total sulphur(IV) oxospecies is kept constant and the pH is varied, an increase in both the initial rate of transfer and in the final amount of sulphur(IV) oxospecies transferred to the cells is observed as pH is decreased. However, when the concentration of free sulphur dioxide in the solution is calculated, the kinetic data are found to obey the rate law:

$$\text{Initial rate} = k[\text{SO}_2]$$

and the close adherence of the experiments to this behaviour is illustrated in Fig. 5.1 for sulphur dioxide concentrations in the range $(0·5–4) \times 10^{-5}$ M. In order to check for effect of variation of pH on the transport mechanism, the uptake of sulphur(IV) oxospecies was measured at constant sulphur dioxide concentration and variable pH but no significant trend was observed. The transport is, therefore, that of sulphur dioxide. The effect of concentration of sulphur dioxide on initial rate of uptake shows saturation kinetics typical of enzyme systems. The apparent temperature dependence is reminiscent of enzyme reactions and the process is inhibited by mercuric chloride, 2,4-dinitrophenol and sodium azide but not by iodoacetamide (Macris, 1972). There is some evidence, therefore, that sulphur dioxide is actively transported across the membrane. Active transport of sulphur dioxide has also been reported for the uptake by rat liver mitochondria where the dicarboxylate carrier is capable of this function (Crompton *et al.*, 1974).

The concentration of free and combined sulphur(IV) oxospecies in the microbial cell is greater than that in the surrounding medium when pH is less than 4·8. For example, when the volume fraction of cells is some 6%, more than 20% of the sulphur(IV) oxospecies present in solution will be transferred to the cells at pH 3·80, increasing to some 55% at pH 3·19 for a total initial sulphur(IV) oxospecies concentration of 1×10^{-3} M. This result does not necessarily represent a concentration gradient opposing diffusion into the cell, since the activity of sulphur dioxide will be low inside the cell at physiological pH and some of the sulphur(IV) oxoanion will undoubtedly combine with cell components.

5.3 TOXICITY

Precautions which must be taken when assessing the effects of sulphur(IV) oxoanions on the growth of yeasts and bacteria have been considered by Carr and coworkers (1976). The medium should not only allow a good growth of the test organism but also bind the minimum quantity of sulphur(IV) oxospecies. For example, replacing glucose with fructose gives a medium with an energy source capable of supporting the growth of most micro-organisms but with a reduced binding capacity towards sulphur(IV) oxospecies. Autoclaving increases this binding capacity (see also Chapter 3) and the alternative procedure which is preferred is to sterile filter the medium. To avoid risk of oxidation of the additive, a problem demonstrated clearly by Postgate (1963), it is necessary to purge the media free of oxygen, using nitrogen, and to maintain a nitrogen filled headspace. Despite such precautions, losses still occur and should be allowed for when calculating the amount of the substance added to the medium (Carr, 1981).

It is stated generally that sulphur dioxide acts as both a biocidal and biostatic agent on micro-organisms in the following order of effectiveness:

bacteria > moulds and yeasts

Gram-negative bacteria > Gram-positive bacteria

(Lueck, 1980; Hammond and Carr, 1976) and the microbial association in sulphited media will be dominated by Gram-positive bacteria. This effect is illustrated in the case of the sulphited sausage: in the absence of the preservative the spoilage is dominated by Gram-negative rods, whilst in the presence of sulphur(IV) oxospecies only Gram-positive rods are found (Dyett and Shelley, 1966; Banks and Board, 1981). Significant differences are also found between organisms of different morphology. In considering the growth of *Lactobacillus* in fermented apple juice, Carr and coworkers (1976) suggest that heterofermentative cocci are more sensitive than heterofermentative rods. The efficacy of a given concentration of sulphur(IV) oxoanion as an inhibitor of the growth of a particular micro-organism increases as pH is reduced and it is accepted that the effective species is molecular sulphur dioxide. Rehm and Wittmann (1963) give the effect of sulphur dioxide as more than 1000 times greater than that of the hydrogen sulphite ion against *Escherichia coli*, the corresponding value for yeasts and *Aspergillus niger* being 500 and 100 respectively. Macris and Markakis (1974) demonstrated further that the effect of pH on survival of *Saccharomyces cerevisiae* in the presence of sulphur(IV) oxoanion does not include an effect of pH on the susceptibility of the yeast to damage by the additive, but results only from the formation of molecular sulphur dioxide. A graph showing the transition from a viable to a non-viable yeast suspension as a function of pH in the presence of $0.0016 M$ sulphur(IV) oxospecies is shown in Fig. 5.2 (Hammond and Carr, 1976), the pH at which growth of this organism is inhibited in the absence of sulphur dioxide being approximately pH 2·4.

    In most cases it is found that adducts of the hydroxysulphonate type are ineffective as antimicrobial agents: for example, Joslyn and Braverman (1954) reported that grape musts containing $0.0375 \, mol \, kg^{-1}$ total (free and reversibly bound) sulphur(IV) oxospecies, of which most was reversibly bound, underwent fermentation. Ingram (1948) demonstrated that glucose hydroxysulphonate did not inhibit yeast. More recently, Carr and coworkers (1976) reported that addition of acetaldehyde overcame the inhibitory effect of sulphur dioxide on certain lactic acid bacteria from ciders, and Banks and Board (1982*a*) also used acetaldehyde to demonstrate the same effect on certain Enterobacteriaceae. There are, however, a small number of reports, notably those of Fornachon (1963), Lafon-Lafourcade and Peynaud (1974) and Mayer and coworkers (1975) which claim that the reverse is true of acetaldehyde hydroxysulphonate with respect to lactic acid bacteria, but the discrepancy may result from differences in experimental methods or choice of micro-organism (Carr *et al.*, 1976).

    The effect of sulphur dioxide concentration on the viability of *Lactobacillus plantarum* appears to be biphasic, as illustrated in Fig. 5.3, and shows a bacteriostatic effect at low concentrations and a bactericidal effect at higher concentrations. There are no comparable detailed data for other organisms, but the lack of an effect on growth of very low concentrations of sulphur dioxide is a feature apparent during inhibition

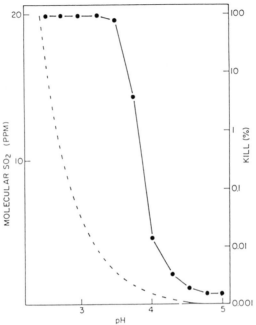

Fig. 5.2. Effect of pH on the viability of a suspension of *Saccharomyces cerevisiae* containing 8×10^6 cells ml^{-1} and 1 mM sulphur(IV) oxospecies, and on the concentration of molecular sulphur dioxide for a total sulphur(IV) oxospecies concentration of 1 mM. Viability, ———; [SO$_2$], – – – –. (Hammond and Carr, 1976.) Reproduced with permission from F. A. Skinner and W. B. Hugo (Eds), *Inhibition and Inactivation of Vegetative Microbes*, Society of Applied Microbiology Symposia Series, 1976, page 93.

studies with the bacteria *Bacillus subtilis* (Freese *et al.*, 1973), *Bacillus cereus*, *E. coli*, *Pseudomonas aeruginosa*, *Serratia marcescens*, and the fungi, *A. niger*, *Botrytis cinerea* and *Trichoderma viride* (Babich and Stotzky, 1978). Schimz (1980) shows colony formation by *S. cerevisiae* to be affected by sulphur dioxide also in a biphasic manner, and the kinetic data of Anacleto and Uden (1982) suggest a very rapid increase in specific death rate at low concentrations of the additive in comparison with the change in specific death rate at higher concentrations, a result consistent with biphasic behaviour. Such an observation implies that the micro-organism is able to accommodate or metabolise a small amount of sulphur(IV) oxospecies, the death phase resulting from the saturation of this mechanism. However, interpretation of the results given above is complicated by the fact that, with the exception of the experiments carried

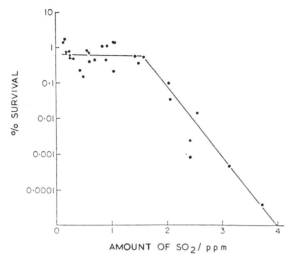

Fig. 5.3. Relationship between the survival of *Lactobacillus plantarum* and the concentration of sulphur dioxide (Carr *et al.*, 1976). Reproduced with permission from *J. Appl. Bact.*, 1976, **40**, page 208.

out by Carr and coworkers (1976), glucose-containing media were employed and heat sterilisation is likely to have been carried out, though this was not stated explicitly by some of the authors referred to above.

Little is known concerning the effects of sulphur(IV) oxoanions on spores. The outgrowth of spores of *Clostridium botulinum* (mixture of type A and B strains) in canned comminuted pork is delayed by the addition of sulphur(IV) oxoanion (Tompkin *et al.*, 1980), an observation which may have practical consequences in the use of sulphite-containing media to enumerate clostridia in foods in the form of black colonies. The inhibition of the germination of spores of *B. cinerea* by sulphur(IV) oxospecies is also reported (Pratella *et al.*, 1979).

The effects of temperature feature prominently in food microbiology. Data for *S. cerevisiae* show clearly the large temperature effect of the inhibitory action of sulphur dioxide, the organism being much more heat labile in the presence of the additive (Schimz, 1980). Similarly, the *D*-value for *Byssochlamys* is halved in the presence of the additive (King *et al.*, 1969).

Evidence concerning the mechanism of the toxic action of sulphur(IV) oxospecies on microbial systems is fragmentary, the main consistent observation being the effect of the additive on ATP levels. Freese and coworkers (1973) report 84% inhibition of ATP formation when *B. subtilis* is placed in a medium containing sulphur(IV) oxospecies at a concentration of 10 mM and at pH 6·5. ATP loss was identified also by Schimz and Holzer (1979) in *S. cerevisiae* and in *Lactobacillus malii* and *Leuconostoc mesenteroides* (Hinze *et al.*, 1981). In the case of *S. cerevisiae* the effect was biphasic with little loss of ATP below a sulphur(IV) oxospecies concentration of 1 mM at pH 3·5, and a rapid increase in loss above this concentration. The reduction in ATP level is accompanied by increase in ADP and, providing cells are not held with a low ATP level for more than, say, 15 min with an ATP level of 1 % of the normal level, complete recovery of ATP and ADP levels is observed. It is concluded that the action of sulphur dioxide on *S. cerevisiae* involves first the activation of an ATP hydrolysing enzyme, presumably located in the cell membrane, followed by ingress of the oxoanions into the cell where reaction with other cell components takes place (Schimz, 1980). Adenosine triphosphatases whose activity is stimulated by sulphite ion are, in fact, known from several microbial sources including *Bacillus megaterium* (Ishida and Mizushima, 1969), *Bacillus stearothermophilus* (Hachimori *et al.*, 1970), thermophilic bacterium PS3 (Yoshida *et al.*, 1975) and *Thiobacillus thiooxidans* (Tomigana, 1978).

It has been suggested that the bacteriostatic and bactericidal effects are, in fact, consequences of interaction with nucleic acids. Shapiro (1977) reports that when *E. coli* is treated with sulphur(IV) oxoanion under conditions of growth inhibition, RNA isolated from the bacterium is found to have combined with the additive leading to interference in protein synthesis, resulting in inhibition of growth. Tanaka and Luker (1978) consider that killing of *B. subtilis* cells by sulphur(IV) oxoanion is a result of DNA breakdown, which was observed as more rapid killing of DNA-repair mutants than their prototrophs. Other suggestions include the interruption of the conversion of glyceraldehyde-3-phosphate to 1,3-diphosphoglycerate in yeasts and the NAD dependent formation of oxalacetate from malate in *E. coli* (Wallnöfer and Rehm, 1965). Evidence exists also (see Section 5.6.3) for the modification of microbial metabolism by sulphur(IV) oxoanions by combination of the latter with metabolic intermediates.

5.4 EFFECTS ON VIRUSES AND BACTERIOPHAGE

Lynt (1966) comments on the uncertain position of the food chain as a vehicle for the distribution of viruses, although a number of cases of poliomyelitis and infectious hepatitis have been linked with food contamination (Jay, 1970). It may be assumed that viruses are as common in foods as are intestinal bacteria and, of the 94 or so types in the enteric virus family, all can be transmitted through the digestive tract. Unlike bacteria, viruses will not multiply in the food, but will remain at a constant level or disappear.

The possibility that viruses are inactivated by sulphur(IV) oxoanions was suggested first by Lynt (1966) in order to explain why type B6 coxsackievirus was rapidly inactivated in coleslaw. More detailed experiments were carried out later by Salo and Cliver (1978) using poliovirus type 1, coxsackievirus type A9 and echovirus type 7, all of which were found to be inactivated with equal efficiency by sulphur(IV) oxoanion. The inactivation caused after 6 h by a 10 mM solution of sulphur(IV) oxoanion at pH 3·6 and room temperature was a reduction by 2–3 orders of magnitude of the 'plaque-forming capacity', and by more than 4–5 orders of magnitude after 24 h, for an initial activity of 10^7 plaque forming units ml^{-1}. The mechanism of action probably does not involve cleavage of disulphide bonds, since neither dithiothreitol nor reduced glutathione caused inactivation of coxsackievirus A9. It is suggested that the nucleic acid component of enteroviruses is the primary target of the inhibitor and any changes to the protein coat are probably secondary effects. It is known that the treatment of bacteriophage MS2 with sulphur(IV) oxoanion produces covalent cross-linkages between the RNA and the protein coat (Budowsky et al., 1976) and the treatment of an RNA-containing plant virus, turnip yellow mosaic virus, with the reagent reduces the cytosine content of the RNA to about 26%, without significant effect on the protein coat (Bouley et al., 1975). The loss of cytosine through deamination is described in Section 1.11.8, and cross-linking occurs as a result of a transamination reaction (Shapiro and Weisgras, 1970) between the cytosine–sulphite adduct and an amine group in close proximity, as follows:

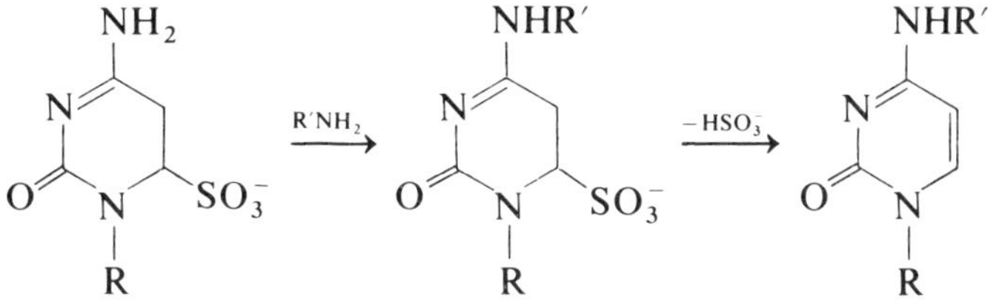

Both deamination of cytosine and the transamination reaction occur normally only with single-stranded nucleic acids *in vitro* but the situation *in*

vivo is not clear, ample evidence being available of reaction with double-stranded DNA under these conditions.

5.5   MUTAGENIC EFFECTS

The deamination of cytosine to uracil is a reaction which, by nature, is expected to be mutagenic, since it leads to the conversion of cytosine–guanine pairs between adjacent DNA strands to adenine–thymine pairs (Shapiro, 1977). Such mutagenicity was first demonstrated using *E. coli*. The test organisms were, in fact, mutants in which the cause of mutation was the presence of a cytosine–guanine pair, and the effect of sulphur(IV) oxoanion studied was that on the reversion of these mutants. It was found that in 13 cases studied the frequency of reversion was significantly higher than in control experiments, the largest increase being by a factor of 20. Corresponding changes were not observed with mutants in which the mutation was a result of replacement of cytosine–guanine by an adenine–thymine pair (Mukai *et al.*, 1970). Furthermore, when mutants are produced by the action of sulphur(IV) oxoanion on *E. coli*, a second treatment with the reagent does not lead to reversion of these mutants (Shapiro, 1977). It is concluded, therefore, that sulphur(IV) oxoanions can cause mutations *in vivo* and that the action is specific to cytosine–guanine pairs as predicted.

Although the mutation has been ascribed to the deamination of cytosine, it is possible that the mechanism may, in fact, involve 5-methylcytosine. This is deaminated directly to thymine:

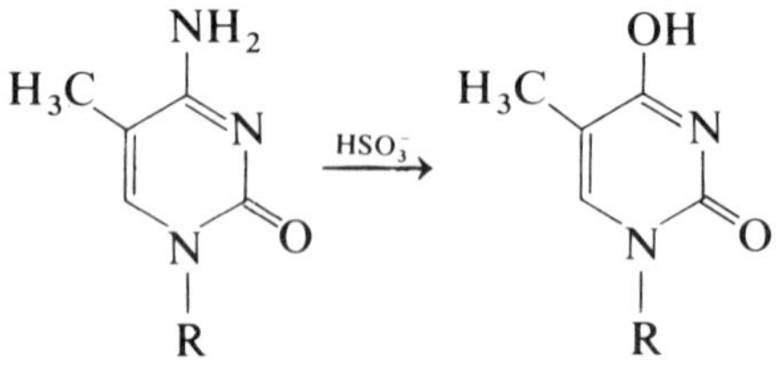

N-substituted thymine

However, the presence of the methyl group renders 5-methylcytosine less reactive in this respect than cytosine, and Wang and coworkers (1980) demonstrated that, under conditions of conversion of more than 96% of cytosine residues in single-stranded DNA to uracil, only 2–3% of 5-methylcytosine residues are converted to thymine. It is known that *E. coli* contains the enzyme uracil–DNA glycosidase, which offers a means of removing uracil residues formed by the deamination of cytosine in order to counteract the mutagenic reaction. This repair will not take place when the base is thymine, and the mutagenic potential of the deamination of 5-methylcytosine is, therefore, greater. Despite the attraction of seeking to explain the mutagenicity of sulphur(IV) oxoanion in terms of the involvement of 5-methylcytosine, there are no quantitative data to support this mechanism, but the greater mutagenic potential of thymine may compensate for the low rate of formation of the latter by this route. The transamination of cytosine, referred to in the previous section, is also likely to contribute to the mutagenic effect, as demonstrated by Hayatsu (1977) using bacteriophage in which the effects of sulphur(IV) oxoanion and nitrogen nucleophile were found to be cooperative. The work of Mukai and coworkers (1970) involved the treatment of cell suspensions with 1 M sulphur(IV) oxoanion at pH 5·2 for 30 min. The effect of low concentrations of sulphur(IV) oxoanion (5 mM, pH 3·6) on mutation in *S. cerevisiae* has been reported by Dorange and Dupuy (1972) who found a two- to twenty-fold increase in the number of mutants in positive cases. Low concentrations of sulphur(IV) oxoanion may be more effective at transforming nucleic acids (Inoue *et al.*, 1972), and Hammond and Carr (1976) suggested that this may be an important consideration in the preservation by sulphur dioxide of products in which legislation restricts the concentration of the preservative. The biphasic dependence of microorganism viability on the concentration of sulphur(IV) oxoanion referred to in Section 5.3 implies that persistent use of low concentration of the preservative with only partial microbial detoxification may be factors in the production of sulphur dioxide resistant strains. In the case of *E. coli*, the resistance acquired after 14 passages is of similar magnitude to the effect of other lipophilic acids and is small (a factor of 1·7–2·0) when compared with certain antibiotics such as oxytetracycline (150) or streptomycin (1670) (Lück and Rickerl, 1959).

Mutation is also reported for bacteriophage. For example, treatment of phage λ with 3 M sulphur(IV) oxoanion at pH 5·6 and 37 °C for 1·5 h leads to a frequency of mutation of the *c* gene by about 10 times that of spontaneous mutation (Hayatsu and Miura, 1970).

5.6 REACTION WITH MICROBIAL METABOLITES

5.6.1 Aflatoxin

Considerable attention has been given to the occurrence of aflatoxins in food. These are carcinogenic metabolites produced by the *Aspergillus flavus–oryzae* group of fungi and their production may be inhibited using sulphur(IV) oxoanions (Davis and Diener, 1967). The formation of toxin is inhibited by the addition of potassium sulphite without measurable inhibition of the growth of the fungus. Complete inhibition of toxin formation was observed at a concentration of 1 % potassium sulphite or more, whilst no inhibition took place when hydrogen sulphite ion was added. No comment on this result is possible, since pH measurements were not made. Studies on individual aflatoxins have been carried out using components labelled, B_1, B_2 and G_1, distinguished as B and G on the basis of blue and green fluorescence, respectively, under ultraviolet light. These aflatoxins, which comprise three out of the four primary aflatoxins, have the following structures:

Trager and Stoloff (1967) reported originally that solutions of 0·5 N hydrogen sulphite or sulphite ion had no effect upon the four primary aflatoxins but, subsequently, Doyle and Marth (1978a, b) demonstrated that aflatoxins B_1 and G_1 were degraded by 0·035 M sulphur(IV) oxoanion at pH 5·5 in 1·3 % methanol at 25 °C, and Moerck and coworkers (1980) showed later that aflatoxin B_2 was also affected.

The kinetics of the degradation of aflatoxins B_1 and G_1 were of first order with respect to aflatoxin and of fractional order (0·8) with respect to sulphur(IV) oxoanion, and the temperature dependence was consistent with an activation energy of $54\,kJ\,mol^{-1}$ (Doyle and Marth, 1978a). The addition of small amounts of methanol, up to 10%, reduces the rate of loss of these aflatoxins indicating, possibly, a free radical mediated reaction. The degradation products appear to be more polar than the aflatoxins (Doyle and Marth, 1978a). On the basis of these results, and consideration of the fact that coumarin reacts readily with sulphur(IV) oxoanion in aqueous solution leading to cleavage of the lactone ring and addition of a sulphonate group to position 4 (Dey and Row, 1924), as follows:

Doyle and Marth (1978b) suggested that one possible mechanism of reaction between aflatoxin and sulphur(IV) oxoanion is attack at the double bond of the lactone ring. However, these authors preferred an alternative mechanism which involved free radical sulphonation at the double bond of the dihydrofuran moiety, since small amounts of aflatoxin B_2 were detected in the degradation products mixture. This evidence was disputed by Moerck and coworkers (1980) who found that aflatoxin B_2, which has no double bond at the position in question, was also readily destroyed by sulphur(IV) oxoanion, and the most likely mechanism is, therefore, addition to and cleavage of the lactone ring, as follows:

A further possible reaction is the addition of hydrogen sulphite ion to the carbonyl group of the cyclopentenone ring.

As yet, the biological safety of the degradation products has not been demonstrated, but a practical demonstration of the removal of aflatoxins B_1 and B_2 from corn contaminated at a level of 235 parts per 10^9 to a 'safe level' after 24 h requires the soaking of the grain in a 2% solution of sodium hydrogen sulphite, presumably added as disulphite (Moerck et al., 1980).

5.6.2 Patulin
The mycotoxin, patulin or clavacin, which has the following structure:

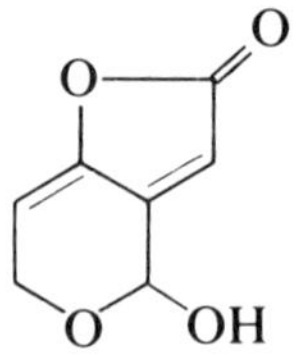

is formed by certain fungi, particularly *Penicillium expansum*, which may cause the rotting of fruit. The reactions of this compound with sulphur(IV) oxoanions are of interest because of toxicological implications of the presence of patulin in food made from contaminated fruit (Austwick, 1975). Burroughs (1977) reports that the affinity of sulphur(IV) oxoanion for patulin is insignificant below a sulphur(IV) oxoanion concentration of 3 mM, which may be used in the processing of apple juice and cider. However, at a concentration of 30 mM sulphur(IV) oxoanion and 90 μM patulin at pH 3·5 and 20 °C, combination was 90% complete in 2 days. Removal of sulphur(IV) oxoanion liberated only part of the patulin, suggesting two mechanisms: one is reversible and is likely to be the opening of the lactone ring, whilst the other is irreversible and involves addition at a double bond. The latter reaction is similar to that of coumarin referred to above.

5.6.3 Intermediate metabolism
The most widely reported effect of sulphur(IV) oxoanion on intermediate metabolism is the combination of the oxoanion with carbonyl-containing metabolites such as acetaldehyde, preventing the further utilisation of this compound. For example, yeast fermentation, in the presence of sulphur(IV) oxoanion, leads to comparatively high levels of acetaldehyde in the product (Otter and Taylor, 1971).

5.7 REACTION WITH ANTIMICROBIAL AGENTS

5.7.1 Spices
A number of spices, including allspice, cinnamon, cloves, horseradish, garlic, onion, oregano and mustard, possess antimicrobial properties, the active principle being eugenol, eugenol acetate or eugenol methyl ether in allspice and cloves; cinnamic aldehyde and eugenol in cinnamon;

allylsulphonyl and allyl sulphide in garlic; aliphatic disulphides such as *n*-propyl allyl and di-*n*-propyl in onion; thymol and carvacrol in oregano; and allyl isothiocyanate in mustard (Jay, 1970; Pivnick, 1980). The ability of sulphur(IV) oxoanions to inhibit the antimicrobial activity of some of these spices is known. Wei and coworkers (1967) demonstrated that the addition of 0·5 % potassium sulphite to an onion–buffer homogenate overcame the bacteriostatic effect of the onion. Hall (1969) showed that *Salmonella* could be recovered from artificially contaminated onion powder, pre-enriched in lactose broth, when 0·5 % potassium sulphite was added but not in the absence of potassium sulphite. The inhibition of the antimicrobial effect of onion was confirmed by Moussa (1973), who found a positive correlation between pungency and the degree to which the addition of sulphur(IV) oxoanion allows the aerobic microbial flora to survive, and by Wilson and Andrews (1976) who demonstrated that *Salmonella* inhibition by garlic powder is also reversible by sulphite ion. However, the antimicrobial agents in allspice, cinnamon, clove and oregano cannot be inactivated in this way and the results have implications to both uses and to the assay of micro-organisms in such spices. It is evident that those antimicrobial agents which contain sulphide or disulphide groups are the most susceptible to modification.

5.7.2 Sorbic acid

Sorbic, or hexadienoic acid ($CH_3CH=CHCH=CHCOOH$) is used frequently as an antimicrobial agent to replace, in part, sulphur dioxide when the use of the latter must be limited. The two preservatives may, therefore, be found mixed together. The reaction between sorbic acid and sulphur(IV) oxoanion is well documented. Hägglund and Ringbom (1926) suggest that it proceeds as a result of 3,4-addition of HSO_3^- to give the following product:

$$CH_3—\underset{\underset{SO_3^-}{|}}{CH}—CH_2—CH=CH—COOH$$

Heintze (1976) reports that, when solutions containing $1·8 \times 10^{-3}$ M sorbic acid and $1·6 \times 10^{-3}$ M sulphur(IV) oxoanion at pH 3·5 are left at room temperature for periods of 65–100 days, a 1:1 addition product is formed in approximately 60 % yield. However, when such products mixtures are analysed for sulphur(IV) oxospecies, all the added sulphur(IV) is recovered. The adduct appears to be labile but such combination between the two substances may reduce their efficacy as preservatives.

TABLE 5.1

Antimicrobial action of combined use of sulphur(IV) oxospecies and other antimicrobial agents

Organism	pH	Formic acid	Sorbic acid	Benzoic acid	p-Hydroxybenzoic acid esters
E. coli	6·0	−	±	−	± to +
S. cerevisiae	5·0		`	± to +	
A. niger	5·0	+	+	+	+

Adapted from tables shown by Lueck (1980). −, Antagonistic action; ±, additive action; +, synergistic action. Reproduced with permission from E. Lueck, *Antimicrobial Food Additives*, pages 38–9. © 1980 Springer-Verlag.

5.7.3 Nitrite ion

The reaction between sulphur(IV) oxoanions and nitrite ion is considered in Section 1.10.6, from which it is evident that these two antimicrobial agents cannot be used together.

5.8 EFFECTS OF COMBINATIONS OF ADDITIVES

Lueck (1980) compares the antimicrobial spectra of sulphur(IV) oxospecies with those of the lipophilic acids commonly used as antimicrobial agents. The latter are more effective against moulds and yeasts than against bacteria and it is evident that sulphur(IV) oxospecies may be used in conjunction with such additives to improve the spectrum of action without introducing an undesirably high concentration of any one antimicrobial agent. In this respect sulphur(IV) oxoanion is used frequently as an antibacterial agent with either benzoic or sorbic acid. In some cases the effect of the combined use of additives does not equal the sum of their individual effects: the action may be antagonistic or synergistic. Examples of such effects for combinations which include sulphur(IV) oxospecies are shown in Table 5.1. It is interesting to note the additive or synergistic effect of sorbic acid, according to the micro-organism used, despite the known reaction of sorbic acid with sulphur(IV) oxoanion.

6

Uses in food

Sulphur(IV) oxospecies are added to food to inhibit non-enzymic browning, to inhibit enzyme catalysed reactions, to inhibit and control micro-organisms and to act as an antioxidant and a reducing agent (Roberts and McWeeny, 1972). The chemical reactions and physical behaviour responsible for this preservative action are considered in the foregoing chapters. The reasons for which sulphur(IV) oxospecies may be added to foods are summarised in Table 6.1, but the appearance of an item in this list does not necessarily imply that the use of the additive is permitted in that application in any given country. Use may also be restricted to specified foods within a given category. The additive may be applied to solid foods as a solution of one of the salts of 'sulphurous acid' or by fumigation with sulphur dioxide gas, usually obtained directly from burning sulphur. Liquid foods are treated with solid salts of 'sulphurous acid', their solutions or the gaseous additive. The most frequently used sulphiting agents are the disulphites of sodium and potassium on account of their good stability towards autoxidation in the solid phase. Gaseous sulphur dioxide is employed where leaching of solids is likely to be a problem, for simplicity, or where the gas may also serve as an acid for the control of pH.

  Not all the sulphur(IV) oxospecies found in food is the result of addition, endogenous formation of the species being the consequence of microbial reduction of sulphate ion. Such endogenous sulphur(IV) oxospecies contribute significantly to the total amount found in fermented products. For example, reduction of sulphate may contribute up to 130 ppm *sulphur*

TABLE 6.1
Preservative effect of sulphur(IV) oxoanion in a wide range of foods

Purpose	Inhibition of non-enzymic browning	Inhibition of enzymes	Inhibition of micro-organisms	Antioxidant	Reducing agent	Other
Meat, fish and protein products						
(a) canned, heat processed	✓				✓	
(b) dehydrated	✓			✓	✓	
(c) raw and pasteurised			✓	✓		
Vegetable products						
(a) canned, heat processed	✓					
(b) dehydrated	✓	✓ᵃ				
(c) frozen		✓ᵃ				
(d) raw		✓				
(e) pickled	✓		✓			
Fruit products						
(a) canned, heat processed	✓	✓				
(b) dehydrated	✓	✓				
(c) frozen		✓				
(d) pasteurised	✓	✓	✓			
(e) raw, for manufacturing purposes			✓			
Beverages						
(a) wine and fruit juices	✓	✓	✓			
(b) beers			✓	✓		
Starch production		✓	✓			
Baked products						✓
Sugar production	✓		✓			✓

Based on information published by Roberts and McWeeny (1972) with minor changes.
ᵃ Inhibition of heat stable enzymes if product is blanched

dioxide in large scale fermentation of grape musts (Würdig and Schlotter, 1971).

Thermal processing operations are often effective in reducing the *sulphur dioxide* content of foods. For example, the boiling of wort in beer manufacture (Thalacker, 1975) and the boiling of sulphited strawberries in the making of jam (McWeeny *et al.*, 1980) lead to respective reductions of more than 94 and 95 % of the measurable additive. At lower temperatures desorption of sulphur dioxide may be assisted with a carrier gas, for example in the desulphiting of fruit juices. The relatively simple removal of the additive renders it useful as a possible processing aid with small residues, perhaps the most widely used application in this respect being the manufacture of sugar, during which process the sulphur(IV) oxospecies are removed by precipitation and by boiling in vacuum pans. The wide uses of *sulphur dioxide* as a processing aid and in ingredients for non-scheduled foods has led some legislating authorities to specify upper limits of *sulphur dioxide* in such foods.

This chapter deals with the interaction of sulphur(IV) oxospecies with food systems as a whole and will rely on the chemical reactions and properties already explained. However, the discussion will begin with a consideration of two specific reactions, hitherto not mentioned. The first, common to a number of fruit products, is the reaction between sulphur(IV) oxoanion and anthocyanins, whilst the importance of the second reaction, the well established sulphur(IV) oxoanion mediated autoxidation of unsaturated organic molecules, is still speculative in food systems.

6.1 GENERAL CONSIDERATIONS

6.1.1 Reaction with anthocyanins

Anthocyanins are derivatives of 2-phenylbenzopyrylium salts with the following substitution pattern:

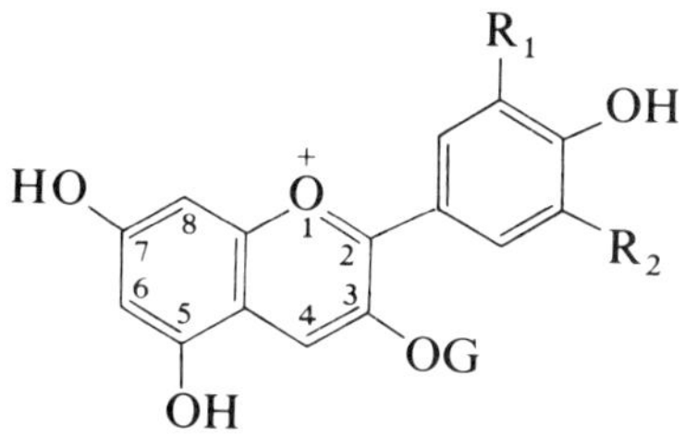

where R_1 and R_2 can be H, OH or OCH_3 according to the particular anthocyanin, and G is a saccharide moiety. The characteristic red colour of anthocyanins results from the presence of the flavylium cation (AH^+) which predominates in acid solution. This species is a moderately strong acid and dissociates to form a blue coloured base (A):

$$AH^+ \rightleftharpoons A + H^+$$

and undergoes a hydration reaction to give the carbinol pseudo base (AHOH):

$$AH^+ + H_2O \rightleftharpoons AHOH + H^+$$

the formation of which is equivalent to the addition of hydroxide ion to position 2 of the benzopyrylium system, with consequent loss of conjugation and, hence, colour. The pseudo base is also the main 'dissociated form' of the acid present below pH 6, the apparent pK_a for this dissociation of AH^+ being 1·91 for cyanin ($R_1 = OH, R_2 = H, G = glucose$) at 25 °C (Brouillard and El Hage Chahine, 1980), and this 'dissociation' accounts for the well known loss of anthocyanin colour as the pH is raised. There is also a small contribution from a colourless chalcone formed by the opening of the pyrylium ring.

The colour of the flavylium cation is also lost when sulphur(IV) oxoanions are added. Timberlake and Bridle (1967) demonstrated that this reaction involves the reversible formation of a 1:1 adduct between sulphur(IV) oxospecies and the cation, with an apparent formation constant of $2·6 \times 10^4 \, M^{-1}$ at pH 3·24 and 20 °C for cyanin, when the sulphur(IV) oxospecies concentration is expressed as the sum of all free forms present. The pH dependence of this equilibrium constant was investigated by Brouillard and El Hage Chahine (1980), who showed that the addition reaction was, in fact, a competitive process involving water and hydrogen sulphite ion, the equilibrium constant for the reaction,

$$AH^+ + HSO_3^- \rightleftharpoons product$$

being $1·1 \times 10^5 \, M^{-1}$ at 25 °C at an ionic strength of 0·2 M. Formation of the adduct takes place with a second order rate constant of $4·2 \times 10^2 \, M^{-1} s^{-1}$ under the same conditions. The significance of the large formation constant is well illustrated by Timberlake (1980) by a calculation of the free hydrogen sulphite ion required for the conversion of half of the anthocyanin to adduct. Using the value of apparent formation constant reported by Timberlake and Bridle (1967), this calculated value is 2·5 ppm expressed as *sulphur dioxide*. The concentration of the free *sulphur dioxide* in an anthocyanin-containing beverage such as blackcurrant juice or wine or in the liquor used for the preservation of fruit is likely to be well in excess of that required for the bleaching of the anthocyanin. The high affinity of hydrogen sulphite ion for anthocyanins is also the basis of a simple method for the determination of anthocyanins in red wine involving the measurement of absorbance before and after addition of sulphur(IV) oxoanion (Ribéreau-Gayon and Stonestreet, 1965).

Nucleophilic attack on the pyrylium cation leads to the formation of a σ-complex of the Meisenheimer type (Section 1.11.6), positions 2 and 4 of anthocyanins being reactive in this way:

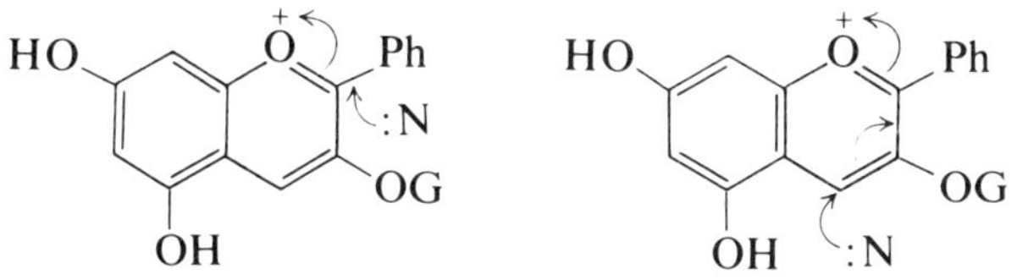

Timberlake and Bridle (1968) found that benzopyrylium compounds with a substituent (methyl or phenyl) in position 4 were less reactive towards sulphur(IV) oxoanion, implying that attack at this position is preferred. There is no evidence of any reaction between the chalcone form and hydrogen sulphite ion (Brouillard and El Hage Chahine, 1980). Anthocyanins are found frequently in combination with other flavanoid compounds, especially in wine, in which such products, which may be polymeric, are formed during the ageing process. The bond to anthocyanin may be to position 4, and compounds of this type are resistant, therefore, to bleaching by hydrogen sulphite ion. Synthetic 4-substituted anthocyanins may also constitute useful food-colouring agents in beverages in which bleaching is troublesome.

Anthocyanins are thermally unstable, the rate-determining step in their degradation being hydrolysis of the glycosidic bond and liberation of the saccharide component (Adams, 1973). When anthocyanins are heated in the presence of sulphur(IV) oxoanion, the three possible degradation steps are:

$$AH^+ \xrightarrow{k_1} products$$

$$AHOH \xrightarrow{k_2} products$$

$$AHSO_3^- \xrightarrow{k_3} products$$

From measurements of the rate of loss of cyanidin-3-rutinoside ($R_1 = OH$, $R_2 = H$, G = rutinose) at 100 °C in the presence of hydrogen sulphite ion, Adams and Woodman (1973) calculated the value of k_3 as zero at pH 3·0 under anaerobic conditions, the half life for the decomposition of anthocyanin being simply related to hydrogen ion concentration by:

$$t_{1/2} = (t_{1/2})_0 \left(1 + \frac{[HSO_3^-]}{A} \right)$$

where $(t_{1/2})_0$ is the value in the absence of hydrogen sulphite ion and A is a combination of equilibrium constants. The value of A is estimated to be $3\cdot5 \times 10^{-3}$ M, illustrating an appreciable stabilising effect by the sulphur(IV) oxoanion: the half life is doubled at a concentration of $3\cdot5 \times 10^{-3}$ M. These authors suggest that stabilisation is associated with the electron-withdrawing effect of the sulphonate group which deactivates the cyanidin–sugar bond, preventing the hydrolysis which would lead to degradation. The possible commercial advantages of the use of sulphur(IV) oxoanions to enhance the thermal stability of anthocyanins are limited by the bleaching action, an effect more serious in this case as the formation constant for the anthocyanin–hydrogen sulphite adduct is much smaller at 100 °C than it is at 20 °C. This disadvantage is not apparent if the product is desulphited after thermal treatment.

The extraction of anthocyanins from fruits may be enhanced by the addition of sulphur(IV) oxoanion. Several examples are cited by Markakis (1982), including the extraction of anthocyanin from apple peel and from grapes.

6.1.2 Sulphite mediated oxidations

The pro-oxidant action of sulphite ion in systems containing unsaturated lipids is discussed in Section 3.3.6 in connection with lipid browning. The essential step in this action is autoxidation of the sulphite ion and interaction between oxidising radical intermediates and the unsaturated compound. Other organic compounds affected in this way include methionine (Yang, 1970), tryptophan (Yang, 1973), chlorophyll (Peiser and Yang, 1977, 1978) and β-carotene (Peiser and Yang, 1979; Wedzicha and Lamikanra, 1983).

The product of oxidation of methionine induced by the autoxidation of sulphur(IV) oxoanion at pH $6\cdot8$ in the presence of a catalytic amount of manganese(II) ion is methionine sulphoxide. The yield of product is more than $0\cdot5$ mol per mole of sulphur(IV) oxoanion oxidised when the amino acid is in excess. Two possible mechanisms have been advanced for this reaction, both involving the same first step, which is based on the belief that the reaction of a sulphide with a one-electron oxidising agent involves transfer of an electron from the sulphur atom of sulphide to give the sulphonium radical, as follows:

$$R—S—R' + \cdot O_2^- + 2H^+ \rightarrow R—\overset{+}{S}—R' + 2 \cdot OH$$

This product may react subsequently with the $\cdot OH$ radical generated during the oxidation of sulphur(IV) oxoanion, as follows:

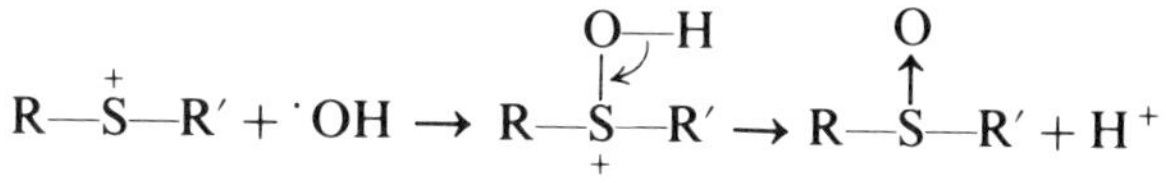

$$R—\overset{+}{S}—R' + \cdot OH \rightarrow R—\underset{+}{\overset{O—H}{S}}—R' \rightarrow R—\overset{O}{\underset{\uparrow}{S}}—R' + H^+$$

or with another superoxide ion to give a persulphoxide which may then oxidise a further methionine molecule, as follows:

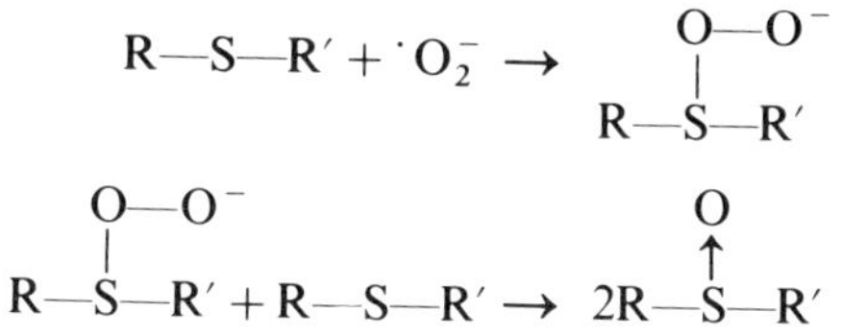

$$R—S—R' + \cdot O_2^- \rightarrow \begin{matrix} O—O^- \\ | \\ R—S—R' \end{matrix}$$

$$\begin{matrix} O—O^- \\ | \\ R—S—R' \end{matrix} + R—S—R' \rightarrow 2R—\overset{O}{\underset{\uparrow}{S}}—R'$$

The involvement of the superoxide ion and hydroxy radical in the autoxidation of sulphur(IV) oxoanion is considered in Section 1.10.1 and the inhibition of this reaction by superoxide dismutase and its enhancement by superoxide-generating systems, such as a mixture of horseradish peroxidase and phenol, add weight to the proposed mechanisms (Yang, 1970).

The nature of the products formed in the other examples of sulphur(IV) oxoanion mediated oxidations is not established. Tryptophan, whose behaviour is similar to that of methionine, is converted to at least four products. One of these has the same mobility as formylkynurenine on paper chromatograms. This compound is considered to be the primary product of photo-oxidation and of photosensitised oxidation of tryptophan (Yang, 1973). In the case of the mediated destruction of β-carotene, the products mixture is more complex. Microanalysis of the oxidation product shows that less than 0.5% sulphur (the limit of detection) and no nitrogen are incorporated into the organic product and there is a large amount ($22–25\%$) of unaccountable mass, presumably due to oxygen. The possibility of formation of sulphonate product is ruled out, since the addition of one sulphur atom per β-carotene molecule would lead to a sulphur content in the region of 5% whereas the observed amounts are at least an order of magnitude less than this. Infrared spectroscopy shows a broad, intense band at 3500 cm^{-1} characteristic of H-bonded OH, and the increase in polarity as β-carotene is converted to products is well demonstrated by mobility on thin layer chromatograms. The products mixture can be resolved into at least 10 components on alkylated dextran with ethanol as the eluting solvent, but there is no evidence from molecular weight determinations of extensive fragmentation or polymerisation reactions. Microanalysis of chromatographic fractions shows that all contain considerable amounts of oxygen, the number of atoms combined per original β-carotene molecule being 10–15. This amount of oxygen is comparable to that found by Teixeira Neto and coworkers (1981) for autoxidation of β-carotene; manometric measurement showed that up to 7 moles of oxygen were consumed per mole of carotenoid, equivalent to 14 atoms of the element. Although the final degree of oxidation is similar to that observed in autoxidation, an apparent difference between the corresponding products is the lack of any clear carbonyl stretching in the products of sulphite mediated autoxidation but observed in direct β-carotene autoxidation (Ouyang *et al.*, 1980). It is likely, therefore, that the oxygen exists mainly in the form of hydroxyl groups and possibly also in the form of stable peroxides. The products still absorb strongly in the ultraviolet region of the spectrum, suggesting that they retain extensive conjugation, and acid–base behaviour observed during titration suggests either enolic OH or the presence of an acid group. Assuming an average molecular weight in the region of 600, the average number of ionisable groups per β-carotene molecule is approximately 0·4. It is possible also to crystallise out a compound with one acid group per molecule (Wedzicha and Lamikanra, 1983).

The experiments involving β-carotene or chlorophyll (Peiser and Yang, 1977) were carried out in homogeneous solution which required the presence of a large proportion of ethanol (75 %) and, in the case of β-carotene, an addition of a small amount of chloroform in order to dissolve the organic and inorganic components. The use of such media is surprising in view of the well known observation that the oxidation of sulphur(IV) oxoanions is effectively inhibited by small amounts of alcohols. An unusual observation is that the addition of catalytic amounts of manganese(II) ion and of glycine or other amine promotes oxidation of sulphur(IV) oxoanion in this system, though the rate of oxidation is somewhat slower than that observed in aqueous solution. Kinetic experiments show that the active constituents with respect to autoxidation are the metal ion–amine complexes and that the reaction will not proceed unless there is a high concentration of ethanol present, that is, it is a general medium effect. An interesting feature is that, if the autoxidation of sulphur(IV) oxoanion is carried out in the presence of nitroso-t-butane in this system, an ESR signal is observed which is most likely to result from the product formed by the trapping the $\cdot SO_2^-$ radical by the reagent (Wedzicha and Lamikanra, 1982). These features of autoxidation in the ethanolic medium are different from those of normal autoxidation of sulphur(IV) oxoanion (Section 1.10.1) and the relevance to aqueous systems of results obtained under such conditions is not clear. Lizada and Yang (1981) successfully oxidised Tween stabilised emulsions of fatty acids, but it is not possible to repeat these experiments with simple dispersions or Tween stabilised emulsions of β-carotene unless some chloroform is also present in the emulsifying agent (Wedzicha and Lamikanra, 1982). However, a simple dispersion of vitamin A alcohol in an aqueous solution of sulphur(IV) oxoanion with a catalytic amount of manganese(II) ion results in rapid oxidation of the vitamin. In this case a comparison between the amount of oxygen incorporated into the vitamin A molecule as a result of oxidation both as a dispersion and in the ethanolic medium, was made. The degree of oxygenation in the ethanolic medium was similar to that reported for β-carotene, whilst that found in the product of heterogeneous oxidation was some 30 % lower. It is concluded that an important factor in the sulphite mediated oxidation of hydrocarbons as dispersion in an aqueous phase is the ability of the free radical intermediates to interact with the substrate. Since these radicals will be solvated, the degree to which the non-polar molecule can interact with the medium, with the help of side chains, is important. A similar effect in the free radical sulphonation of dodecane is described in Section 1.11.2.

Thus, the sulphite mediated oxidation of unsaturated food components is possible both in aqueous solution and in dispersions, and the probability of such reactions taking place in food systems may now be considered. Oxidation of sulphur(IV) oxoanion to sulphate ion in food undoubtedly takes place. For example, some 20–30 % of the additive present in dehydrated vegetables may be converted to sulphate ion (Mangan and Doak, 1949). Despite this fact, it is generally accepted that the presence of sulphur(IV) oxospecies is beneficial to the retention of carotenoids, for example in dehydrated vegetables (Weier and Stocking, 1946; Nutting et al., 1970; Arya et al., 1979), dehydrated fruits (Woodroof and Luh, 1975) and in model dehydrated systems using sulphited cellulose as an adsorbent for β-carotene (Baloch et al., 1977). The additive is used also to prevent the oxidation of essential oils in citrus products whilst, in model systems, it may enhance the oxidation of orange oil (Beard et al., 1972). Although sulphur(IV) oxoanion mediates the oxidation of emulsified lipids in model systems (Section 3.3.6), there are no reports in the literature of the oxidative deterioration of fats as a result of the use of this additive: dehydrated potato is potentially vulnerable in this respect.

It is likely that the success of sulphur(IV) oxoanion as an antioxidant arises from the physical state of the substrate in contact with the aqueous medium and the presence of natural or added antioxidants in the aqueous and non-aqueous phases. The second possibility is demonstrated by the observation that, when β-carotene is solvent-extracted from carrots, the crude product is less easily oxidised in the presence of sulphur(IV) oxoanion than is pure β-carotene (Wedzicha and Lamikanra, 1982). When sulphited foods are packaged in containers impermeable to oxygen, residual oxygen in the headspace may be removed by oxidation of sulphite ion. If, during the course of this oxidation, the unsaturated components of the food are protected in some way, the effect of the additive will be that of an antioxidant with respect to the oxidation of these components by oxygen.

6.1.3 Metal corrosion

When tin plate is placed in contact with most food electrolytes, the electrochemical cell shown in Fig. 6.1 is set up, with tin acting as the anode, and access to the iron (as steel) is through imperfections in the tin coating. The net result is dissolution of tin and evolution of hydrogen at the iron electrode. Since the area of tin greatly exceeds that of iron, the rate of dissolution of tin is controlled by the small area of iron. Tin, on the other hand, offers cathodic protection to the iron. The rate of dissolution of the tin may be increased by the addition of electron acceptors which act as depolarisers and whose reduction provides the necessary electron sink for the formation of tin ions. Examples of suitable oxidising agents include oxygen, nitrate ion, organosulphur compounds and sulphur(IV) oxospecies. The sulphur-containing compounds are reduced to hydrogen

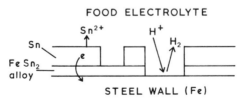

Fig. 6.1. Diagrammatic representation of tin plate in contact with a typical food electrolyte.

sulphide, the reaction in the case of sulphur(IV) oxospecies proceeding presumably through dithionite ion (Section 1.10.7). This hydrogen sulphide may react with the tin surface to leave a black stain of insoluble sulphide (Andrae, 1969) which may also reduce the cathodic protection of the iron afforded by tin.

The effect of low concentrations of sulphur(IV) oxospecies in grapefruit juice on the rate of de-tinning of cans is shown in Fig. 6.2 (Saguy *et al.*, 1973). The shapes of the corrosion curves in juice with and without added

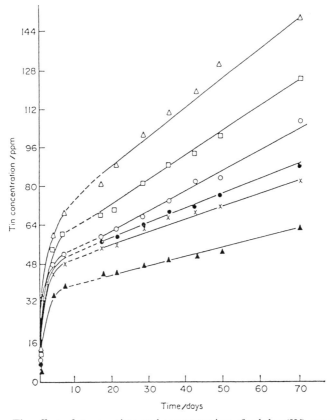

Fig. 6.2.   The effect of storage time and concentration of sulphur(IV) oxospecies (expressed as *sulphur dioxide*) on de-tinning of tin plate cans (A2 size) containing grapefruit juice at 35 °C and with 3 mm head space. Sulphur(IV) oxospecies added as $Na_2S_2O_5$. ▲, No additive; ×, 1 ppm; ●, 2 ppm; ○, 3 ppm; □, 4 ppm; △, 5 ppm (Saguy *et al.*, 1973). Reprinted with permission from *J. Food Technol.*, 1973, **8**, page 150. © 1973 Institute of Food Science and Technology.

sulphur(IV) oxospecies are similar, the differences being the slope of the linear portion and the intercept. Whilst the latter does not appear to show any relationship to the concentration of the additive, there is a linear correlation between slope and concentration for the duration of the experiments. The lowest concentration of sulphur(IV) oxospecies, 1 ppm ($1 \cdot 6 \times 10^{-5}$ M), shows a clear de-tinning effect and it is likely, therefore, that the threshold value for this additive is below this concentration, particularly when it is considered that some of the additive in grapefruit juice will be in reversible combination with carbonylic components. No pH control was attempted in this work but it is unlikely that the addition of sodium disulphite at the concentrations used will give rise to significant changes in pH. If the maximum oxidising capacity of sulphur(IV) oxospecies is given by:

$$SO_2 + 6e^- + 6H^+ \rightarrow H_2S + 2H_2O$$

then one mole of sulphur(IV) oxospecies will allow the liberation of three ions of tin. Therefore, at the highest concentration of the additive, 5 ppm ($7 \cdot 8 \times 10^{-5}$ M), the maximum stoichiometric increase in concentration of tin is by 28 ppm. It is evident, however, that the extent of de-tinning after 70 days is approximately four times greater than this amount. The addition of 10 ppm *sulphur dioxide* reduces the life of a can of grapefruit juice from approximately 1 year to some 21 weeks; in the case of orange juice the reduction is from 2 years to some 40 weeks, can failure corresponding to the observation of hydrogen swells. At such concentrations of *sulphur dioxide* improvement in the quality of tinplate does not prevent its corrosion (Semel and Saguy, 1974). An interesting case of electrolytic reduction is the formation of hydrogen sulphide in crown-capped bottles containing sulphited beverages, an effect attributed to the presence of sulphur(IV) oxospecies and damage to the surface of the cap (D. Hicks, personal communication).

Aqueous solutions of sulphur dioxide are known to corrode stainless steel also, an effect which may be controlled by the use of molybdenum-containing stainless steels (Brown, 1977). Bauer (1972) compares two types of stainless steel for wine storage tanks. One type containing $\geq 18\%$ chromium, $\geq 8\%$ nickel and occasionally sulphur is less resistant to corrosion than another type containing 18% chromium, 10–12% nickel and 2–3% molybdenum. Experiments at room temperature have shown that tanks made from the first type of steel could withstand the action of 50–70 ppm free sulphur dioxide, but tanks of the molybdenum-containing stainless steel should be used at higher concentrations.

TABLE 6.2

Reactivity of flavour components of foods towards sulphur(IV) oxospecies

Food	Compound	Group	Reactivity towards S(IV)
Citrus fruits	Terpenes	C=C	Free radical?
	Terpenoid alcohols and esters		
	Citral	C=C / CHO	Observed (Erman et al., 1979)
	Citronellal	C=C	Free radical?
		CHO	Expected
	Aliphatic aldehyde	CHO	Expected
Apple	Hexanal	CHO	Expected
	Hex-2-enal	CHO	Expected
		C=C	Expected
Carrot	2-Alkyl-3-methoxypyrazines	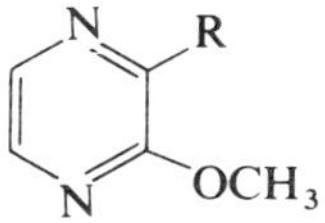 (see structure)	Unlikely
Potato			
Cabbage	Isothiocyanates	RNCS	Expected
	Sulphides	R_2S	Free radical?
	Disulphides	R—S—S—R'	Expected
	Hydrogen sulphide	H_2S	Expected
Onion	Aldehydes	CHO	Expected
Garlic	Thiols	RSH	Free radical?
	Disulphides	R—S—S—R'	Expected
	Trisulphides	R—S—S—S—R'	Expected
	Hydrogen sulphide	H_2S	Expected
Mustard	Allylisothiocyanate	RNCS	Observed (Frijters et al., 1981)

2-Alkyl-3-methoxypyrazine structure: pyridazine ring with R and OCH₃ substituents.

6.1.4 Effects on flavour

The most direct effect of *sulphur dioxide* on flavour is that of the taste of the additive. This is generally detrimental to the quality of sulphited foods, giving rise to a metallic taste or pungent odour. However, since the amount of the additive present in a food at the time of consumption is normally below its sensory threshold, the most important effects on flavour will be the result of interactions of the additive with the volatile components responsible for flavour. The wide range of functional groups encountered in these substances is shown in Table 6.2 with an indication of known and possible reactivity towards sulphur(IV) oxospecies. Little is known concerning the implications of these reactions to perceived food flavour. The suppression of oxidised flavour in beer by reaction of offending volatile aldehydes with hydrogen sulphite ion (Hashimoto, 1972) illustrates the sensory consequences of the carbonyl reaction. The only other reaction studied in connection with sensory characteristics is that involving allyl isothiocyanate in mustard paste (Griffiths *et al.*, 1980; Frijters *et al.*, 1981). The reaction between isothiocyanates and sulphur(IV) oxoanion is shown in Section 1.11.5, the product being an involatile and odourless sulphonate. Sulphur dioxide is required for the processing of certain mustard pastes in order to inhibit microbial growth and prevent discoloration. Although the storage of mustard paste in the absence of the additive leads to loss of allyl isothiocyanate, the addition of sulphur dioxide at a level of 1600 ppm to a mixture of mustard flour:water (1:4 w/w) leads to a considerable acceleration of this loss. For example, after 10 days of storage the concentration of allyl isothiocyanate in a sulphited mustard paste had decreased by 85% compared with 55% in the non-sulphited paste. In another paste the corresponding values were 79% and 48%, and the effect appeared to be independent of isothiocyanate concentration. These changes in concentration were reflected in changes to headspace composition and were accompanied by an increase with time in the amount of allylaminothiocarbonylsulphonate which could be measured in the sulphited paste. Over this storage period the sulphited paste decreased in odour intensity but developed a garlic quality so that after 13 days its odour was equally mustard and garlic. Further storage led to loss of this garlic odour, which was not an attribute of the non-sulphited system. The source of the garlic component is not clear but it is possible that it may arise from further reactions of the sulphonate, since solutions of the adduct in water were shown to possess a garlic odour.

The implication of the information shown in Table 6.2 and the examples referred to above suggest that many sulphited foods may be organoleptically distinct on account of the presence of the additive. This view may be reinforced by considering two further situations. First, the composition of fermented beverages with respect to the products and by-products of fermentation is affected by the presence of sulphur(IV) oxospecies. The trapping of acetaldehyde as the hydroxysulphonate during the conversion of carbohydrate to ethanol leads to its accumulation. In the case of beer, acetaldehyde has little effect on palate unless it is present in amounts greater than 25 ppm, but it is a precursor of other compounds which have pronounced tastes. Ethanol reacts with acetaldehyde to produce acetal which can be detected by taste at 1 ppm. Acetaldehyde may be a precursor of diacetyl and will also react with hydrogen sulphide to produce thioacetaldehyde, which results in a particularly unpleasant flavour at very low concentrations (Otter and Taylor, 1971). Under conditions where acetaldehyde is not available, dihydroxyacetone phosphate may become the substrate for further reduction leading to the formation of glycerol (Nord and Weiss, 1958), as follows:

$$
\begin{array}{ccc}
\mathrm{CH_2OH} & \mathrm{CH_2OH} & \mathrm{CH_2OH} \\
| & | & | \\
\mathrm{C{=}O} \xrightarrow[\mathrm{H^+}]{\mathrm{NADH}} & \mathrm{CHOH} \xrightarrow{\alpha\text{-glycerophosphatase}} & \mathrm{CHOH} \\
| & | & | \\
\mathrm{CH_2OPO_3H_2} & \mathrm{CH_2OPO_3H_2} & \mathrm{CH_2OH}
\end{array}
$$

The increase in glycerol formation in the presence of sulphur(IV) oxospecies is well known and may have organoleptic consequences. The effects of the additive on other enzyme systems responsible for the development of flavour are not known. Secondly, a more subtle possible effect of sulphur(IV) oxospecies is that on non-enzymic development of flavour. Whereas the additive is essential for inhibition of browning and of flavour during processing and storage, it is also capable of reducing the development of certain cooked flavours during the preparation of the processed food for consumption. There is clearly a need here for further investigation.

6.1.5 Reaction with artifical food colouring agents

The most common food colouring agents are azo compounds. Triphenylmethyl compounds, as well as a small number of dyes based on quinoline, xanthene and indigoid structures, constitute the remainder. Azo compounds may be reduced by sulphur(IV) oxoanion to colourless hydrazo compounds (Section 1.11.11), whilst the triarylmethyl cation may be decolorised by combination with sulphite ion (Sections 1.9 and 2.3.3.1).

TABLE 6.3

Reactivity of artificial food colouring agents towards sulphur(IV) oxospecies
(Society of Dyers and Colourists, 1971)

Common name	Colour Index name	Chemical class	Stability towards S(IV)
—	Yellow 2	Monoazo	Poor
Sunset Yellow FCF	Yellow 3	Monoazo	Fair
Tartrazine	Yellow 4	Monoazo	Excellent
Yellow 2G	Yellow 5	Monoazo	Fair
Chrysoin	Yellow 8	Monoazo	Excellent
Quinoline Yellow	Yellow 13	Quinoline	Excellent
Orange RN	Orange 1	Monoazo	Fair (slightly yellower)
—	Orange 2	Monoazo	Poor
Orange G	Orange 4	Monoazo	Fair
—	Red 1	Monoazo	Very good
—	Red 2	Monoazo	Good
Carmoisine	Red 3	Monoazo	Fair
Fast Red E	Red 4	Monoazo	Poor
Ponceau MX	Red 5	Monoazo	Good
Ponceau 4R	Red 7	Monoazo	Fair
—	Red 8	Monoazo	Good
Amaranth	Red 9	Monoazo	Fair
Red 2G	Red 10	Monoazo	Good
Red 10B	Red 12	Monoazo	Good
Red FB	Red 13	Monoazo	Fair
Erythrosine BS	Red 14	Xanthene	Good
Violet BNP	Violet 3	Triarylmethane	Poor
Indigo Carmine	Blue 1	Indigoid	Poor
—	Blue 2	Triarylmethane	Very good
Green S	Green 4	Triarylmethane	Good
Brown FK	Brown 1	Complex azo	Good
Chocolate Brown FB	Brown 2	Monoazo	Good
Chocolate Brown HT	Brown 3	Disazo	Fair
Black PN	Black 1	Disazo	Fair
Black 7894	Black 2	Disazo	Fair

Compiled from *Colour Index*, 3rd edition, Vol. 2, pages 2775–80.

The stability of possible food colouring agents with respect to sulphur(IV) oxospecies is shown in Table 6.3 (Society of Dyers and Colourists, 1971); the reader is advised that the presence of an entry in this list does not imply that the use of a given dye-stuff in food is permitted at the present time. The regulations concerning this use vary internationally and from time to time.

6.1.6 Packaging of foods containing *sulphur dioxide*

The headspace of foods containing sulphur(IV) oxospecies always contains a low concentration of gaseous sulphur dioxide and, therefore, the permeability to sulphur dioxide of packaging materials needs to be taken into account. Since autoxidation of the additive is possible, the permeability of the packaging material towards oxygen may also be of significance.

The main considerations in the choice of polymer films for the packaging of foods containing *sulphur dioxide* are explained by Davis and coworkers (1975). Experiments using low density polyethylene, polyamide (nylon 11) and polycarbonate (bisphenol-A polycarbonate) films show that all three types are more permeable to sulphur dioxide than to oxygen, and that their behaviour towards sulphur dioxide is non-ideal since the permeability depends upon the partial pressure of the gas. It is not possible, therefore, to calculate permeabilities at low concentrations from data obtained at high concentrations. For a concentration of 600 ppm sulphur dioxide at a total pressure of gas of 1 atm the permeability was found to vary by a factor of 2000 when the characteristics of common packaging films were compared. The most permeable films are low density polyethylene, polycarbonate and polystyrene, whilst the least permeable are films coated with PVDC (a copolymer of vinylidene chloride and vinyl chloride). Between these extremes the order of increasing permeability is PVC (rigid), polyester, polypropylene, polyamide (nylon 11) and high density polyethylene. Davis and coworkers also estimate that the contribution of sulphur dioxide permeation to the loss of additive from sulphited apricots packaged in low density polyethylene is in the region of 15 %. The permeation of oxygen into the pack is responsible for some 30 % of the lost additive.

6.2 SPECIFIC APPLICATIONS

6.2.1 Food dehydration

6.2.1.1 Introduction

Food dehydration is an effective method of preservation, since all forms of spoilage, with the exception of lipid oxidation, may be arrested by appropriate reduction of water activity. The most widely practised method of dehydration is that of air drying, which necessitates the use of a heated air stream. As water is removed, the concentration of solutes increases and the temperature of the food rises. The majority of foods which are dehydrated in this way contain the reactants of non-enzymic browning reactions (Chapter 3), and the conditions in the early stages of the drying operation are conducive to Maillard and ascorbic acid browning. The good supply of oxygen also renders dehydroascorbic acid browning likely. As the water activity is reduced further, the rate of non-enzymic browning will begin to fall and the success of a dehydration operation, with respect to browning, depends upon the extent to which browning had been suppressed in the intermediate range of water activities. The only known method of controlling these undesirable reactions is the addition of *sulphur dioxide*. The additive is not effective at inhibiting the formation of early intermediates in browning, and the sulphonates (such as the 3,4-dideoxy-4-sulpho osuloses), formed as a result of the inhibition of Maillard and ascorbic acid browning, are themselves capable of further reaction to coloured products in the absence of free additive (Sections 3.3.4 and 3.3.5.1 and Wedzicha, 1984). Since it is not practical to reduce the water activity to such an extent as to stop all chemical reactions, it is necessary for there to be residual *sulphur dioxide* present in the food to prevent subsequent reactions of these intermediates and products. The foods most commonly dehydrated in the presence of sulphur(IV) oxospecies are fruits and vegetables and their juices. Roberts and McWeeny (1972) draw attention to two other areas of food dehydration in which the use of *sulphur dioxide* may be beneficial. In one case the appearance and acceptability of dehydrated, light-coloured protein products such as pork and chicken may be improved by the addition of sulphur(IV) oxoanion or by storage in an atmosphere containing sulphur dioxide. The other instance is that of reduction of browning and the development of off-flavour during the storage of milk powder, and inhibition of onset of insolubility.

6.2.1.2 Sulphiting practice
The methods available for the application of the additive to fruits and vegetables lead to wide variation in uptake. Vegetables are blanched prior to dehydration. Most commonly, the additive is incorporated at the blanching stage, either in the blanch liquor if the vegetable is to be dipped, or as a spray in the case of steam blanching. Uptake is expected to be a function of concentration of sulphur(IV) oxospecies, length of treatment and time allowed for draining, size and geometry of the food being treated and the pH of the blanch liquor or spray. Such treatments may be applied to peas, beans, cabbage, swede, carrot, potato and tomato. In the case of potato granule manufacture, a solution of the additive is conveniently metered into cooked mashed potato and the amount is, therefore, more predictable. Fruits for dehydration are normally treated with gaseous

sulphur dioxide from burning sulphur. Such treatment is used in the manufacture of dried apricots, peaches and bananas, raisins and sultanas. On the other hand, apple slices are generally dipped in solutions of the additive and may receive an additional treatment with gaseous sulphur dioxide during the drying operation. Dipping of apricots in solutions containing hydrogen sulphite ion is said to offer considerable advantages over the uses of the gaseous additive in better control over the amount taken up, greatly reduced time for the treatment and decreased desorption losses during drying (Stafford *et al.*, 1972). The immersion process demonstrates that the mean rate of uptake of additive is proportional to its concentration in solution and that this rate of absorption increases as pH is reduced. The pH dependence could be indicative of selective transport of molecular sulphur dioxide, as is the case in microbial cell wall transport (Section 5.2) or, since mean rates were measured, may be explained by the low activity of sulphur dioxide within the fruit. Fruit and vegetable juices are sulphited by direct addition prior to dehydration. The practices of commercial fruit and vegetable dehydration are described in detail by Van Ardsel and coworkers (1973) and Woodroof and Luh (1975).

Dehydration times in excess of 12 h for fruits and vegetables and of several days in the sun-drying of fruits necessitate the use of large amounts of *sulphur dioxide*. Suggested residual amounts in a variety of dehydrated products are shown in Table 6.4 and these represent some of the highest levels of use of the additive in foods. Legislation may, in fact, permit the addition of greater amounts of *sulphur dioxide*: for example, the legislation for the UK stipulates a uniform limit of 2000 ppm for dehydrated fruit and

TABLE 6.4

Residual levels of *sulphur dioxide* in dehydrated fruits and vegetables required for their successful preparation and storage

Vegetable	SO$_2$ (ppm)	Fruit	SO$_2$ (ppm)
Green beans[a]	500	Apple[c]	1 000–2 000
Cabbage[a]	1 000–2 000	Apricot[c]	2 000–4 000
Carrot[a]	500–1 000	Peach[c]	2 000–4 000
Celery[a]	500–1 000	Pear[c]	1 000–2 000
Peas[a]	300–500	Raisin[c]	1 000–1 500
Potato granule[b]	250		

[a] Van Ardsel and coworkers (1973).
[b] Sahasrabudhe and coworkers (1976).
[c] Woodroof and Luh (1975).

vegetables other than cabbage or potato, which have limits of 2500 and 550 ppm respectively, in the final product. The reason for the lower requirement in dehydrated potato is its relatively low reducing sugar content (0.5%—*c.* 2% per dry weight), which may be compared with 15–25% per dry weight expected for fresh cabbage. In their studies on the loss of *sulphur dioxide* during the dehydration of cabbage, potato and carrot, Gilbert and McWeeny (1976) found that only 35–45% of the sulphur(IV) oxoanion incorporated into the vegetable was measurable after drying. Therefore, in order to achieve say 2000 ppm *sulphur dioxide* in the dehydrated product, at least 4400 ppm *sulphur dioxide* per dry weight is required at the time of blanching.

6.2.1.3 Reactions during dehydration

The available information is limited to a small number of dehydrated foods. The evidence which suggests that the organic products derived from sulphur(IV) oxospecies during the dehydration of cabbage are the result of interactions with intermediates of the Maillard reaction is the following series of observations (Panahi, 1982):

(a)　if freeze-dried cabbage is fractionated into components of varying polarity by successive solvent extractions (petroleum spirit, ether, acetone, ethanol and water) and the extracts tested for their reactivity towards sulphur(IV) oxospecies, only the aqueous extract shows any reactivity;

(b)　if dehydrated cabbage is prepared using ^{35}S-labelled sulphite as the additive and the distribution of radioactive components is analysed by means of chromatography on DEAE Sephacel (Section 2.7.3.2 and Fig. 2.16), the chromatogram shows the same components as one obtained for the reaction products in a mixture of aqueous extract of non-sulphited cabbage and ^{35}S-labelled sulphite which have been allowed to react for 6 h at 80 °C;*

(c)　the aqueous extract of non-sulphited cabbage may be rendered unreactive towards sulphur(IV) oxospecies by passing it through a column of cation exchange resin in the H^+ or Na^+ forms. Such treatment removes amino compounds and peptides which would participate in Maillard browning.

---

* The most abundant component, characterised by a sharp peak at an elution volume of 130 ml in Fig. 2.16, is sulphate ion. There is no evidence of a significant amount of free sulphur(IV) oxospecies. It is presumed that the latter is oxidised to sulphate during work-up and in the early stages of chromatography.

Further evidence is available from studies involving ^{14}C-labelling of glucose and fructose in the aqueous extracts of non-sulphited cabbage. When such labelled extracts are allowed to react with inactive sulphur(IV) oxospecies and the distribution of ^{14}C-labelled products analysed by the chromatographic method referred to in (b), the sulphur-containing products running earlier than sulphate (Fig. 2.16) match ^{14}C components very well. An approximate calculation of the composition of the main component using ^{14}C and ^{35}S activities in the respective experiments shows that the ratio of glucose residues to sulphur is close to 1:1. Experiments involving ^{35}S-labelled products and inactive sulphur(IV) oxoanions show that the sulphonates present in this fraction do not exchange sulphur with inactive sulphite ion in the medium and are, therefore, not simply hydroxysulphonate adducts of carbonyl components. The later running products in Fig. 2.16 could originate from a combination of non-enzymic browning reactions of glucose and fructose in the presence of sulphur(IV) oxospecies. However, the correspondence between retention data in the ^{35}S and ^{14}C experiments is less satisfactory in this region of the chromatogram.

The only quantitative analysis of specific sulphonates known to be formed in foods has been carried out for the dehydration of sulphited swede, in which case the combined yields of 3,4-dideoxy-4-sulphohexosulose and 3,4-dideoxy-4-sulphopentosulose on dehydration represent a conversion of only 27 ppm *sulphur dioxide* for a residual *sulphur dioxide* content of 1140 ppm (Wedzicha and McWeeny, 1975).

6.2.1.4 Reactions on storage

The loss of *sulphur dioxide* from dehydrated vegetables continues on storage and determines the practical shelf life with respect to spoilage through non-enzymic browning. The loss of the additive in dehydrated apricots (Stadtman *et al.*, 1946), carrot, potato and cabbage (Legault *et al.*, 1949) appears to be kinetically of overall first order. This disappearance of *sulphur dioxide* is sensitive to temperature and the moisture content of the product (as determined by vacuum oven drying at 60 °C), as illustrated in Fig. 6.3 for dehydrated carrot; the composition of the atmosphere with respect to oxygen in the headspace is relatively unimportant at higher temperatures (37 °C: Mangan and Doak, 1949; 49 °C: Legault *et al.*, 1949) but is significant at lower temperatures (24 °C: Legault *et al.*, 1949), when packs containing oxygen show a greater rate of loss of the additive.

The known reaction products are sulphate ion (Mangan and Doak, 1949) and the organic sulphonates indicative of the inhibition, by sulphur(IV) oxospecies, of Maillard and ascorbic acid browning (see

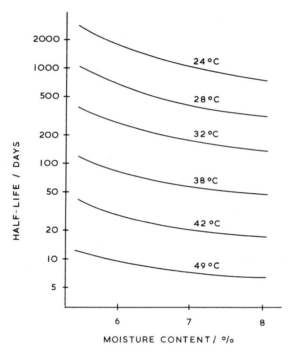

Fig. 6.3. The effect of moisture content and temperature on the half-life of *sulphur dioxide* in nitrogen-packed dehydrated carrot (Legault *et al.*, 1949). Reproduced with permission from *Ind. Eng. Chem.*, 1949, **41**, page 1450. © 1949 American Chemical Society.

Sections 2.7.3.2 and 3.3.3). No dithionate is formed. The only reported quantitative analysis of these sulphonates is that for stored dehydrated cabbage (Wedzicha and McWeeny, 1975). After 5 years of storage the combined amount of 3,4-dideoxy-4-sulphohexosulose and 3,4-dideoxy-4-sulphopentosulose was equivalent to the reaction of some 800 ppm *sulphur dioxide*.

6.2.1.5 Preservative action
Although much attention has been given to the uses of *sulphur dioxide* for the control of non-enzymic browning reactions in dehydrated foods, it is important not to lose sight of its role as an antioxidant. This is discussed in relation to the stability of carotenoids and other unsaturated food components in Section 6.1.2 and is referred to in connection with oxidation of ascorbic acid in Section 3.3.5.2.

6.2.2 Preservation of fresh fruit and vegetables

6.2.2.1 Microbial decontamination

Sulphur dioxide is very effective as a fumigant in the control of post-harvest decay of grapes caused by the fungus *Botrytis cinerea*. The fumigant may initially be applied at a concentration of 1 % by volume in the air of a fruit store for 20 min, followed by applications at one-quarter of this level for 30–60 min every 7–10 days. It is interesting to note that grapes held at a relative humidity of 95% require 15–20% more additive for a given exposure than those held at 85% relative humidity (Nelson and Baker, 1963). Raspberries may similarly be treated for the control of *Botrytis* and *Cladosporium* (Cappellini *et al.*, 1961), though at lower concentrations of the additive. At an initial concentration of 0·5 % sulphur dioxide, a slight bleaching and softening of the berries is noted but, when this concentration is reduced to 0·25%, the only noticeable effect is a delay in post-harvest ripening.

6.2.2.2 Storage as pulp

The term 'pulp' is used to describe fruit stored in a sulphite liquor and applies to all qualities ranging from whole fruit to completely disintegrated fruit. The liquor consists of a solution of *sulphur dioxide* at a concentration of up to 3000 ppm and is saturated with slaked lime in order to improve the texture of the product, giving a final pH of approximately 3·0. Such a medium allows fruit such as blackberries, blackcurrants, gooseberries, loganberries, raspberries and strawberries to be stored for several months. Indeed, after 8 months storage the retention of ascorbic acid content of blackcurrants and strawberries is 69 % and 88 % respectively, and is 43 % in blackcurrants stored for 20 months (Baker, 1947). Fruit whose colour depends upon the presence of anthocyanin is bleached under these conditions. The bleaching of cherries in sulphite liquor followed by leaching and colouring leads to a product which is uniform and brightly coloured, as well as having improved texture (Woodroof and Luh, 1975). The main application of storage in pulp is for subsequent jam manufacture, enabling processing to take place all the year round, despite a short fruit season. It is economically advantageous over frozen storage and, in the UK for example, 70 % of the strawberries to be used for jam are stored in this way.

The high concentration of sulphur(IV) oxospecies and the prolonged contact time with the fruit renders important the assessment of the extent of reaction between the additive and the components of the fruit. The uptake of *sulphur dioxide* by strawberries and its distribution between free and reversibly bound and irreversibly bound forms has been investigated by

Fig. 6.4. The effect of sulphur dioxide concentration on the breakdown of sulphited fruit by *R. sexualis* (Harris and Dennis, 1979). Reproduced with permission of C. Dennis from *J. Sci. Food Agric.*, 1979, **30**, page 708. © The authors.

McWeeny and coworkers (1980). When 530 g berries are placed in 76 ml liquor (2000 ppm *sulphur dioxide* + 6700 ppm CaO), 78 % of the *sulphur dioxide* is transferred to the fruit after 6 months at room temperature. This transfer corresponds to a final level of 225 ppm in the fruit. However, Monier-Williams analysis revealed that only 140 ppm *sulphur dioxide* was in a recoverable form, corresponding to a conversion of 85 mg *sulphur dioxide* per kilogram of fruit to products in which it is irreversibly combined. McWeeny and coworkers also considered the changes in the amounts of recoverable and non-recoverable *sulphur dioxide* brought about by the conversion of the fruit to jam. The recipe was based on fruit (45 parts), sugar (50 parts), glucose syrup (20 parts), water (5 parts) and pectin (0·7 % solution, 10 parts) and required heating conditions which would boil off 20 parts of water in approximately 20 min. The process led to a physical loss of 111 mg *sulphur dioxide* per kilogram of fruit, giving a total (free + reversibly bound + irreversibly bound) *sulphur dioxide* content of 114 ppm, of which 110 ppm were irreversibly bound. The boiling operation has, therefore, led to the conversion of a further 25 mg *sulphur dioxide* per kilogram of fruit to a form in which it is irreversibly combined. The nature of these reactions has yet to be established. However, the presence of glucose from both the added syrup and the inversion of sucrose during boiling renders possible the formation of 3,4-dideoxy-4-sulphohexosulose (Section 3.3.3).

The success of the storage of strawberry pulp depends undoubtedly upon the broad spectrum of action of sulphur(IV) oxospecies in the inhibition of microbial action, the action of enzymes, non-enzymic browning and oxidative spoilage. Occasionally, however, sulphited strawberries are found to disintegrate, a phenomenon which has been investigated in detail in a series of papers (Dennis and Harris, 1979; Archer, 1979; Harris and Dennis, 1979; Archer and Fielding, 1979; Dennis *et al.*, 1979; Harris and Dennis, 1981). It is evident that the problem arises through contamination of the liquor by *Rhizopus sexualis* which, in particular, is capable of producing an endopolygalacturonase isoenzyme with considerable stability in the sulphite liquor. Although the effect may be prevented by increasing the concentration of *sulphur dioxide* as demonstrated in Fig. 6.4, the reason for the improvement in product at higher concentrations of the additive is not the specific action of sulphur(IV) oxospecies, but the effect of pH on the isoenzyme in question.

6.2.2.3 Inhibition of oxidising enzymes
The most common use of *sulphur dioxide* for the control of oxidising enzymes is the inhibition of enzymic browning of the polyphenoloxidase type (Section 4.2.6) when plant tissues are damaged by cutting. An example is the use of a 'sulphite dip' to protect apple slices. A solution containing 300 ppm *sulphur dioxide* and 1000 ppm calcium ion at either pH 3·7 or pH 7 is adequate to permit the storage of slices of Golden Delicious apples for several weeks at 1 °C in sealed containers, although flavour and texture are better after treatment at pH 7. The main detrimental effect is that, as *sulphur dioxide* concentration is increased to 1000 ppm or more, the apples become very soft, especially at the natural pH. The actions of calcium ion and *sulphur dioxide* appear to be synergistic, the calcium ion allowing a much lower concentration of *sulphur dioxide* to be used. For example, after 9 weeks of storage at 1 °C, apple slices treated with 1000 ppm *sulphur dioxide* alone were unsatisfactory, whilst those treated with only 100 ppm *sulphur dioxide* but with 1000 ppm calcium ion added were still acceptable (Ponting *et al.*, 1972). It should be emphasised, however, that whilst only enzymic browning is likely to be the cause of colour development on a short time scale, the situation which exists after several weeks of storage is by no means clear. Potatoes are frequently dipped in a solution containing sulphur(IV) oxospecies after peeling or cutting up, thereby providing a convenience food with a storage life of a few days. Francis and Amla (1961) found that the amount of *sulphur dioxide* taken up by potatoes depended, not unexpectedly, upon the concentration of the additive in solution and also upon its pH, the amount incorporated into the potato increasing with decreasing pH. The sign of the latter effect is the same as that for uptake by

microbes (Section 5.2) and by apricots (Section 6.2.1.2), but there is no reason to expect similar transport properties in all these cases. For potato the amount taken up is independent of the period of immersion for times of 2–10 min in the case of abrasion peeled potatoes and 10–30 s for hand peeled potatoes. The observed pH dependence renders retention of the additive in the form of a simple surface film an unlikely explanation and it is more probable that the process involves rapid transport of *sulphur dioxide* into the surface layers of cells under a concentration gradient, but further transport into the potato is slow on the time scale of the immersion. Abrasion peeling leads to a greater uptake than hand peeling, probably on account of the larger surface area exposed. The actual level of use of *sulphur dioxide* in any particular situation depends upon the activity of the enzyme and, clearly, no hard and fast rules can be made. Since oxygen is required for polyphenoloxidase activity, the demand on *sulphur dioxide* may be reduced by vacuum packaging. Anderson and Zapsalin (1957) suggest that vacuum packed potato strips require less than one fifth of the amount of *sulphur dioxide* used in air packs, thereby reducing the undesirable effects of tissue-softening by the additive. The mechanism of this effect on plant tissues is not known.

The enzyme, ascorbate oxidase, is present widely in plant tissues and could be responsible for the loss of ascorbic acid in macerated tissue which had not been blanched. Haisman (1974) reviewed some early work on the inactivation of this enzyme by sulphur(IV) oxospecies, the results in question being shown in Table 6.5. Pumpkin extract is a good source of the enzyme and the effectiveness of the additive is clearly demonstrated. Haisman considers that it is possible that the destruction of the vitamin in the other systems is a result of a coupled oxidation involving polyphenoloxidase and phenolic substrate, in which case sulphur dioxide exerts its effect by acting on the polyphenoloxidase.

Haisman (1974) also reports two further instances in which the inactivation of plant enzymes is implied. Delay off-flavours in vined peas have been attributed to the action of lipoxygenase and have been associated with the level of hexanal in the peas. Dipping or blanching in a solution containing approximately 700 ppm *sulphur dioxide* has been shown to eliminate the off-flavour and the appearance of hexanal. It is not known, however, whether this is a result of the inactivation of the enzyme or simply the formation of a hydroxysulphonate adduct between hydrogen sulphite ion and the carbonyl compounds responsible for the off-flavour. The second instance relates to the production of off-flavour in high-temperature–short-time sterilised peas and is attributed to the regeneration

TABLE 6.5
The influence of vegetable extracts and sulphur dioxide
on the decomposition of ascorbic acid at pH 6·0
(Haisman, 1974)

Catalytic agent	Sulphur dioxide content (ppm)	Decomposition rate constant (min^{-1})
Pumpkin extract	0	26
	10	5·5
	100	0·01
	1 000	0·02
	5 000	0·05
Cauliflower extract	0	0·09
	50	0·05
	100	0·04
	1 000	0·03
Potato extract	0	1·15
	1 000	0·10
Potato juice	0	7·3
	1 000	0·26

Reproduced with permission from *J. Sci. Food Agric.*,
1974, **25**, page 806. © D. R. Haisman.

of peroxidase. The use of a solution containing 700–1100 ppm *sulphur dioxide* prevented the formation of off-flavour but had little or no effect on regeneration of peroxidase. It is not known how enzyme activity was measured in this case, but it is presumed that the addition of hydrogen peroxide for the assay removed all the remaining sulphur(IV) oxospecies. It is possible that, when present in the vegetable, the sulphur(IV) oxospecies are removing any hydrogen peroxide, as explained in Section 4.2.5, thereby effectively preventing the action of the enzyme.

6.2.3 Meat and fish products
6.2.3.1 Meat products
The use of *sulphur dioxide* in meat is generally limited to comminuted products in which the additive serves as an antimicrobial agent. When present in amounts of 450 ppm, as is the case in the British fresh sausage, the additive owes its success to the inhibition of Enterobacteriaceae and particularly Salmonellae. The latter are not only more sensitive to *sulphur dioxide* than other Enterobacteriaceae, but do not recover when the

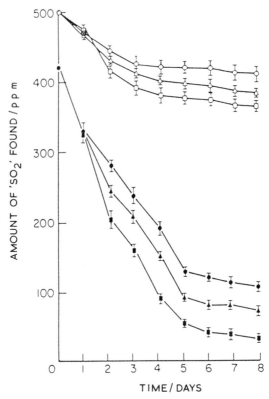

Fig. 6.5. The effect of temperature on the loss of free and total sulphur dioxide from British fresh sausages: ○, total SO_2 at 4°C; △, total SO_2 at 10°C; □, total SO_2 at 22°C; ●, free SO_2 at 4°C; ▲, free SO_2 at 10°C; ■, free SO_2 at 22°C (Banks and Board, 1982b). Reproduced with permission of J. G. Banks from *J. Sci. Food Agric.*, 1982, **33**, page 200. © The authors.

concentration of *sulphur dioxide* falls to below its inhibitory value (Banks and Board, 1982a). Being confined to Gram-positive microflora consisting of Lactobacilli and *Microbacterium thermosphactum* (Section 5.3), the spoilage of sulphited meat tends to be associated with a sour odour, unlike the spoilage of untreated meat which leads to a 'putrid' smell (Christian, 1963; Dyett and Shelley, 1966).

The manufacture of pork sausage with a target of 450 ppm recoverable *sulphur dioxide* requires the addition of some 600 ppm of the additive, since approximately 25 % is lost irreversibly during the process. Since there is no reason to expect reversibly bound additive to exert an antimicrobial effect

(Section 5.3), the extent to which the free additive is retained during storage is significant. The amounts of free and total recoverable *sulphur dioxide* at various storage temperatures are shown as a function of time in Fig. 6.5 (Banks and Board, 1982*b*). The rapid loss of free *sulphur dioxide* compared with the loss of total recoverable additive suggests that free *sulphur dioxide* is being converted to a reversibly bound form. The nature of the binding agent, and whether its source is of meat or microbial origin or is present in the other components of the sausage (cereal, spices) is not known. The nature of the antimicrobial sulphur(IV) oxospecies at around neutral pH is uncertain. It is evident that, at pH 6–7, the concentration of molecular sulphur dioxide is negligible at the concentrations used here and the ionic forms are, therefore, the most likely active species. Irreversible combination of the additive is said to be the result of oxidation (Anderton and Locke, 1956; Hibbert, 1970).

The reaction of sulphur(IV) oxospecies with thiamine is discussed in Section 4.2.3.5. Hermus (1969) has found that the extent of destruction of the vitamin in sulphited meat increases linearly with increasing *sulphur dioxide* content to 1000 ppm, the loss occurring essentially within the first hour after addition. Thus, the addition of 400 ppm and 1000 ppm *sulphur dioxide* depletes thiamine content by 55 % and 95 % respectively. The effect is independent of temperature (at 4 °C and 15 °C), but the presence of *sulphur dioxide* during frying of meat increases losses considerably. Whereas the loss on frying pork is 44 % in the absence of the additive, this is increased to 80 % and 97 % in the presence of 100 ppm and 400 ppm *sulphur dioxide* respectively.

Sulphur(IV) oxospecies have a beneficial effect on meat colour. Krol and Moerman (1959) demonstrated that the additive can preserve the colour of meat, probably by inhibition of the oxidation of myoglobin to metmyoglobin. It does not, however, restore the colour of discoloured meat.

6.2.3.2 *Fish products*
Sulphur dioxide may be used to preserve shrimps and lobsters against discoloration during iced storage. The treatment of shrimps for 10 min with solutions of increasing concentrations of sodium sulphite, 0, 1, 3, 5 %, leads to the formation of increasing amounts of dimethylamine, 1·8, 6, 10, 34 ppm respectively (Yoshida, 1981). A possible source of this compound is the reaction of sulphur dioxide with trimethylamine oxide, which is abundant to the extent of 600–800 ppm in the sulphited shrimps. Spinelly and Koury (1979) demonstrated the feasibility of this reaction in model

systems at pH 5·6, and the mechanism suggested in earlier work by Lecher and Hardy (1948) is as follows:

$$(CH_3)_3NO + SO_2 \rightarrow (CH_3)_3NOSO_2$$

$$(CH_3)_3NOSO_2 \rightarrow (CH_3)_3N^{2+} + SO_3^{2-}$$

$$(CH_3)_3N^{2+} \rightarrow CH_2\!\!=\!\!\overset{+}{N}(CH_3)_2 + H^+$$

$$CH_2\!\!=\!\!\overset{+}{N}(CH_3)_2 \rightarrow CH_2O + (CH_3)_2\overset{+}{N}H_2$$

with formaldehyde as the by-product. This reaction does not lead to loss of sulphur dioxide although, in practice, some sulphate ion is also formed through an alternate intermediate, $(CH_3)_3NSO_3$.

6.2.4 Addition to wheat flour

Sodium disulphite is added during the preparation of wheat flour doughs intended for biscuit manufacture. The principal technical advantages of the use of the additive are:

1.   It greatly reduces the time required to mix a batch of dough.
2.   The elasticity of the dough is reduced, thereby eliminating the need for a standing time to allow stress relaxation to occur and facilitating production of a satisfactory dough sheet in modern continuous biscuit plants.
3.   The additive reduces the variation in product attributable to variations in the flour used (Wade, 1974).

Although it is possible to make measurements of the viscoelastic behaviour of dough in relation to the amount of sulphur(IV) oxoanion incorporated, the interpretation of such results is not yet sufficiently refined to provide unambiguous information on the effect of the additive on the molecular structure of dough components. For this reason, the discussion here is confined to a consideration of empirical observations which provide adequate indication of the sign of specific changes in physical characteristics.

In bakery practice, the dough is mixed and sheeted using fixed operating procedures. The sheet is finally passed through two power driven gauge rollers. Owing to the elastic properties of the dough, the thickness of the final dough sheet is greater than the gap between the rollers. If the gap is maintained constant, then the thickness of the dough will be a function (though complex) of the viscoelastic behaviour of the dough. Biscuits are formed by stamping oval shapes from the final dough sheet. The weight of

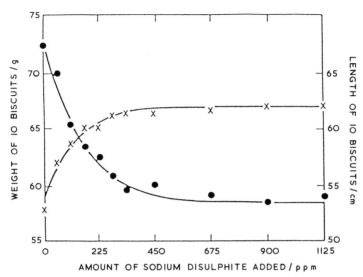

Fig. 6.6. Effect of the amount of sodium disulphite added to dough on the weight and length of a fixed number of biscuits: ●, weight; ×, length (Wade, 1972). Reproduced with permission from *J. Sci. Food Agric.*, 1972, **23**, page 334. © P. Wade.

the biscuits will, therefore, also be a function of the viscoelastic behaviour of the dough. After stamping, the oval shapes of dough shrink anisotropically to circular shapes, owing to tension in the sheet. The final size of the biscuit is, therefore, also a function of the viscoelastic properties of the sheet (Wade, 1971). The effect of changing the amount of sodium disulphite added to the dough on the weight and size of biscuits is shown in Fig. 6.6 (Wade, 1972). Doughs containing no additive are hard, tough and rubbery. The presence of only 56 ppm sodium disulphite renders the dough markedly more extensible while still firm and elastic. Further increase in the level of treatment increases the softening and reduces the elasticity until optimum processing conditions are attained for an addition of 225 ppm. When 1125 ppm sodium disulphite is added, the dough rapidly loses cohesion after removal from the mixer.

The modification of the rheological behaviour of dough after treatment with sodium disulphite is attributed to its cleavage of disulphide bonds in the flour. Indeed, in their study of the fate of sulphur(IV) oxospecies in biscuit doughs, Thewlis and Wade (1974) found that 93 % of the sulphur added to the dough remained in the biscuit. This is made up of some 30 % sulphate ion and 63 % organo-sulphur compounds. The evidence is consistent with the latter being protein-S-sulphonates. Less than 0·2% of the sulphur remains as sulphur(IV) oxospecies in the biscuit. Using the known disulphide content of the flour, Wade (1972) estimated that, under optimum processing conditions, some 25% of the disulphide groups present in the flour are cleaved; the elastic properties are essentially destroyed when sufficient sodium disulphite is added to cleave approximately 50% of the bonds.

Stevens (1966, 1973) found that the susceptibility of disulphide groups in flour and in a protein extract (gliadin) from flour to cleavage by sulphite ion depended upon the presence of a denaturing agent (urea). In the case of flour, approximately 50% of the groups which reacted in the presence of 8 M urea were found to react in the absence of urea. The disulphide bonds which are cleaved in the absence of urea were regarded as inter-chain links, whilst those which require denaturation of the protein for reaction with sulphite ion were regarded as intra-chain. A simple test with sulphite ion may, therefore, distinguish these two types of disulphide group in wheat flour.

6.2.5 Use in beverages

6.2.5.1 Non-fermented beverages

Sulphur(IV) oxospecies are used to prevent enzymic and non-enzymic browning in fruit juices. Browning is, however, not always detrimental: the characteristic golden yellow colour of apple juice is a result of polyphenoloxidase activity during the milling and pressing of the fruit, whilst the colour of white wine results mainly from enzymic oxidation of catechins (Burroughs, 1974).

Non-enzymic browning is important in pasteurised, lightly coloured juices such as lemon juice. The pH of the latter is 2·5 and, although it may contain nitrogen (0·07%, mainly as amino acids) and sugars (3%), Maillard browning is not significant, probably as a result of the low pH (Section 3.2). Instead, the main cause of browning is the degradation of ascorbic acid; lemon juice may be fortified (0·1%) with respect to this vitamin (Clegg, 1964). The browning of lemon juice is complicated by the presence of other components. Glucose and fructose could exert a synergistic effect on ascorbic acid browning, and citric acid is clearly a promoter of colour production (Clegg, 1966). Although the browning of lemon juice is successfully controlled by the addition of 350 ppm *sulphur dioxide* which is, therefore, presumed to interrupt ascorbic acid browning as explained in Section 3.3.5, Wedzicha and McWeeny (1975) could find no significant amounts of 3,4-dideoxy-4-sulpho osuloses, indicative of this inhibition, in commercial, fortified, sulphited lemon juice.

6.2.5.2 *Fermented beverages*

Sulphur(IV) oxospecies are added to grape musts to prevent polyphenol-oxidase mediated browning and to prevent the growth of acetic and lactic acid bacteria, so that the fermentation of wine is dominated by a desirable strain of yeast. In red wines a small amount may be added to assist in the extraction of the colour from the grape skins and to the finished wine to stabilise the colour (Burroughs, 1974). In storage of wine, the additive serves as an antimicrobial agent, an antioxidant, a trapping agent for acetaldehyde and an inhibitor of non-enzymic browning. Sulphur dioxide is added in the manufacture of fermented apple juice to control similarly the microflora in fermentation and to act as a preservative on storage (Beech and Carr, 1977).

A significant reaction of sulphur(IV) oxospecies with components of fermented beverages is that with carbonyl components. This reaction is of practical significance, since the antimicrobial and antioxidant action of the additive is related to the amount of the free additive which is present. It is possible to account quantitatively for this binding using measured dissociation constants for the specific hydroxysulphonate adducts and the concentrations of the respective carbonyl compounds in solution. This is demonstrated by Burroughs and Sparks (1973a, b, c) who took the principal *sulphur dioxide* binding components, with respective hydroxy-sulphonate dissociation constants at pH 3·8 and 20 °C, to be the following: D-*threo*-2,5-hexodiulose ($3·3 \times 10^{-4}$ M), L-xylosone ($1·4 \times 10^{-3}$ M), acetaldehyde ($1·4 \times 10^{-6}$ M), galacturonic acid ($2·0 \times 10^{-2}$ M), pyruvic acid ($2·0 \times 10^{-4}$ M), 2,5-diketogluconic acid ($4·3 \times 10^{-4}$ M) and 2-ketoglutaric acid ($6·6 \times 10^{-4}$ M). The amount of each varies with the quality of the grapes and, particularly, with the extent to which grapes have been affected by moulds, there being, in general, higher levels of carbonyl components if such microbial growth has taken place. The high quality of predictions obtained by these workers is demonstrated for two wines in Fig. 6.7. That made chiefly from the cultivar Müller-Thurgau used grapes which were in fairly good condition and contained considerably smaller amounts of carbonyl components than the other wine, prepared from the cultivar Madeleine-Sylvaner, which had been badly affected by mould at the time of harvest. An examination of the changes in the concentration of individual carbonylic components as the amount of free *sulphur dioxide* is increased shows that, even at the lowest concentration of free additive used for calculation (6·4 ppm), essentially all the acetaldehyde present is in the form of hydroxysulphonate on account of its very low dissociation constant. The increase in reversible binding of *sulphur dioxide* with increasing

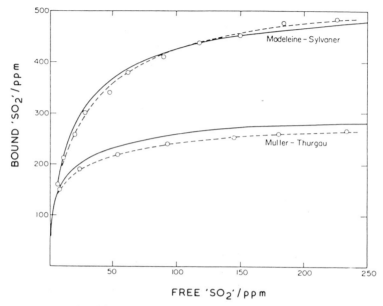

Fig. 6.7. Reversible binding of sulphur dioxide in wines: ———, calculated values; --O--O--, experimental data (Burroughs and Sparks, 1973c). Reproduced with permission of L. F. Burroughs from *J. Sci. Food Agric.*, 1973, **24**, page 216. © The authors.

concentration results, therefore, from the progressive formation of hydroxysulphonates with the other carbonyl components.

Whereas in wine and cider making *sulphur dioxide* plays a vital role in restricting the growth of unwanted micro-organisms, the importance of the additive in brewing is, possibly, not so great. Addition during malting increases the hot water extractability of the malt. Sulphur dioxide gas is frequently applied in the drying of hops, causing them to lose their green colour and assume a more uniform bright yellow tinge, which is rated highly in hand evaluation. This treatment also prevents formation of 'raw' aromas in the hops, but reduces their bittering potential. Most of the sulphur dioxide found in fermented beer arises from the action of yeast, and perhaps the best defined preservative action on beer is that of an antioxidant (Hough *et al.*, 1971). The presence of trace amounts of *N*-nitrosodimethylamine in certain beers is a well known fact which has been attributed to the use of direct-fire heating in the kilning of malt. It is presumed that oxides of nitrogen present in the drying air nitrosate the amines in the malt. A significant observation is that the amount of *N*-nitroso compound may be

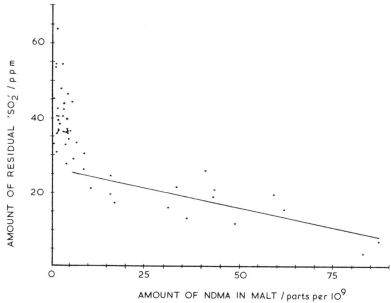

Fig. 6.8. Relationship between residual sulphur dioxide and the *N*-nitrosodimethylamine content of production malts (Lukes *et al.*, 1980). Reproduced with permission from *J. Am. Soc. Brew. Chem.*, 1980, **38**, page 147. © 1980 American Society of Brewing Chemists.

reduced by the addition of sulphur dioxide to the drying air (O'Brien *et al.*, 1980), the relationship between residual *sulphur dioxide* in the malt and its *N*-nitroso content being shown in Fig. 6.8 (Lukes *et al.*, 1980). The mechanism of this effect is not understood. An interesting observation concerning the effect of addition of *sulphur dioxide* during the beer making process on the quality of the final product is the relationship between foam quality and the presence of the additive at the stage of wort boiling. The presence of approximately 25 ppm *sulphur dioxide* in the kettle results in a substantial loss of foam properties of both wort and beer. It is believed that the foam destruction results from the cleavage of disulphide groups in the wort proteins which, in turn, reduces the quality of higher molecular weight proteinaceous materials believed to be necessary for sustaining good foam (Brenner and Bernstein, 1975).

6.2.6 Other uses

6.2.6.1 Sugar manufacture

The addition of *sulphur dioxide* is an essential part of the processes of manufacture of sugar from beets and from cane. In beet sugar manufacture, the sucrose is extracted from sliced beet by counter-current diffusion and a process involving treatment with lime and carbon dioxide is used to remove the protein, pectin and some organic acids. *Sulphur dioxide* is added to this clarified extract before evaporation, and sometimes additional *sulphur dioxide* is introduced before crystallisation (Oldfield, 1974). The additive has a slight bleaching effect on preformed colour, but when present in amounts of 250–500 ppm its function is to prevent colour formation in the evaporation and crystallisation stages. The pH of a concentrated syrup is typically 7·8 and may contain of the order of 0·3 % reducing sugar. Heating of such a syrup to 85 °C results in colour formation, presumably due to alkaline degradation of the reducing sugar component. Under these conditions the effect of low concentrations of *sulphur dioxide* (50 ppm) is to delay the start of colour formation but, once started, the colour development is more rapid than in mixtures to which no *sulphur dioxide* has been added (compare with the mechanisms at pH 5·5 discussed in Section 3.3.4) and after 12 h there is no difference between the sample containing the additive and one to which none had been added. On the other hand, the presence of 300 ppm *sulphur dioxide* or more leads to optimum inhibition of colour.

An interesting observation is the tendency for the production of imidodisulphonate ion after the addition of *sulphur dioxide*. This compound is the result of reaction between nitrite ion and sulphite ion and is described in Section 1.10.6. While no nitrite is found in beets, amounts of 2–45 ppm nitrite may occur in the extract. Since diffusion extraction at a temperature high enough to suppress microbial activity produces nitrite-free extracts, the nitrite is formed by the reduction of nitrate by bacteria; the organism is believed to be a strain of *Bacillus stearothermophilus* (Carruthers *et al.*, 1958). Potassium imidodisulphonate is able to crystallise out with sucrose increasing the ash content and it is reported that in one instance the effect reached such proportions that the process had to be shut down (McGinnis, 1971). An indicator of the possible presence of nitrite ion is an unexpectedly low level of *sulphur dioxide* in the sucrose solution.

The uses of *sulphur dioxide* in cane sugar manufacture have been considered in detail by Honig (1953). The manufacturing process involves the milling of cane and extraction with water followed by treatment with lime and carbon dioxide or sulphur dioxide. The clarified extract is evaporated and crystallised. The functions of the additive are summarised as follows:

1. The formation of insoluble complexes with dissolved silica and metal oxides.
2. Removal, as insoluble calcium sulphite, of calcium ion added as lime. The precipitate has a coarse structure and acts as a filter-aid.
3. Reaction with calcium salts of organic acids to form insoluble calcium sulphite and free acid.
4. Inhibition of the oxidation of polyphenols.
5. Inhibition of non-enzymic browning reactions.

6.2.6.2 Pectin manufacture

The first step in the manufacture of pectin is acid hydrolysis of protopectin in the raw material. This is conveniently carried out using hot dilute aqueous sulphur dioxide at pH 1·8–2·7. The use of sulphur dioxide as an acidulant has the advantage over other acids that it can be conveniently removed by aeration, thereby avoiding the formation of salts by neutralisation with alkali. Subsequent steps in production involve concentration of the extract and precipitation (Joseph and Havighorst, 1952).

7

Toxicology

The known reactivity of sulphur(IV) oxospecies with components of biological systems, including structural components, enzymes and nucleic acids *in vitro* and in a limited number of cases *in vivo*, has led to concern over the uses of this substance as a food additive, despite a long history of use. There has also been an awareness of potential hazards associated with exposure to atmospheric pollution by sulphur dioxide and much epidemiological evidence links this pollutant with mortality and several respiratory diseases. The toxicity of inhaled sulphur dioxide is concerned mainly with its effects on the tissues of the respiratory system and will not be considered here, where the emphasis is on ingested sulphur(IV) oxospecies. However, inhaled sulphur dioxide is absorbed and will contribute to the overall body burden.

The toxicology of sulphur(IV) oxospecies has been reviewed by Gunnison (1981), Shapiro (1977), by the Institute of Food Technologists' Expert Panel on Food Safety and Nutrition and the Committee on Public Information (Institute of Food Technologists and Committee on Public Information, 1976), and by the Scientific Committee for Food (1981). All these reports consider the toxicology of free sulphur(IV) oxospecies, there being essentially no reported work on the consequences of products of reactions of sulphur(IV) oxospecies with food components, at the time of publication of the reports. It is evident from the data presented in Chapter 6 that a consequence of the use of sulphur(IV) oxospecies in food systems is reaction with food components to form products in which the additive is combined reversibly and irreversibly, the latter case accounting

312

often for more than 50 % of the added sulphur(IV) oxospecies. Some toxicological data on such products are now available.

7.1 BODY BURDEN

The body burden with respect to sulphur(IV) oxoanion is the result of ingestion, inhalation and the presence of metabolic pathways which lead to the formation of sulphite ion as an intermediate.

The level of ingested sulphur(IV) oxospecies may be deduced in various ways (Quattrucci, 1981):

(i)   balance studies which monitor food production and the import and export of food
(ii) shopping basket surveys which monitor only what is purchased
(iii) consumption surveys of selected groups of the population
(iv) trade information of the total production and imports or total sales.

As expected, these surveys reveal wide variations in ingestion of sulphur(IV) oxospecies, even within a given population, since the consumption of sulphited foods is subject to personal preferences. Larger differences, reflecting eating habits, are apparent when populations are compared and, hence, the need for legislation appropriate to each country concerning maximum permitted levels in particular foods, in order to maintain an acceptable overall intake. Any assessment of intake is hindered by losses of the additive during food preparation; for example, a large proportion of the sulphur(IV) oxospecies in dehydrated vegetables is lost in the water used for reconstitution.

Bigwood (1973) gives the daily intake of sulphur(IV) oxospecies in Belgium as 0·08–1·1 mmol *per capita*, which is expected to be representative of consumption in other developed countries, although the amount consumed by an individual could exceed this amount by as much as an order of magnitude (Institute of Food Technologists and Committee on Public Information, 1976). The Scientific Committee for Food (1981) gives the estimates *per capita* for daily consumption for Europe as 0·25 mmol from food and non-alcoholic beverages, and 0·63 mmol and 1·41 mmol if additional quantities of 300 ml and 700 ml, respectively, of wine are consumed. In the US the usual assessment for daily intake from solid foods and non-alcoholic beverages is in the region of 0·03 mmol *per capita*.

The amount of sulphur dioxide inhaled is also subject to considerable uncertainty and depends upon the concentration of the gas and also the sex and level of activity of the exposed person. Shapiro (1977) quotes that, at a sulphur dioxide concentration of 0·02 ppm, a male engaged in light work would inhale 0·033 mmol of the gas in 24 h, whilst an industrial worker exposed to an environment containing 4 ppm sulphur dioxide could inhale 2·2 mmol after 8 h of light work. At normal levels of exposure to sulphur dioxide (less than 0·1 ppm, long term average), the amount of gas inhaled will be small compared with the amount ingested as sulphur(IV) oxospecies in food.

The turnover of endogenous sulphur(IV) oxoanion produced as a result of the catabolism of sulphur-containing amino acids is greater than that from exogenous sources. The average amount of sulphate normally excreted daily is approximately 25 mmol *per capita* and, if it is assumed that all this sulphate is derived from the oxidation of sulphur(IV) oxoanion, then this value represents the daily turnover of the latter. The result puts into perspective the amount of exogenous sulphur(IV) oxoanion, the distribution of which will, of course, be different from that produced in normal metabolism.

7.2 POTENTIAL TOXIC EFFECTS

The most sensitive biological components with respect to sulphur(IV) oxoanion are enzymes, and the principles underlying their inactivation are considered in detail in Chapter 4. In addition to the possible inhibition of NAD^+- and flavin-dependent enzymes and cytochrome oxidase, and the activation of α-glucan phosphorylase, Gunnison (1981) cites the action of sulphur(IV) oxoanion on the activity of 2,3-diphosphoglyceric acid phosphatase and the sulphatase group of enzymes. The former is thought to be the physiological phosphatase catalysing the conversion:

2,3-diphosphoglyceric acid → 3-phosphoglyceric acid + phosphate

and has the unusual property of increased activity in the presence of sulphur(IV) oxoanion. There are relatively large amounts of 2,3-diphosphoglyceric acid in human erythrocytes, where it is of importance in the regulation of oxygen affinity for haemoglobin. Therefore, any factor which affects the concentration of 2,3-diphosphoglyceric acid may be of importance. Manyai and Varady (1958) found that human erythrocytes could be selectively depleted of the compound by the action of sulphur(IV) oxoanion, leaving ATP content and glycolysis intact; the concentration required to halve the 2,3-diphosphoglyceric acid content of erythrocytes after 4 h of exposure to sulphur(IV) oxoanion is approximately 4 mM (Parker, 1969). The effect of sulphur(IV) oxoanion on the phosphatase in question was investigated by Harkness and Roth (1969). At the lowest concentration of sulphur(IV) oxoanion used (2·5 mM) and at pH 7·8 and 37 °C, incubation of the enzyme for 1 h resulted in a fifteen-fold increase in activity. Aryl sulphatases are known to catalyse the hydrolysis of the ester bond in various aromatic sulphates, but more recently it has been suggested that the physiological substrates for the sulphatases include lipids containing galactosyl-3-sulphate residues (such as cerebroside sulphate), mucopolysaccharides containing N-acetylgalactoseamine-4-sulphate residues (such as dermatan sulphate), heparin sulphate, chondroitin sulphates and steroid sulphates. The reactions and properties of the aryl sulphatases are reviewed by Roy (1976) and it is evident that sulphur(IV) oxoanions strongly inhibit these enzymes. The K_i value for sulphite ion towards human liver aryl sulphatase A is in the region of 1 μM (James, 1979).

The possible enhancement of adenosine triphosphatase activity by sulphur(IV) oxoanions has been considered in Section 5.3 in relation to the inhibition of microbial growth.

The risk of mutagenic effects in animal tissues containing sulphur(IV) oxoanions was discussed by Shapiro (1977). One possible cause of mutation is the transformation in DNA of cytosine–guanine pairs to adenine–thymine pairs by deamination of cytosine (Section 1.11.8), which is the main cause of mutation in micro-organisms (Section 5.5). Using the value of the rate constant for the deamination of cytosine, extrapolated to pH 7·4 and 37 °C, and converting this to the number of transformations per generation, the estimated concentration of hydrogen sulphite ion which would double the spontaneous mutation rate is of the order of 3×10^{-7} M. The calculation of this estimate was carried out assuming that the reaction of cytosine in chromosomes proceeds at the same rate as for single-stranded DNA. Since the deamination reaction *in vivo* could proceed more slowly than that observed for single-stranded DNA *in vitro*, the estimate given above may be low, but the potential for mutagenicity is clearly illustrated. Other reactions which interfere with the pairing of bases in DNA include the reversible formation of adducts between sulphur(IV) oxoanion and thymine or uracil (Section 1.11.8) and the transamination of cytosine described in Section 5.4 (Shapiro, 1977). The extent to which free radical reactions are involved in the damaging of genetic material is unknown. Sulphite mediated oxidation and free radical sulphonation reactions are described in connection with lipid oxidation in Section 3.3.6 and other unsaturated food components in Chapter 6. The possibility of such reactions within cells depends upon autoxidation or other free radical mediated oxidations of sulphur(IV) oxoanion. The presence of numerous radical inhibitors would tend to disfavour such reactions. However, it is known that a portion of the sulphur(IV) oxoanion which binds to RNA from *Escherichia coli*, *in vivo*, is not released by alkali, and Shapiro suggests that this binding may be the result of free radical addition to 4-thiouridine, a known component of bacterial transfer RNA, which is also known to undergo such an irreversible addition, as follows:

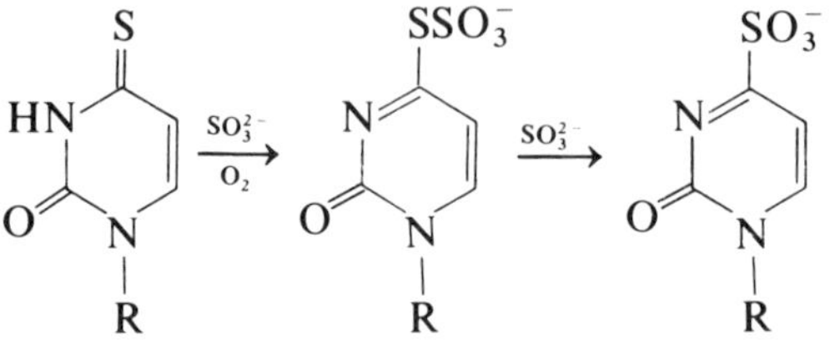

The existence of numerous proteins with disulphide groups favours the possibility of disulphide cleavage. The mechanism of this reaction is explained in Section 1.11.9 and it is expected that at physiological pH the position of equilibrium will lie well to the side of the cleaved protein. The formation of such sulphitolysis products is considered in Section 7.4.

7.3   SULPHITE OXIDASE

It is known that, when rats ingest ^{35}S-labelled sulphite ion, 70–95 % of the radioactivity is absorbed through the intestines and excreted as sulphate, with no detectable sulphite in the urine (Gibson and Strong, 1973). This observation is attributed to the presence of the enzyme, sulphite oxidase, which is responsible for the final step in the catabolism of sulphur-containing amino acids. The enzyme has been assayed in a number of animal tissues; in rats the greatest activity is found in the liver, kidney containing approximately half the activity, and less is found in the terminal intestinal ileum and heart and least in the lung (Cohen *et al.*, 1973). The importance of this enzyme as a means of detoxifying sulphite ion in humans is illustrated by the only reported case of sulphite oxidase deficiency. The patient described in this report was a $2\frac{1}{2}$ year old boy who showed signs of extreme brain damage, mental retardation and dislocated ocular lenses.

The main chemical disturbance included urinary excretion of S-sulpho-L-cysteine, sulphite ion and thiosulphate ion, whilst the excretion of sulphate ion was markedly lower than in normal controls. The amount of sulphate ion excreted was also unaffected by the administration of L-cysteine (Irreverre *et al.*, 1967).

The properties and mode of action of sulphite oxidase are described in Section 4.3.2. The discovery of the presence of molybdenum as an essential prosthetic group in sulphite oxidase represented a breakthrough in the study of this enzyme in animals since tungsten, which is a competitive antagonist of molybdenum, can be used to produce experimental sulphite oxidase deficiency. The large surplus of sulphite oxidase activity in rats is illustrated by the observation that animals whose enzyme activity is less than 3% of that of controls remain healthy, but are more susceptible to sulphur(IV) oxoanion toxicity. The median lethal dose for intraperitoneal sulphur(IV) oxoanion injection was found to be $181\,mg\ NaHSO_3$ per kilogram for animals deficient in sulphite oxidase compared with $473\,mg$ per kilogram for normal rats (Cohen *et al.*, 1973). The symptoms associated with toxicity suggest that the target of sulphite may be the central nervous system. A feature of the systemic sulphite exposure of normal and sulphite oxidase deficient rats is the markedly different level of tissue S-sulphonates produced by sulphitolysis in the two types. In general, relatively low intakes of exogenous sulphur(IV) oxoanion (0–$3\cdot5\,mmol$ $kg^{-1}\,day^{-1}$) by enzyme deficient rats produced the same systemic sulphite exposure as in normal rats after ingestion of heavily sulphited diets (13–$25\,mmol\ kg^{-1}\,day^{-1}$). Thus, there is evidence that sulphite oxidase deficiency is associated with greater tissue protein damage, presumably as a result of the presence of free sulphur(IV) oxoanion in the circulatory system (Gunnison *et al.*, 1981).

The administration of certain drugs and toxic agents causes the increase in concentration of enzymes responsible for their metabolism. No increase in sulphite oxidase activity may be detected when rats are exposed to sulphur(IV) oxoanions.

The apparent capacity of the sulphite oxidase system in mammals is high compared with the endogenous load, and the system seems to be well adapted to large variations in exogenous sulphur(IV) oxospecies. Indeed, Cohen and coworkers (1973) estimate that, on the basis of the amount of sulphite oxidase present in a rat, the capacity for oxidation is around $0\cdot75\,mol\ kg^{-1}\,day^{-1}$. It is possible, however, that, owing to large species variations, this high figure may be misleading in connection with toxicity in man. Gunnison and coworkers (1977) measured the rate of clearance of injected sulphite in rats, rabbits and rhesus monkeys in order to estimate the *in vivo* sulphite oxidase and found that the order of clearance was rat > rabbit > rhesus monkey, with relative rates of $1:0·34:0·20$ respectively. *In vitro* measurements of the sulphite oxidase activity for these three species indicated the following order: rat > rhesus monkey > rabbit, with relative rates of $1:0·12:0·05$ respectively. It seems, therefore, that factors other than the absolute amount of sulphite oxidase are important in the function of the enzyme. Irrespective of the reasons for the different orders of reactivity, the rate of metabolism of sulphur(IV) oxoanion by the rat is clearly far greater than that of the other two species and is, presumably, also greater, if metabolism in the rhesus monkey is a valid model, than that in man. The kinetic studies of Gunnison and coworkers (1977) show that the elimination of sulphur(IV) oxoanion is of first order, with a half-life in the region of 1 min in the case of the rat.

7.4   OBSERVED TOXICITY

The actual toxicity of sulphur(IV) oxoanions will depend upon the relative rates of potential undesirable reactions, described in Section 7.2, and that of the detoxifying mechanism. If the *in vivo* oxidation of sulphur(IV) oxoanion takes place in the organs in which sulphite oxidase activity is greatest, that is in the liver and kidneys, absorption and transport of the ions will be necessary and the components of blood and of blood vessel walls are, therefore, likely species for interaction with sulphur(IV) oxoanions. The absorption of the oxoanions from the digestive tract is demonstrated by a dose-dependent formation of S-sulphonates in blood plasma (Gunnison and Palmes, 1978) and with the walls of the aorta (Gunnison *et al.*, 1981). Plasma S-sulphonates are formed from sulphur(IV) oxoanion in high yield (Gunnison and Benton, 1971) and are metabolised only slowly, the half-life for their elimination being of the order of days. The amount of S-sulphonate formed depends upon the rate at which the sulphur(IV) oxoanion is absorbed and the rate at which it is oxidised; the latter point is illustrated by elevated S-sulphonate levels in both plasma and aorta in sulphite oxidase deficient rats (Gunnison *et al.*, 1981). The availability of disulphide-containing proteins must also be a primary factor in S-sulphonate formation, the proteins of the aorta being particularly susceptible in this respect (Gunnison and Farrugella, 1979). The factors controlling S-sulphonate formation lead also to variations in plasma S-sulphonate formation in different species of test animal

(Gunnison and Palmes, 1978). The rat, for example, shows very little exogenous plasma S-sulphonate compared with rabbit and rhesus monkey, and it is suggested that an inverse relationship exists between cleavage of disulphide bonds by exogenous sulphur(IV) oxoanion and sulphite oxidase activity. However, although a significant level of endogenous plasma S-sulphonate is present in some animals, for example the rat, these levels do not show any relationship to sulphite oxidase activity.

Gunnison and coworkers (1977) have demonstrated that, in the normal rat, the sulphite oxidase activity is rather high, impairing the suitability of this species as a test animal upon which a recommended acceptable daily intake (ADI) of sulphur(IV) oxoanion for man may be based. They consider that the sulphite oxidase deficient rat, produced artificially by the administration of tungsten, which allows higher levels of systemic doses, is a better test animal. High levels of exposure to sulphur(IV) oxoanions lead to toxicity as a result of decreased food consumption, reactions with constituents of the diet and irritation of the gut. Exposure more closely allied to that of the human population is achieved by conditions of low sulphite oxidase level and low exogenous sulphur(IV) oxoanion concentration (Gunnison et al., 1981). However, much of the reported work involves the use of rats.

The most widely recognised adverse effect of sulphur(IV) oxoanion is that upon the thiamine content of diets. Bhagat and Lockett (1964), for example, found that, when a standard diet containing thiamine (approximately $140 \mu g/100 g$) was stored with added sodium disulphite $(0.6\% w/w)$ at room temperature, half of the thiamine was lost after 5–6 days. The toxicity to rats of fresh sulphited diets is related to this loss of thiamine, and the toxic effects may be prevented or cured by the administration of thiamine or cocarboxylase. The deficiency of thiamine results only from destruction of the vitamin in the diet, since Gunnison and coworkers (1981) conclude from their experiments on sulphite oxidase deficient rats that there is no evidence of systemic destruction of thiamine, even under conditions of extreme systemic sulphur(IV) oxoanion exposure. In addition to thiamine deficiency, Bhagat and Lockett (1964) found that the storage of diets at room temperature led to the slow development of a different form of toxicity, which was not apparent at 31 days but readily detectable at 75 days and which could not be reversed by administration of thiamine. A detailed study of the toxicity of stored sulphited diets was carried out by Til and coworkers (1972a, b). They overcame the problem of thiamine deficiency by drastic supplementation of sulphited diets with the vitamin and report the level at which there was no apparent toxic effect as $1·13$ mmol sulphur(IV) oxospecies per kilogram of body weight per day for rats, and $1·44$ mmol kg^{-1} day^{-1} for pigs. The most sensitive criteria for damage by sulphur(IV) oxospecies in rats were the presence of occult blood in the faeces and changes in gastric morphology.

The requirement for the allocation of an unconditional ADI to substances is the availability of adequate short term and long term toxicological data, or information on the biochemistry and metabolic fate of the compound, or both (Bigwood, 1973). The toxicity data reported by Til and coworkers (1972a) fulfil this requirement and the 'no effect' level in animal studies is converted to an ADI by introducing a safety factor of 100, that is the ADI is $0·011$ mmol sulphur(IV) oxoanion kg^{-1} day^{-1}, which is the level recommended by the United Nations FAO/WHO. The Scientific Committee for Food (1981) noted the relevance of toxicity data published by Til and coworkers (1972b) for pigs in view of the greater relevance of this species to man.

There are very few reported instances of adverse effects on man of sulphur(IV) oxoanions consumed in foods, the most notable being a systemic allergic reaction evident in a patient a few minutes after eating a meal treated with the additive. The patient showed both positive skin reaction and positive systematic response to oral doses of sulphur(IV) oxoanion and the symptoms, which were itching of ears and legs, nausea, warmness, coughing and tightness in the throat were terminated immediately after injection of epinephrine. The formation of specific antibodies to sulphur(IV) oxoanions was suggested as the cause of this allergic reaction (Prenner and Stevens, 1976).

It is known that sulphur(IV) oxospecies present in animal diets and in human food react with the components of those diets to form products in which the additive is reversibly and irreversibly combined (Chapter 6). Gibson and Strong (1976) used the hydroxysulphonate of acetaldehyde as a model of reversibly bound sulphur(IV) oxoanion and administered this to rats in their drinking water as a single dose equivalent to $0·78$ mmol sulphur(IV) oxoanion per kilogram. After 48 h, $72–85\%$ of the sulphur had been excreted in the urine and $10–15\%$ in the faeces. Elimination took place at virtually the same rate as ingested free sulphur(IV) oxoanion, and less than $1·5\%$ of the exogenous sulphur remained in the body after two weeks. The test animals showed no obvious signs of toxicity and, presumably, the hydroxysulphonate adducts are decomposed in the alkaline environment of the gut, the released sulphur(IV) oxoanion being metabolised. The report of the development of a toxic factor in diets stored for long periods of time (Bhagat and Lockett, 1964) emphasises the need for tests to be carried out on the sulphonates which contribute to the irreversible combination of sulphur(IV) oxoanion in foods. The LD_{50} for acute toxicity of 3,4-dideoxy-4-sulphohexosulose in rats and mice exceeds 5 g per kilogram body weight (Walker et al., 1983a). Metabolic studies (Walker et al., 1983b) indicate that this compound is only partly absorbed from the gastrointestinal tracts of rats and mice, 50–60 % remaining in the faeces. Absorption probably occurs in the stomach and upper small intestine, but that which is absorbed is excreted rapidly in urine with no evidence of the degradation of the carbon chain to products which can enter the pathways of intermediary metabolism, or of the release of sulphur into the body sulphur pool. The only organ with an appreciable concentration of 3,4-dideoxy-4-sulphohexosulose after dosing is the pancreas, but, overall, the picture is one of a metabolically inert compound which is rapidly cleared by the kidney. Investigation for mutagenicity against four strains of Salmonella typhimurium using the Ames test showed negative results (Walker et al., 1983a).

7.5 MUTAGENICITY

The possible mutagenic action of sulphur(IV) oxoanion is fertile ground for speculation but there has been no unequivocal evidence of genetic damage resulting from the presence of the oxoanion to cells in vivo in normal animals. Combinations of sulphur(IV) oxoanions and other mutagens such as benzo-α-pyrene (Laskin et al., 1976) or ultraviolet light (Mallon and Rossman, 1981) show that the oxoanion could act as a co-mutagen in rats and in Chinese hamster cells respectively, but that, alone, it is not mutagenic. Gunnison and coworkers (1981) suggested tentatively that sulphur(IV) oxoanion may be the cause of mammary adenocarcinoma observed at a very low incidence rate in young sulphite oxidase deficient female rats, although no such observations had been made previously in studies of chronic sulphur(IV) oxoanion toxicity. The systemic sulphur(IV) oxoanion levels in other studies were, however, lower than those in sulphite oxidase deficient rats used by Gunnison and coworkers (1981) and the lack of previous observations does not rule out, therefore, the reported possible carcinogenicity, but the authors themselves, suggest that further work is necessary to assess the possible risks in this respect.

7.6 CONCLUSION

The unconditional ADI for sulphur(IV) oxoanion *per capita* is in the region of $0.5–0.8$ mmol day^{-1}, which is well above the daily consumption in food and non-alcoholic beverages but is comparable to the amount of additive consumed if, in addition, 300 ml wine is drunk daily. It is likely, therefore, that this ADI is frequently exceeded, particularly by groups of the population who consume large amounts of wine. The toxicological and metabolic data are very favourable towards the continued use of sulphur(IV) oxoanion as a food additive, but the consequences of exceeding the ADI by large amounts are unclear. It is evident, however, that the use of sulphur(IV) oxoanions is unacceptable in foods which act as important sources of thiamine.

References

Abel, E. (1951). Theory of the oxidation of sulphite to sulphate by oxygen. *Monatsh.*, **82**, 815–34.

Abrams, A., Lowy, P. H. and Borsook, H. (1955). Preparation of 1-amino-1-deoxy-2-ketohexoses from aldohexoses and α-amino acids. I. *J. Amer. Chem. Soc.*, **77**, 4794–6.

Adams, C. A., Warnes, G. M. and Nicholas, D. J. D. (1971). A sulphite dependent nitrate reductase from *Thiobacillus denitrificans*. *Biochim. Biophys. Acta*, **235**, 398–406.

Adams, E. (1969). Reaction of dithionite and bisulphite with pyridoxal and pyridoxal enzymes. *Anal. Biochem.*, **31**, 484–92.

Adams, J. B. (1973). Thermal degradation of anthocyanins with particular reference to the 3-glycosides of cyanidin. I. In acidified solution at 100 °C. *J. Sci. Food Agric.*, **24**, 747–62.

Adams, J. B. and Woodman, J. S. (1973). Thermal degradation of anthocyanins with particular reference to the 3-glycosides of cyanidin. II. The anaerobic degradation of cyanidin-3-rutinoside at 100 °C and pH 3·0 in the presence of sodium sulphite. *J. Sci. Food Agric.*, **24**, 763–8.

Adams, R. and Garber, J. D. (1949). Aminebisulphites. II. Their use as resolving agents for aldehydes and ketones. *J. Amer. Chem. Soc.*, **71**, 522–6.

Al-Abachi, M. Q., Belcher, R., Bogdanski, S. L. and Townshend, A. (1976). Molecular emission cavity analysis. Part IX. The simultaneous determination of sulphur anions in admixture. *Anal. Chim. Acta*, **86**, 139–46.

Albers, H., Mueller, E. and Dietz, H. (1963). Dimeric dehydroascorbic acid. *Z. Physiol. Chem.*, **334**, 243–58.

Albu, H. W. and Schweinitz, H. D. G. von (1932). Autoxidations. V. Formation of dithionate by the oxidation of aqueous sulphite solutions. *Ber.*, **65B**, 729–37.

Anacleto, J. and Uden, N. van (1982). Kinetics and activation energetics of death in *Saccharomyces cerevisiae* induced by sulphur dioxide. *Biotech. Bioeng.*, **24**, 2477–86.

Anbar, M., Hefter, H. and Kremer, M. L. (1962). Reaction of organic hydroperoxides with sodium sulphite. *Chem. Ind. (London)*, 1055–7.

Anderson, E. E. and Zapsalin, C. (1957). Technique ups quality, shelf-life of pre-peeled potatoes. *Food Eng.*, **29**, 114–16.

Anderton, J. I. and Locke, D. T. (1956). Studies in the manufacturing and keeping qualities of suasages. Research report No. 72, British Food Manufacturers Research Association, Leatherhead.

Andrae, W. (1969). Mottled stains in tin cans. 2. Reaction of sulphur compounds. *Nahrung*, **13**, 549–57.

Anet, E. F. L. J. (1957). Chemistry of non-enzymic browning II. Some crystalline amino acid-deoxy-sugars. *Aust. J. Chem.*, **10**, 193–7.

Anet, E. F. L. J. (1958). A new reaction of aldoses with primary amines and its significance for non-enzymic browning reactions. *Chem. Ind. (London)*, 1438–9.

Anet, E. F. L. J. (1959a). Chemistry of non-enzymic browning. VII. Crystalline di-D-fructose-glycine and some related compounds. *Aust. J. Chem.*, **12**, 280–7.

Anet, E. F. L. J. (1959b). Chemistry of non-enzymic browning. X. Difructose-amino acids as intermediates in browning reactions. *Aust. J. Chem.*, **12**, 491–6.

Anet, E. F. L. J. (1960a). 3-Deoxyhexosones. *J. Amer. Chem. Soc.*, **82**, 1502.

Anet, E. F. L. J. (1960b). Degradation of carbohydrates. I. Isolation of 3-deoxyhexosones. *Aust. J. Chem.*, **13**, 396–403.

Anet, E. F. L. J. (1961). Degradation of carbohydrates. II. The action of acid and alkali on 3-deoxyhexosones. *Aust. J. Chem.*, **14**, 295–301.

Anet, E. F. L. J. (1962a). Degradation of carbohydrates. III. Unsaturated hexosones. *Aust. J. Chem.*, **15**, 503–9.

Anet, E. F. L. J. (1962b). Formation of furan compounds from sugars. *Chem. Ind. (London)*, 262.

Anet, E. F. L. J. (1963). Unsaturated sugars: enols of 3-deoxy-D-'glucosone'. *Chem. Ind. (London)*, 1035–6.

Anet, E. F. L. J. (1970). Proton exchange during the conversion of hexose into 5-hydroxymethyl-2-furaldehyde. *Aust. J. Chem.*, **23**, 2383–5.

Anet, E. F. L. J. and Ingles, D. L. (1964). Mechanism of inhibition of non-enzymic browning by sulphite. *Chem. Ind. (London)*, 1319.

Araiso, T., Miyoshi, K. and Yamazaki, I. (1976). Mechanism of electron transfer from sulphite to horseradish peroxidase–hydroperoxide compounds. *Biochem.*, **15**, 3059–63.

Archer, S. A. (1979). Pectolytic enzymes and degradation of pectin associated with breakdown of sulphited strawberries. *J. Sci. Food Agric.*, **30**, 692–703.

Archer, S. A. and Fielding, A. H. (1979). Polygalacturonase isoenzymes of fungi involved in the breakdown of sulphited strawberries. *J. Sci. Food Agric.*, **30**, 711–23.

Arkhipova, G. P. and Chistyakova, I. I. (1971). Spectrophotometric study of aqueous bisulphite and pyrosulphite solutions. *Zh. Prikl. Khim.*, **44**, 2193–8. Through *Chemical Abstracts*, **76**, 38033v.

Arya, S. S., Natesan, V., Parihar, D. B. and Vijayaraghavan, P. K. (1979). Stability of carotenoids in dehydrated carrots. *J. Food Technol.*, **14**, 579–86.

Atkinson, G. and Petrucci, S. (1966). Ion association of magnesium sulphate in water at 25 °C. *J. Phys. Chem.*, **70**, 3122–8.

Attari, A. and Jaselskis, B. (1972). Spectrophotometric determination of sulphur dioxide by reduction of iron(III) in the presence of ferrozine. *Anal. Chem.*, **44**, 1515–17.

Attari, A., Igielski, T. P. and Jaselskis, B. (1970). Spectrophotometric determination of sulphur dioxide. *Anal. Chem.*, **42**, 1282–5.

Audrieth, L. F., Sveda, M., Sisler, H. H. and Butler, M. J. (1940). Sulphamic acid, sulphamide and related aquo-ammonosulphuric acids. *Chem. Rev.*, **26**, 49–94.

Austwick, P. K. C. (1975). Mycotoxins. *Br. Med. Bull.*, **31**, 222–9.

Axelrod, H. D., Bonelli, J. E. and Lodge Jr, J. P. (1970). Fluorimetric determination of sulphur dioxide as sulphite. *Anal. Chem.*, **42**, 512–15.

Babich, H. and Stotzky, G. (1978). Influence of pH on inhibition of bacteria, fungi and coliphages by bisulphite and sulphite. *Environ. Res.*, **15**, 405–17.

Backer, H. J. and Mulder, H. (1933). Acyclic derivatives of aminomethane-sulphonic acid. *Rec. Trav. Chim.*, **52**, 454–68.

Backer, H. J. and Mulder, H. (1934). The acetylation of 1,1-aminosulphonic and 1,1-hydrazinosulphonic acids. *Rec. Trav. Chim.*, **53**, 1120–7.

Backer, H. J., Mulder, H. and Froentjes, W. (1935). The addition of bisulphite to isothiocyanates. *Rec. Trav. Chim.*, **54**, 57–61.

Bäckström, H. (1934). The chain mechanism in the autoxidation of sodium sulphite solutions. *Z. Phys. Chem.*, **B25**, 122–38.

Bailey, P. L. and Riley, M. (1975). Performance characteristics of gas-sensing membrane probes. *Analyst (London)*, **100**, 145–56.

Baker, L. C. (1947). Nutritional aspects of the chemical preservation, colouring and flavouring of fruit and vegetables. *Brit. J. Nutr.*, **1**, 258–62.

Baloch, A. K., Buckle, K. A. and Edwards, R. A. (1977). Stability of β-carotene in model systems containing sulphite. *J. Food Technol.*, **12**, 309–16.

Baltes, W., Franke, K., Hörtig, W., Otto, R. and Lessig, U. (1981). Investigations on model systems of Maillard reactions. *Prog. Food Nutr. Sci.*, **5**, 137–45.

Banks, J. G. and Board, R. G. (1981). Sulphite: the elective agent for the microbial associations in British fresh sausage. *J. Appl. Bact.*, **51**, pIX.

Banks, J. G. and Board, R. G. (1982*a*). Sulphite—inhibition of Enterobacteriaceae including Salmonella in British fresh sausage and in culture systems. *J. Food Prot.*, **45**, 1292–7.

Banks, J. G. and Board, R. G. (1982*b*). Comparison of methods for the determination of free and bound sulphur dioxide in stored British fresh sausage. *J. Sci. Food Agric.*, **33**, 197–203.

Barnett, D. (1975). Electrodes for estimating sulphur dioxide. *CSIRO Food Res. Quart.*, **35**, 68–9.

Barrie, L. A. and Georgii, H. W. (1976). An experimental investigation of the absorption of sulphur dioxide by water droplets containing heavy metal ions. *Atmospheric Environment*, **10**, 743–9.

Bassett, H. and Henry, A. J. (1935). The formation of dithionate by the oxidation of sulphurous acid and sulphites. *J. Chem. Soc.*, 914–29.

Bassett, H. and Parker, W. (1951). The oxidation of sulphurous acid. *J. Chem. Soc.*, 1540–60.

Bauer, A. E. (1972). Stainless Cr/Ni steel for wine storage tanks. *Schweiz. Z. Obst-Weinbau*, **108**, 499–503.

Baumgarten, P. (1936). The reaction mechanism of sulphite autoxidation. II. The oxidation of aqueous sulphite solutions in the presence of pyridine. *Ber.*, **69B**, 229–42.

Beard, J. H., Fletcher, B. C. and Beresteyn, E. C. H. van (1972). The enhancement of orange oil oxidation by sulphur dioxide. *J. Sci. Food Agric.*, **23**, 207–13.

Beech, F. W. and Carr, J. G. (1977). Cider and perry. In *Economic Microbiology. Vol. 1. Alcoholic Beverages* (A. H. Rose, Ed.), Academic Press, New York.

Beetch, E. B. and Oetzel, L. I. (1957). Colorimetric determination of sulphur dioxide from malt and beer by complexing with sodium tetrachloromercurate(II). *J. Agric. Food Chem.*, **5**, 951–2.

Beilke, S. and Lamb, D. (1975). Remarks on the rate of formation of bisulphite ions in aqueous solution. *A.I.Ch.E. Journal*, **21**, 402–4.

Beilke, S., Lamb, D. and Müller, J. (1975). On the catalysed oxidation of atmospheric SO_2 by oxygen in aqueous systems. *Atmospheric Environment*, **9**, 1083–90.

Belcher, R., Bogdanski, S. L., Knowles, D. J. and Townshend, A. (1975). Molecular emission cavity analysis—a new flame analytical technique. Part V. The determination of some sulphur anions. *Anal. Chim. Acta*, **77**, 53–63.

Belcher, R., Bogdanski, S. L. and Townshend, A. (1973). Molecular emission cavity analysis—a new flame analytical technique. Part I. Description of the technique and the development of a method for the determination of sulphur. *Anal. Chim. Acta*, **67**, 1–16.

Bell, R. P. (1973). *The Proton in Chemistry*, 2nd edn, Cornell University Press, Ithaca, New York.

Bennet, A. H. and Donovan, F. K. (1943). Determination of sulphur dioxide in citrus juices. *Analyst (London)*, **68**, 140–2.

Berger, S. (1977). Vitamin C—a ^{13}C magnetic resonance study. *Tetrahedron*, **33**, 1587–9.

Bergner, K. (1947). The behaviour of L-ascorbic acid-oxidising enzymes and copper ions towards sulphur dioxide. *Deut. Lebensm. Rundschau*, **43**, 95–101.

Bernasconi, C. F. and Bergstrom, R. G. (1973). Intermediates in nucleophilic aromatic substitution. IX. 1:1 and 1:2 sulphite complexes of 1,3,5-trinitrobenzene. *Cis–trans* isomerism in the 1:2 complex. *J. Amer. Chem. Soc.*, **95**, 3603–8.

Beroza, M. and Coad, R. A. (1967). Reaction gas chromatography. In *The Practice of Gas Chromatography* (L. S. Ettre & A. Zlatkis, Eds), Interscience Publishers, New York.

Bethea, R. M. and Meador, M. C. (1969). Gas chromatographic analysis of reactive gases in air. *J. Chromatog. Sci.*, **7**, 655–64.

Betts, R. H. and Voss, R. H. (1970). The kinetics of oxygen exchange between sulphite ion and water. *Can. J. Chem.*, **48**, 2033–41.

Bhagat, B. and Lockett, M. F. (1964). The effect of sulphite in solid diets on the growth of rats. *Food Cosmet. Toxicol.*, **2**, 1–13.

Bhatty, M. K. and Townshend, A. (1971). An improved ultraviolet spectrophotometric method for the determination of sulphur dioxide. *Anal. Chim. Acta*, **55**, 263–5.

Bigeleisen, J. (1949). The validity of the use of tracers to follow chemical reactions. *Science*, **110**, 14–16.

Bigwood, E. J. (1973). The acceptable daily intake of food additives. *CRC Crit. Rev. Toxicol.*, **2**, 41–93.

Biilman, E. and Blom, J. H. (1924). Electrometric studies on azo- and hydrazo-compounds. *J. Chem. Soc.*, **125**, 1719–31.

Birch, T. W. and Harris, L. J. (1933). LXXIX. The titration curve and dissociation constants of vitamin C. *Biochem. J.*, **27**, 595–9.

Birk, J. R., Larsen, C. M. and Wilbourn, R. G. (1970). Gas chromatographic determination of sulphide, sulphite and carbonate in solidified salts. *Anal. Chem.*, **42**, 273–5.

Bischof, P., Eckert-Maksic, M. and Maksic, Z. B. (1981). On the relative stability of ascorbic acid tautomers. *Z. Naturforsch.*, **36a**, 502–4.

Blackadder, D. H. and Hinshelwood, Sir Cyril (1958). The kinetics of the decomposition of the addition compounds formed by sodium bisulphite and a series of aldehydes and ketones. *J. Chem. Soc.*, 2720–7.

Blasius, E. and Ziegler, K. (1974). Photometric determination of sulphite with trinitrobenzenesulphonic acid. *Z. Anal. Chem.*, **269**, 15–18.

Boedeker, C. (1861). A new reagent for sulphurous acid. *Ann.*, **117**, 193–5.

Bouley, J. P., Briand, J. P., Jonard, G., Witz, J. and Hirth, L. (1975). Low pH RNA–protein interactions in turnip yellow mosaic virus. III. Reassociation experiments with other viral RNAs and chemically modified TYMV-RNA. *Virology*, **63**, 312–19.

Bourne, D. W. A., Higuchi, T. and Pitman, I. H. (1974). Chemical equilibria in solutions of bisulphite salts. *J. Pharm. Sci.*, **63**, 865–8.

Brekke, J. E. (1963). Collaborative study on the assay of sulphur dioxide in dried apricots. *J. Ass. Off. Anal. Chem.*, **46**, 618–22.

Brenner, M. W. and Bernstein, L. (1975). Effect of sulphites on beer foam quality. *Proc. Am. Soc. Brew. Chem.*, **33**, 171–3.

Brimblecombe, P. and Spedding, D. J. (1974). The catalytic oxidation of micromolar aqueous sulphur dioxide. I. Oxidation of dilute solutions of iron(III). *Atmospheric Environment*, **8**, 937–45.

Britton, H. T. S. and Robinson, R. A. (1932). The use of the glass electrode in titrimetric work and precipitation reactions. The application of the principle of the solubility product to basic precipitates. *Trans. Faraday Soc.*, **28**, 531–45.

Brouillard, R. and El Hage Chahine, J. M. (1980). Chemistry of anthocyanin pigments. 6. Kinetic and thermodynamic study of hydrogen sulphite addition to cyanin. Formation of a highly stable Meisenheimer-type adduct derived from a 2-phenylbenzopyrylium salt. *J. Amer. Chem. Soc.*, **102**, 5375–8.

Brown, M. H. (1977). Stainless steels in the chemical and process industries. In *Handbook of Stainless Steels* (D. Peckner & I. M. Bernstein, Eds), McGraw-Hill Book Co., New York.

Brun, P. C., Gasquet, C., Stoutz, S. de and Nicoli, A. (1961). Amperometric determination of sulphur dioxide in wine. *Ann. Fals. Expert. Chim.*, **54**, 412–20.

Bruno, P., Caselli, M., Fano, A. Di and Traini, A. (1979). Fast and simple polargraphic method for the determination of free and total sulphur dioxide in wines and other common beverages. *Analyst* (*London*), **104**, 1083–7.

Budowsky, E. I., Simukova, N. A., Turchinsky, M. F., Boni, I. V. and Skoblov, Y. M. (1976). Induced formation of covalent bonds between nucleoprotein components. V. UF or bisulphite induced polynucleotide–protein cross linkage in bactiophage MS. *Nucleic Acid Res.*, **3**, 261–76.

Bünau, G. and Eigen, M. (1962). On the kinetics of the iodine–sulphite reaction. *Z. Phys. Chem.*, **32**, 27–50.

Bunte, H. (1874). The constitution of hyposulphurous acid. *Ber.*, **7**, 646–8.

Burroughs, L. F. (1974). Browning control in fruit juices and wines. *Chem. Ind.*, 718–20.

Burroughs, L. F. (1977). Stability of Patulin to sulphur dioxide and to yeast fermentation. *J. Ass. Off. Anal. Chem.*, **60**, 100–3.

Burroughs, L. F. and Sparks, A. H. (1963). The determination of the total sulphur dioxide content of ciders. *Analyst (London)*, **88**, 304–9.

Burroughs, L. F. and Sparks, A. H. (1964). The determination of the free sulphur dioxide content of ciders. *Analyst (London)*, **89**, 55–60.

Burroughs, L. F. and Sparks, A. H. (1973a). Sulphite-binding power of wines and ciders. I. Equilibrium constants for the dissociation of carbonyl bisulphite compounds. *J. Sci. Food Agric.*, **24**, 187–98.

Burroughs, L. F. and Sparks, A. H. (1973b). Sulphite-binding power of wines and ciders. II. Theoretical consideration and calculation of sulphite-binding equilibria. *J. Sci. Food Agric.*, **24**, 199–206.

Burroughs, L. F. and Sparks, A. H. (1973c). Sulphite-binding power of wines and ciders. III. Determination of carbonyl compounds in a wine and calculation of its sulphite-binding power. *J. Sci. Food Agric.*, **24**, 207–17.

Burton, H. S., McWeeny, D. J. and Biltcliffe, D. O. (1962). Sulphites and aldose–amino reactions. *Chem. Ind. (London)*, 219–21.

Burton, H. S., McWeeny, D. J. and Biltcliffe, D. O. (1963). Non-enzymic browning: the role of unsaturated carbonyl compounds as intermediates and of SO_2 as an inhibitor of browning. *J. Sci. Food Agric.*, **14**, 911–20.

Burton, K. W. C. and Nickless, G. (1968). Amido- and imido-sulphonic acids. In *Inorganic Sulphur Chemistry* (G. Nickless, Ed.), Elsevier Publishing Co., Amsterdam.

Campbell, A. D., Hubbard, D. P. and Tioh, N. H. (1975). Some observations on the coulometric determination of sulphur dioxide. *Anal. Chim. Acta*, **76**, 483–6.

Cappellini, R. A., Stretch, A. W. and Walton, G. S. (1961). Effects of sulphur dioxide on the reduction of post-harvest decay of latham red raspberries. *Pl. Dis. Rep.*, **45**, 301–3.

Carlyle, D. W. and Zeck Jr, O. F. (1973). Electron transfer between sulphur(IV) and hexaaquoiron(III) ion in aqueous perchlorate solution. Kinetics and mechanism of uncatalysed and copper(II) catalysed reactions. *Inorg. Chem.*, **12**, 2978–83.

Carr, J. G. (1981). Antimicrobial activity of sulphur dioxide. *J. Sci. Food Agric.*, **32**, 1140.

Carr, J. G., Davies, P. A. and Sparks, A. H. (1976). The toxicity of sulphur dioxide towards certain lactic acid bacteria from fermented apple juice. *J. Appl. Bact.*, **40**, 201–12.

Carruthers, A., Gallaher, P. J. and Oldfield, J. F. T. (1958). Nitrate reduction by thermophilic bacteria in sugar beet diffusion systems. *Int. Sugar J.*, **60**, 335.

Carruthers, A., Heaney, R. K. and Oldfield, J. F. T. (1965). Determination of sulphur dioxide in white sugar. *Int. Sugar J.*, **67**, 364–8.

Carswell, D. R. (1977). Sulphur dioxide—a literature review of analytical methods. Scientific and Technical Surveys, No. 103, The British Food Manufacturing Industries Research Association, Leatherhead, Surrey.

Cecil, R. and Loening, U. E. (1957). The reaction of the disulphide bonds of insulin with sodium sulphite. *Biochem. J.*, **66**, 18P–19P.

Cecil, R. and McPhee, J. R. (1955). A kinetic study of the reactions of some disulphides with sodium sulphite. *Biochem. J.*, **60**, 496–506.

Cecil, R. and Wake, R. G. (1962). Reactions of inter- and intra-chain disulphide bonds in proteins with sulphite. *Biochem. J.*, **82**, 401–6.

Cermak, V. and Smutek, M. (1975). Mechanisms of decomposition of dithionite in aqueous solutions. *Coll. Czech. Chem. Comm.*, **40**, 3241–63.

Chantry, G. W., Horsfield, A., Morton, J. R., Rowlands, J. R. and Whiffen, D. H. (1962). The optical and electron resonance spectra of SO_3^-. *Mol. Phys.*, **5**, 233–9.

Charbonnier, F., Faure, R. and Loiseleur, H. (1977). Refinement of the structure of tetraaquabis(methanesulphonato)copper(II) $[Cu(CH_3SO_3)_2(H_2O)_4]$. Evidence for a plane of symmetry in the methanesulphonato coordinate. *Acta Cryst.*, **B33**, 1845–8.

Charon, D. and Szabo, L. (1973). Synthesis and acid degradation of 3-deoxy-ketoaldonic acids. *J. Chem. Soc., Perkin Trans.*, **1**, 1175–9.

Chen, R. Y., Yokota, K. and Sakaguchi, O. (1972). Food hygienic studies of sulphite. Inhibition of amylase, in particular salivary amylase. *Nippon Eiseigaku Zasshi*, **26**, 486–91. Through *Chemical Abstracts*, **78**, 67630j.

Cheronis, N. D. and Ma, T. S. (1964). *Organic Functional Group Analysis by Micro and Semimicro Methods*, Interscience Publishers, New York.

Chio, K. S. and Tappel, A. L. (1969). Synthesis and characterisation of the fluorescent products derived from malonaldehyde and amino acids. *Biochem.*, **8**, 2821–7.

Chmielnicka, J. (1963). Influence of sulphurous acid on oxidation of L-ascorbic acid by peroxidase in the presence of hydrogen peroxide. *Roczniki Zakladu Hig.*, **14**, 393–402.

Christian, J. H. B. (1963). Preservation of minced meat with sulphur dioxide. *CSIRO Food Pres. Quart.*, **23**, 30–3.

Clark, D. R. and Miles, P. (1978). Addition of sulphur dioxide to Schiff's bases. *J. Chem. Res. (S)*, 124; *J. Chem. Res. (M)*, 1752–63.

Clark, W. M. (1960). *Oxidation–Reduction Potentials of Organic Systems*, Baillere, Tindall & Cox Ltd, London.

Clegg, K. M. (1964). Non-enzymic browning of lemon juice. *J. Sci. Food Agric.*, **15**, 878–85.

Clegg, K. M. (1966). Citric acid and the browning of solutions containing ascorbic acid. *J. Sci. Food Agric.*, **17**, 546–9.

Coggiola, I. M. (1963). 2,5-Dihydro-2-furoic acid: a product of the anaerobic decomposition of ascorbic acid. *Nature*, **200**, 954–5.

Cohen, H. J. and Fridovich, I. (1971a). Hepatic sulphite oxidase. Function and properties. *J. Biol. Chem.*, **246**, 359–66.

Cohen, H. J. and Fridovich, I. (1971b). Hepatic sulphite oxidase. The nature and functions of the haem prosthetic groups. *J. Biol. Chem.*, **246**, 367–73.

Cohen, H. J. and Fridovich, I. (1971c). Hepatic sulphite oxidase. A functional role for molybdenum. *J. Biol. Chem.*, **246**, 374–82.

Cohen, H. J., Betcher-Lange, S., Kessler, D. L. and Rajagopalan, K. V. (1972). Hepatic sulphite oxidase. Congruency in mitochondria of prosthetic groups and activity. *J. Biol. Chem.*, **247**, 7759–66.

Cohen, H. J., Drew, R. T., Johnson, J. L. and Rajagopalan, K. V. (1973). Molecular basis of the biological function of molybdenum. The relationship between sulphite oxidase and the acute toxicity of bisulphite and SO_2. *Proc. Nat. Acad. Sci.*, **70**, 3655–9.

Colowick, S. P., Kaplan, N. O. and Ciotti, M. M. (1951). The reaction of pyridine nucleotide with cyanide and its analytical use. *J. Biol. Chem.*, **191**, 447–59.

Connant, J. B. and Pratt, M. F. (1926). The irreversible reduction of organic compounds. III. The reduction of azo dyes. *J. Amer. Chem. Soc.*, **48**, 2468–84.

Conway, E. J. (1962). *Microdiffusion Analysis and Volumetric Error*, 5th edn, Crosby Lockwood & Sons Ltd, London.

Conway, E. J. and Byrne, A. (1933). LXI. An absorption apparatus for the micro-determination of certain volatile substances. 1. The micro-determination of ammonia. *Biochem. J.*, **27**, 419–29.

Cooperstein, S. J. (1963). Reversible inactivation of cytochrome oxidase by disulphide bond reagents. *J. Biol. Chem.*, **238**, 3606–10.

Cordingly, R. H. (1959). An investigation of the sulphonic acids derived from xylose and arabinose, PhD Thesis, Institute of Paper Chemistry, Lawrence College, Appleton, Wisconsin.

Cotton, F. A. and Wilkinson, G. W. (1966). *Advanced Inorganic Chemistry*, 2nd edn, Interscience Publishers, London.

Crampton, M. R. (1967). A spectroscopic study of complex formation between aromatic nitro-compounds and sodium sulphite. *J. Chem. Soc. B*, 1341–6.

Crampton, M. R. (1969). Meisenheimer complexes. *Adv. Phys. Org. Chem.*, **7**, 211–57.

Cresser, M. S. and Isaacson, P. J. (1976). The analytical potential of gas-phase molecular absorption spectrometry for the determination of anions in solution. *Talanta*, **23**, 885–8.

Critchfield, F. E. and Johnson, J. B. (1956). Determination of alpha- and beta-unsaturated compounds by reaction with sodium sulphite. *Anal. Chem.*, **28**, 73–5.

Crompton, M., Palmeri, F., Capano, M. and Quagliariello, E. (1974). The transport of sulphate and sulphite in rat liver mitochondria. *Biochem. J.*, **142**, 127–37.

Cruickshank, D. W. J. (1961). The role of $3d$-orbitals in π bonds between (a) silicon, phosphorus, sulphur or chlorine and (b) oxygen or nitrogen. *J. Chem. Soc.*, 5486–504.

Danilczuk, E. and Swinarski, A. (1961). The complex ion $[Fe(III)(SO_3)_n]^{3-2n}$. *Roczniki Chem.*, **35**, 1563–71.

Dasgupta, P. K., Cesare, K. De and Ullrey, J. C. (1980). Determination of atmospheric sulphur dioxide without tetrachloromercurate(II) and the mechanism of the Schiff reaction. *Anal. Chem.*, **52**, 1912–22.

Davies, A. G. (1961). *Organic Peroxides*, Butterworths, London.

Davies, A. G. and Feld, R. (1956). Organic peroxides. Part VI. A stereochemical investigation of the preparation and reactions of 1-phenylethylhydroperoxide. *J. Chem. Soc.*, 665–70.

Davies, C. W. (1962). *Ion Association*, Butterworths, London.

Davis, A. R. and Chatterjee, R. M. (1975). A vibrational–spectroscopic study of the SO_2–H_2O system. *J. Solution Chem.*, **4**, 399–412.

Davis, N. D. and Diener, U. L. (1967). Inhibition of aflatoxin synthesis by *p*-aminobenzoic acid, potassium sulphite and potassium fluoride. *Appl. Microbiol.*, **15**, 1517–18.

Davis, R. E. (1968). Mechanisms of sulphur reactions. In *Inorganic Sulphur Chemistry* (G. Nickless, Ed.), Elsevier Scientific Publishing Co., Amsterdam.

Davis, R. E., Cohen, A. and Louis, J. A. (1963). A novel correlation of activation energy with sulphur–sulphur bond distance. *J. Amer. Chem. Soc.*, **85**, 3050–1.

Davis, E. G., McBean, D. McG. and Rooney, M. L. (1975). Packaging foods that contain sulphur dioxide. *CSIRO Food Res. Quart.*, **35**, 57–62.

Dawes, I. W. and Edwards, R. A. (1966). Methyl substituted pyrazines as volatile reaction products of heated aqueous aldose–amino acid mixtures. *Chem. Ind. (London)*, 2203.

Dennis, C. and Harris, J. E. (1979). The involvement of fungi in the breakdown of sulphited strawberries. *J. Sci. Food Agric.*, **30**, 687–91.

Dennis, C., Davis, R. P., Harris, J. E., Calcutt, L. W. and Cross, D. (1979). The relative importance of fungi in the breakdown of commercial samples of sulphited strawberries. *J. Sci. Food Agric.*, **30**, 959–73.

Deveze, D. and Rumpf, P. (1964). Spectrophotometric study of aqueous solutions of SO_2 in various acid buffers. *Compt. Rend.*, **258**, 6135–8.

Dey, B. B. and Row, K. K. (1924). Action of sodium sulphite on coumarins. *J. Chem. Soc.*, **125**, 554–64.

Dharmatti, S. S. and Weaver Jr, H. E. (1951). The magnetic moment of [33]S. *Phys. Rev.*, **83**, 845.

Diemair, W., Koch, J. and Hess, D. (1961). The determination of bound and free sulphurous acid in wine. *Z. Anal. Chem.*, **178**, 321–30.

Dietz, H. (1970). New preparation of dimer dehydro-L-ascorbic acid and its dissociation in solution. *Ann.*, **738**, 206–8.

Dobbs, H. E. (1966). New oxygen flask apparatus for the assay of tritium and carbon-14. *J. Appl. Rad. Isotop.*, **17**, 363–4.

Doddi, G., Fornarini, S., Illuminati, G. and Stegel, F. (1979). Rates and equilibria for the addition of methoxide ion to 2,6-diphenyl- and 4-methoxy-2,6-diphenylpyrylium cations. *J. Org. Chem.*, **44**, 4496–500.

Doerge, D. R. and Ingraham, L. L. (1980). Kinetics of thiamine cleavage by bisulphite ion. *J. Amer. Chem. Soc.*, **102**, 4828–30.

Dorange, J. L. and Dupuy, P. (1972). Evidence of the mutagen action of sodium sulphite on yeast. *Compt. Rend. Hebd. Seances Acad. Sci., Ser. D, Sci. Nat.*, **274**, 2798–800.

Downer, A. W. E. (1943). The preservation of citrus juices with sulphurous acid. *J. Soc. Chem. Ind.*, **62**, 124–7.

Doyle, M. P. and Marth, E. H. (1978a). Bisulphite degrades aflatoxin: effect of citric acid and methanol and possible mechanism of degradation. *J. Food Prot.*, **41**, 891–6.

Doyle, M. P. and Marth, E. H. (1978b). Bisulphite degrades aflatoxin: effect of temperature and concentration of bisulphite. *J. Food Prot.*, **41**, 774–80.

Drago, R. S. (1962). *Free Radicals in Inorganic Chemistry* (Advances in Chemistry Series, No. 36), American Chemical Society, Washington.

Duborg, J. and Devillers, P. (1957). Maillard reaction. *Bull. Soc. Chim. France*, 333–6.

Dyett, E. J. and Shelley, D. (1966). The effects of sulphite preservative in British fresh sausages. *J. Appl. Bact.*, **29**, 439–46.

Edwards, J. O. (1952). Rate laws and mechanisms of oxyanion reactions with bases. *Chem. Rev.*, **50**, 455–82.

Edwards, J. O. (1954). Correlation of relative rates and equilibria with a double basicity scale. *J. Amer. Chem. Soc.*, **76**, 1540–7.

Edwards, J. O. and Pearson, R. G. (1962). The factors determining nucleophilic reactivities. *J. Amer. Chem. Soc.*, **84**, 16–24.

Egorov, A. M., Popov, V. O., Berezin, I. V. and Rodionov, U. V. (1980). Kinetic and structural properties of NAD-dependent bacterial formate dehydrogenase. *J. Solid Phase Biochem.*, **5**, 19–33.

Eigen, M. and Tamm, K. (1962a). Sound absorption in electrolytes as a consequence of chemical relaxation. I. Relaxation theory of stepwise dissociation. *Z. Electrochem.*, **66**, 93–107.

Eigen, M. and Tamm, K. (1962b). Sound absorption in electrolytes as a consequence of chemical relaxation. II. Result of measurement and relaxation mechanism for 2:2 electrolytes. *Z. Electrochem.*, **66**, 107–21.

Eigen, M., Kruse, W., Maass, G. and Maeyer, L. De (1964). Rate constants of protolytic reactions in aqueous solution. In *Progress in Reaction Kinetics*, Vol. 2 (G. Porter, Ed.), Pergamon Press, Oxford.

Eigen, M., Kustin, K. and Maass, G. (1961). Rate of the hydration of SO_2 in aqueous solution. *Z. Phys. Chem.*, **30**, 130–6.

Eisner, U. and Kuthan, J. (1972). The chemistry of the dihydropyridines. *Chem. Rev.*, **72**, 1–42.

Elbeih, I. I. M. and Abou-Elnaga, M. A. (1960). A new scheme of analysis for the common anions based on paper chromatography. *Anal. Chim. Acta*, **23**, 30–5.

Eldick, R. van and Harris, G. M. (1980). Kinetics and mechanism of the formation, acid-catalysed decomposition and intramolecular redox reaction of oxygen bonded (sulphite) pentamminecobalt(III) ions in aqueous solution. *Inorg. Chem.*, **19**, 880–6.

Ellis, A. J. and Anderson, D. W. (1961). Effect of pressure on the first acid dissociation constants of 'sulphurous' and phosphoric acids. *J. Chem. Soc.*, 1765–7.

Ellis, G. P. (1959). The Maillard reaction. *Adv. Carbohydr. Chem.*, **14**, 63–134.

Ellis, G. P. and Honeyman, J. (1955). Glycosylamines. *Adv. Carbohydr. Chem.*, **10**, 95–168.

Embs, R. J. and Markakis, P. (1965). The mechanism of sulphite inhibition of browning caused by polyphenol oxidase. *J. Food Sci.*, **30**, 753–8.

Embs, R. J. and Markakis, P. (1969). Bisulphite effect on the chemistry and activity of horseradish peroxidase. *J. Food Sci.*, **34**, 500–1.

Enders, C. and Theis, K. (1938). Melanoidins and their relationship to the humic acids. *Brennstoff. Chem.*, **19**, 360–5, 402–7, 439–49.

Erdey, L., Simon, J., Gal, S. and Liptay, G. (1966). Thermoanalytical properties of analytical grade reagents. IVA. *Talanta*, **13**, 67–80.

Eriksen, T. E. (1972). Sulphur isotope effects. II. The isotopic exchange coefficients for the sulphur isotopes ^{34}S–^{32}S in the system SO_2–aqueous solutions of SO_2. *Acta Chem. Scand.*, **26**, 581–4.

Eriksen, T. E. and Lind, J. (1972). Spectrophotometric determination of sulphur dioxide and thiosulphate in aqueous solutions of hydrogen sulphite ion. *Acta Chem. Scand.*, **26**, 3325–32.

Erman, M. B., Shmelev, L. V., Pribytkova, I. M. and Aul'chenko, I. S. (1979). Structure and properties of the bisulphite adducts of citral. *Zh. Org. Khim.*, **15**, 1598–605.

Euler, H. and Eistert, B. (1957). *Chemie und Biochemie der Reduktone und Reduktonate*, Ferdinand Enke Verlag, Stuttgart.

Falk, M. and Giguere, P. A. (1958). On the nature of sulphurous acid. *Can. J. Chem.*, **36**, 1121–5.

Faure, R., Vincent, E. J., Ruiz, J. M. and Lena, L. (1981). Sulphur-33 nuclear magnetic resonance studies of sulphur compounds with sharp resonance lines. ^{33}S NMR spectra of some sulphones and sulphonic acids. *Org. Magnetic Resonance*, **15**, 401–3.

Fava, A. and Iliceto, A. (1958). Kinetics of displacement reactions at the sulphur atom. II Stereochemistry. *J. Amer. Chem. Soc.*, **80**, 3478–9.

Fava, A. and Pajaro, G. (1956). Kinetics of displacement reactions at the sulphur atom. I. Isotopic exchange between sulphite and alkythiosulphates. *J. Amer. Chem. Soc.*, **78**, 5203–8.

Feather, M. S. (1970). The conversion of D-xylose and D-gluconic acid to 2-furaldehyde. *Tetrahedron Lett.*, 4143–7.

Feather, M. S. (1981). Amine-assisted sugar dehydration reactions. *Prog. Food Nutr. Sci.*, **5**, 37–45.

Feather, M. S. and Harris, J. F. (1968). The absence of proton exchange during the conversion of hexoses to 5-(hydroxymethyl)-2-furaldehyde. *Tetrahedron Lett.*, 5807–10.

Feather, M. S. and Harris, J. F. (1970). On the mechanism of conversion of hexoses into 5-(hydroxymethyl)-2-furaldehyde and metaroccharinic acid. *Carbohydr. Res.*, **15**, 340–9.

Feather, M. S. and Harris, J. F. (1973). Dehydration reactions of carbohydrates. *Adv. Carbohydr. Chem. Biochem.*, **28**, 161–224.

Feather, M. S. and Russell, K. R. (1969). Deuterium incorporation during conversion of 1-amino-1-deoxy-D-fructose derivates to 5-(hydroxymethyl)-2-furaldehyde. *J. Org. Chem.*, **34**, 2650–2.

Feather, M. S., Harris, D. W. and Nichols, S. B. (1972). Routes of conversion of D-xylose, hexuronic acids and L-ascorbic acid to 2-furaldehyde. *J. Org. Chem.*, **37**, 1606–8.

Feigl, F. and Anger, V. (1972). *Spot Tests in Inorganic Analysis*, 6th edn, Elsevier Publishing Co., Amsterdam.

Fernelius, W. C., Loening, K. and Adams, R. M. (1973). Notes on nomenclature. *J. Chem. Educ.*, **50**, 341–2.

Filippi, L. J. De, Toler, L. S. and Hultquist, D. E. (1979). Reactivities of hydroxylamine and sodium bisulphite with carbonyl-containing haems and with the prosthetic groups of the erythrocyte green haem proteins. *Biochem. J.*, **179**, 151–60.

Finch, H. D. (1962). Bisulphite adducts of acrolein. *J. Org. Chem.*, **27**, 649–51.

Fischer, D. W. and Baun, W. L. (1965). Effect of chemical combination on the soft X-ray L emission spectra of potassium, chlorine and sulphur using a stearate soap film crystal. *Anal. Chem.*, **37**, 902–6.

Fischer, K. (1935). A new method for the analytical determination of the water content of liquids and solids. *Angew. Chem.*, **48**, 394-6.

Flockhart, B. D., Ivin, K. J., Pink, R. C. and Sharma, B. D. (1971). The nature of the radical intermediates in the reactions between hydroperoxides and sulphur dioxide and their reaction with alkene derivatives. Electron spin resonance study. *Chem. Comm.*, 339–40.

Foe, L. G. and Trujillo, J. L. (1980). Activation of phosphofructokinase by divalent anions. *Arch. Biochem. Biophys.*, **199**, 1-6.

Foerster, F. and Hamprecht, G. (1926). Contribution to the knowledge of sulphurous acid and its salts. V. The behaviour of pyrosulphite on heating. *Z. Anorg. Allgem. Chem.*, **158**, 277-315.

Foerster, G., Weiss, W. and Staudinger, H. (1965). Electron spin resonance determination on semihydroascorbic acid. *Ann.*, **690**, 166-9.

Fogg, A. G., Moser, W. and Chalmers, R. A. (1966). The Boedeker reaction. Part III. The detection of sulphite ion. *Anal. Chim. Acta*, **36**, 248–51.

Fornachon, J. C. M. (1963). Inhibition of certain lactic acid bacteria by free and bound sulphur dioxide. *J. Sci. Food Agric.*, **14**, 857-62.

Francis, F. J. and Amla, B. L. (1961). The effect of residual sulphur dioxide on the quality of pre-peeled potatoes. *Amer. Potato J.*, **38**, 89–94.

Freese, E., Sheu, C. W. and Galliers, E. (1973). Function of lipophilic acids as antimicrobial food additives. *Nature (London)*, **241**, 321-5.

Fridovich, I. and Handler, P. (1958). Xanthine oxidase. III. Sulphite oxidation as an ultrasensitive assay. *J. Biol. Chem.*, **233**, 1578-80.

Fridovich, I. and Handler, P. (1959). Utilisation of the photosensitised oxidation of sulphite for manometric actinometry. *Biochim. Biophys. Acta*, **35**, 546-7.

Fridovich, I. and Handler, P. (1961). Detection of free radicals generated during enzymic oxidations by the initiation of sulphite oxidation. *J. Biol. Chem.*, **236**, 1836–40.

Frijters, J. E. R., Griffiths, N. M., Mather, A. M., Reynolds, J. and Fenwick, G. R. (1981). Effect of sulphur dioxide on the chemical composition and odour of mustard paste. *Chem. Senses*, **6**, 33–43.

Fujita, K., Ikuzawa, M., Izumi, T., Hamano, T., Mitsuhashi, Y., Matsuki, Y., Adachi, T., Nonogi, H., Fuke, T., Suzuki, M.,Toyoda, M., Ito, Y. and Iwaida, M. (1979). Establishment of a modified Rankine method for the separate determination of free and combined sulphites in foods. III. *Z. Lebensm. Untersuch.-Forsch.*, **168**, 206–11.

Fuller, E. C. and Crist, R. H. (1941). The rate of oxidation of sulphite ions by oxygen. *J. Amer. Chem. Soc.*, **63**, 1644-50.

Garber, R. W. and Wilson, C. E. (1972). Determination of atmospheric sulphur dioxide by differential pulse polarography. *Anal. Chem.*, **44**, 1357–60.

Garnier, Y. and Duval, C. (1959). Paper chromatographic separation of inorganic ions containing sulphur and their identification by infrared spectroscopy. *J. Chromatog.*, **2**, 72-5.

Geneste, P., Lamaty, G. and Roque, J. P. (1971). Nucleophilic addition reactions to ketones. Sulphite ion addition. Detection of hyperconjugation by measuring the deuterium secondary isotope effect. *Tetrahedron*, **27**, 5539-59.

Geneste, P., Lamaty, G. and Roque, J. P. (1972). Hammett ρ constant in the formation of oximes and bisulphite compounds. *Rec. Trav. Chim.*, **91**, 188–94.

Gibson, W. B. and Strong, F. M. (1973). Metabolism and elimination of sulphite by rats, mice and monkeys. *Food Cosmet. Toxicol.*, **11**, 185–98.

Gibson, W. B. and Strong, F. M. (1976). Metabolism and elimination of α-hydroxyethanesulphonate by rats. *Food Cosmet. Toxicol.*, **14**, 41–3.

Gilbert, J. and McWeeny, D. J. (1976). The reactions of sulphur dioxide in dehydrated vegetables. *J. Sci. Food Agric.*, **27**, 1145–6.

Gmelin, L. (1963). *Gmelins Handbuch Der Anorganischen Chemie. Achte Völlig Neu Bearbeitete Auflage. Schwefel*, Part B, No. 3, Verlag Chemie GmbH, Weinheim/Bergstrasse.

Goldberg, R. N. (1981). Evaluated activity and osmotic coefficients for aqueous solutions: thirty six uni-bivalent electrolytes. *J. Phys. Chem. Ref. Data*, **10**, 671–764.

Golding, R. M. (1960). Ultraviolet absorption studies of the bisulphite–pyrosulphite equilibrium. *J. Chem. Soc.*, 3711–16.

Gomez-Moreno, C. and Palacian, E. (1974). Nitrate reductase from *Chlorella fusca*. Reversible inactivation by thiols and by sulphite. *Arch. Biochem. Biophys.*, **160**, 269–73.

Gomez-Moreno, C., Choy, M. and Edmondson, D. E. (1979). Purification and properties of the bacterial flavoprotein: thiamine dehydrogenase. *J. Biol. Chem.*, **254**, 7630–5.

Gomyo, T., Kato, H., Udaka, K., Horikoshi, M. and Fujimaki, M. (1972). Effects of heating on chemical properties of melanoidin prepared from glycine–xylose system. *Agric. Biol. Chem.*, **36**, 125–32.

Goshima, K., Maezono, N. and Tokuyama, K. (1973). A novel degradation pathway of L-ascorbic acid under non-oxidative conditions. *Bull. Chem. Soc. Japan*, **46**, 902–4.

Grassini, G. and Lederer, M. (1959). Paper electrophoresis of inorganic anions in 0·1 M NaOH. *J. Chromatog.*, **2**, 326.

Green, L. R. and Hine, J. (1974). The pH independent equilibrium constants and rate constants for formation of the bisulphite addition compound of isobutyraldehyde in water. *J. Org. Chem.*, **39**, 3896–901.

Greenberg, F. H., Leung, K. K. and Leung, M. (1971). The reaction of vitamin K_3 with sodium bisulphite. *J. Chem. Educ.*, **48**, 632–4.

Greenstock, C. L. and Miller, R. W. (1975). The oxidation of tiron by superoxide anion. Kinetics of the reaction in aqueous solution and in chloroplasts. *Biochim. Biophys. Acta*, **396**, 11–16.

Griffiths, N. M., Mather, A. M., Fenwick, G. R., Frijters, J. E. R. and Merz, J. H. (1980). An effect of sulphur dioxide on the odour of mustard paste. *Chem. Ind. (London)*, 239–40.

Gringras, L. and Sjoestedt, G. (1961). Paper chromatography of some sulphinic acids and sulphonyl chlorides. *Acta Chem. Scand.*, **15**, 435–6.

Guggenheim, E. A. (1935). The specific thermodynamic properties of aqueous solutions of strong electrolytes. *Phil. Mag.*, **19**, 588–643.

Guilbault, G. G., Brignac Jr, P. and Zimmer, M. (1968). Homovanillic acid as a fluorometric substrate for oxidative enzymes. Analytical applications of the peroxidase, glucose oxidase and xanthine oxidase systems. *Anal. Chem.*, **40**, 190–6.

Gunnison, A. F. (1981). Sulphite toxicity: a critical review of *in vitro* and *in vivo* data. *Food Cosmet. Toxicol.*, **19**, 667–82.

336   CHEMISTRY OF SULPHUR DIOXIDE IN FOODS

Gunnison, A. F. and Benton, A. W. (1971). Sulphur dioxide: sulphite. Interaction with mammalian serum and plasma. *Arch. Environ. Health*, **22**, 381–8.

Gunnison, A. F. and Farrugella, T. J. (1979). Preferential S-sulphonate formation in lung and aorta. *Chem. Biol. Interact.*, **25**, 271–9.

Gunnison, A. F. and Palmes, E. D. (1978). Species variability in plasma S-sulphonate levels during and following sulphite administration. *Chem. Biol. Interactions*, **21**, 315–29.

Gunnison, A. F., Bresnahan, C. A. and Palmes, E. D. (1977). Comparative sulphite metabolism in the rat, rabbit and rhesus monkey. *Toxicol. Appl. Pharmacol.*, **42**, 99–109.

Gunnison, A. F., Dulak, L., Chiang, G., Zaccardi, J. and Farrugella, T. J. (1981). A sulphite oxidase deficient rat model: subchronic toxicology. *Food Cosmet. Toxicol.*, **19**, 221–32.

Gürtler, O., Muller, G. and Holzapfel, H. (1968). Determination of equilibrium coefficients and affinities for 28 anions towards WOFATIT SBK. *Z. Naturforsch.*, **23B**, 1382–3.

Guthrie, P. J. (1979). Tautomeric equilibria and pK_a values for 'sulphurous acid' in aqueous solution. A thermodynamic analysis. *Can. J. Chem.*, **57**, 454–7.

Hachimori, A., Muramatsu, N. and Nosoh, Y. (1970). Studies on an ATPase of thermophilic bacteria. I. Purification and properties. *Biochim. Biophys. Acta*, **206**, 426–37.

Hägglund, E. and Ringbom, A. (1926). On the addition of sulphite to unsaturated compounds. *Z. Anorg. Allgem. Chem.*, **150**, 231–53.

Hägglund, E., Johnson, T. and Urban, H. (1930). Influence of sulphite and bisulphite solutions on sugars at high temperatures. *Ber.*, **63B**, 1387–95.

Haight, G. P., Perchonock, E., Emmenegger, F. and Gordon, G. (1965). The mechanism of the oxidation of sulphur(IV) by chromium(VI) in acid solution. *J. Amer. Chem. Soc.*, **87**, 3835–40.

Haisman, D. R. (1974). The effect of sulphur dioxide on oxidizing enzymes in plant tissues. *J. Sci. Food Agric.*, **25**, 803–10.

Hall, J. R. (1969). Enhanced recovery of *Salmonella montevideo* from onion powder by the addition of potassium sulphite to lactose broth. *J. Ass. Off. Anal. Chem.*, **52**, 940–2.

Halperin, J. and Taube, H. (1950). Oxygen atom transfer in the reaction of chlorate with sulphite in aqueous solution. *J. Amer. Chem. Soc.*, **72**, 3319–20.

Halperin, J. and Taube, H. (1952). The transfer of oxygen atoms in oxidation reduction reactions. *J. Amer. Chem. Soc.*, **74**, 380–2.

Hamano, T., Mitsuhashi, Y., Matsuki, Y., Ikuzawa, M., Fujita, K., Izumi, T., Adachi, T., Nonogi, H., Fuke, T., Suzuki, H., Toyoda, M., Ito, Y. and Iwaida, M. (1979). Application of gas chromatography for the separate determination of free and combined sulphite in foods. I. *Z. Lebensm. Untersuch.-Forsch.*, **168**, 195–9.

Hammond, S. M. and Carr, J. G. (1976). The antimicrobial activity of SO_2—with particular reference to fermented and non-fermented fruit juices. In *Inhibition and Inactivation of Vegetative Microbes* (F. A. Skinner & W. B. Hugo, Eds), Society of Applied Microbiology Symposia Series, Academic Press, New York.

Hansen, E. H., Filho, H. B. and Růžička, J. (1974). Determination of the hydrogen sulphite content of wine by means of the air-gap electrode. *Anal. Chim. Acta*, **71**, 225–7.

Hardell, H. and Theander, O. (1971). Treatment of D-erythrose and D-xylose with sulphite. *Acta Chem. Scand.*, **25**, 877–82.

Harkness, D. R. and Roth, S. (1969). Purification and properties of 2,3-diphosphoglyceric acid phosphatase from human erythrocytes. *Biochem. Biophys. Res. Comm.*, **34**, 849–56.

Harlow, G. A. and Morman, D. H. (1964). Automatic ion exclusion–partition chromatography of acids. *Anal. Chem.*, **36**, 2438–42.

Harris, J. E. and Dennis, C. (1979). The stability of pectolytic enzymes in sulphite liquor in relation to breakdown of sulphited strawberries. *J. Sci. Food Agric.*, **30**, 704–10.

Harris, J. E. and Dennis, C. (1981). Susceptibility of strawberry varieties to breakdown in sulphite by endopolygalacturonase from zygomycete spoilage fungi. *J. Sci. Food Agric.*, **31**, 87–95.

Harris, R. K. (1978). Introduction. In *NMR and the Periodic Table* (R. K. Harris & B. E. Mann, Eds), Academic Press, New York.

Hashimoto, N. (1972). Oxidation of higher alcohols by melanoidins in beer. *J. Inst. Brew.*, **78**, 43–51.

Hashmi, M. H. and Chughtai, N. A. (1968). Semiquantitative determination of anions by circular thin-layer chromatography. *Mikrochim. Acta*, 1040–4.

Haworth, D. T. and Ziegert, R. M. (1968). The thin-layer and column chromatographic separation of some inorganic anions on microcrystalline cellulose. *J. Chromatog.*, **38**, 544–7.

Hayashi, T., Manou, F., Namiki, M. and Tsuji, K. (1981). Structure and interrelation of the free radical species and the precursor produced by the reaction of dehydro-L-ascorbic acid and amino acid. *Agric. Biol. Chem.*, **45**, 711–16.

Hayashi, T., Ohta, Y. and Namiki, M. (1977). Electron spin resonance spectral study on the structure of the novel free radical products formed by the reactions of sugars with amino acids or amines. *J. Agric. Food Chem.*, **25**, 1282–7.

Hayatsu, H. (1977). Cooperative mutagenic actions of bisulphite and nitrogen nucleophiles. *J. Mol. Biol.*, **15**, 19–32.

Hayatsu, H. and Inoue, M. (1971). The oxygen-mediated reaction between 4-thiouracil derivatives and bisulphite. Isolation and characterisation of 1-methyluracil-4-thiosulphate as an intermediate in the formation of 1-methyluracil-4-sulphonate. *J. Amer. Chem. Soc.*, **93**, 2301–6.

Hayatsu, H. and Miura, A. (1970). The mutagenic action of sodium bisulphite. *Biochem. Biophys. Res. Comm.*, **39**, 156–60.

Hayatsu, H. and Shiragami, M. (1979). Reaction of bisulphite with the 5-hydroxymethyl group in pyrimidines and in phage DNA's. *Biochem.*, **18**, 632–7.

Hayatsu, H., Wataya, Y. and Kai, K. (1970). Addition of sodium bisulphite to uracil and to cytosine. *J. Amer. Chem. Soc.*, **92**, 724–6.

Hayon, E., Treinin, A. and Wilf, J. (1972). Electronic spectra, photochemistry and autoxidation mechanism of sulphite–bisulphite–pyrosulphite systems. The SO_2^-, SO_3^-, SO_4^- and SO_5^- radicals. *J. Amer. Chem. Soc.*, **94**, 47–57.

Hegg, D. A. and Hobbs, P. A. (1978). Oxidation of sulphur dioxide in aqueous systems with particular reference to the atmosphere. *Atmospheric Environment*, **12**, 241–53.

Heilbron, Sir Ian, Cook, A. H., Bunbury, H. M. and Hey, D. H. (1965). *Dictionary of Organic Compounds*, 4th edn, Eyre & Spottiswoode Publishers, London.

Heintze, K. (1962). Behaviour of sulphur dioxide in fruit preserves. *Deut. Lebensm. Rundschau*, **58**, 224–7.

Heintze, K. (1970). Determination of sulphurous acid in foods by different methods of analysis. *Ind. Obst-Gemüseverwert.*, **55**, 185–6.

Heintze, K. (1976). Mutual effects of sorbic acid and sulphurous acid. *Ind. Obst-Gemüseverwert.*, **61**, 555–6.

Herbert, R. W., Hirst, E. L., Percival, E. G. V., Reynolds, R. J. W. and Smith, F. (1933). The constitution of ascorbic acid. *J. Chem. Soc.*, 1270–90.

Hermus, R. J. J. (1969). Sulphite-induced thiamine cleavage. *Int. J. Vit. Res.*, **39**, 175–81.

Herrera, J. C. and Wedzicha, B. L. (1981). Unpublished results.

Heunisch, G. W. (1977). Stoichiometry of the reaction of sulphites with hydrogen sulphide ion. *Inorg. Chem.*, **16**, 1411–13.

Hevesi, L. and Bruice, T. C. (1973). Reaction of sulphite with isoalloxazines. *Biochem.*, **12**, 290–7.

Heyns, K. and Hauber, R. (1970). Determination of the structure of specific ^{14}C labelled brown polymerisates of sorbose by thermal fragmentation. *Ann.*, **733**, 159–69.

Hibbert, H. R. (1970). White spot in sausages. Research report No. 159, British Food Manufacturers Research Association, Leatherhead.

Higginson, W. C. E. and Marshall, J. W. (1957). Equivalence changes in oxidation–reduction reactions in solution: some aspects of the oxidation of sulphurous acid. *J. Chem. Soc.*, 447–58.

Hine, J. (1971). Rate and equilibrium in the addition of bases to electrophilic carbon and in S_N1 reactions. *J. Amer. Chem. Soc.*, **93**, 3701–8.

Hine, J., Green, L. R., Meng Jr, P. C. and Thiagarajan, V. (1976). Rates and equilibria in hydration and bisulphite addition to 1,3-dimethoxyacetone. *J. Org. Chem.*, **41**, 3343–9.

Hinze, H., Maier, K. and Holzer, H. (1981). Rapid decrease of the adenosine triphosphate content in *Lactobacillus malii* and *Leuconostoc mesenteroides* after incubation with low concentrations of sulphite. *Z. Lebensm. Untersuch.-Forsch.*, **172**, 389–92.

Hinze, W. L., Elliott, J. and Humphrey, R. E. (1972). Spectrophotometric determination of sulphite with mercuric thiocyanate and ferric ion. *Anal. Chem.*, **44**, 1511–13.

Histaune, I. C. and Heicklen, J. (1975). Infrared spectroscopic study of ammonia–sulphur dioxide–water solid state system. *Can. J. Chem.*, **53**, 2646–56.

Hodge, J. E. (1953). Dehydrated foods. Chemistry of browning reactions in model systems. *J. Agric. Food Chem.*, **1**, 928–43.

Hodge, J. E. (1955). The Amadori rearrangement. *Adv. Carbohydr. Chem.*, **10**, 169–205.

Hodge, J. E. and Rist, C. E. (1952). *N*-Glycosyl derivatives of secondary amines. *J. Amer. Chem. Soc.*, **74**, 1494–7.

Hodge, J. E., Fisher, B. E. and Nelson, E. C. (1963). Dicarbonyls, reductones and heterocyclics produced by reactions of reducing sugars with amine salts. *Proc. Am. Soc. Brew. Chem.*, 84–8.

Hodge, J. E., Mills, F. D. and Fisher, B. E. (1972). Compounds of browned flavour derived from sugar–amine reactions. *Cereal Sci. Today*, **17**, 34–40.

Hoffman, K. A. (1900). Pentacyanoiron compounds. *Ann.*, **312**, 1–33.

Hoffman, M. R. and Edwards, J. O. (1975). Kinetics of the oxidation of sulphite by hydrogen peroxide in acidic solution. *J. Phys. Chem.*, **79**, 2096–8.

Honig, P. (1953). Inorganic nonsugars. In *Principles of Sugar Technology* (P. Honig, Ed.), Vol. 1, Elsevier Publishing Company, Amsterdam.

Horowitz, W. (1975). *Official Methods of Analysis of the Association of Official Analytical Chemists*, 12th edn, Association of Official Analytical Chemists, Washington.

Hough, J. S., Briggs, D. E. and Stevens, R. (1971). *Malting and Brewing Science*, Chapman and Hall Ltd, London.

Huelin, F. E., Coggiola, I. M., Sidhu, G. S. and Kennett, B. H. (1971). The anaerobic decomposition of ascorbic acid in the pH range of foods and in more acid solutions. *J. Sci. Food Agric.*, **22**, 540–2.

Hug, D. H., O'Donnell, P. S. and Hunter, J. K. (1978). Photoactivation of urocanase in *Pseudomonas putida*. Role of sulphite in enzyme modification. *J. Biol. Chem.*, **253**, 7622–9.

Huitt, H. A. and Lodge Jr, J. P. (1964). Equilibrium effects in determination of sulphur dioxide with pararosaniline and formaldehyde. *Anal. Chem.*, **36**, 1305–8.

Humphrey, R. E. and Hinze, W. (1971). Spectrophotometric determination of cyanide, sulphide and sulphite with mercuric chloranilate. *Anal. Chem.*, **43**, 1100–2.

Humphrey, R. E. and Laird, C. E. (1971). Polarographic determination of chloride, cyanide, fluoride, sulphate and sulphite with metal chloranilates. *Anal. Chem.*, **43**, 1895–7.

Humphrey, R. E. and Sharp, S. W. (1976). Polarographic determination of chloride, cyanide, fluoride, sulphate and sulphite ions by an amplification procedure employing metal iodates. *Anal. Chem.*, **48**, 222–3.

Humphrey, R. E., Ward, M. H. and Hinze, W. (1970). Spectrophotometric determination of sulphite with 4,4-dithiopyridine and 5,5′-dithiobis(2-nitrobenzoic acid). *Anal. Chem.*, **42**, 698–702.

Hurd, C. D. (1970). The acidities of ascorbic and sialic acids. *J. Chem. Educ.*, **47**, 481–2.

Husbands, M. J. and Scott, G. (1979). Mechanism of antioxidant action: the behaviour of sulphur dioxide in autoxidising systems. *Eur. Polymer J.*, **15**, 249–53.

Huss Jr, A. and Eckert, C. A. (1977). Equilibria and ion activities in aqueous sulphur dioxide solutions. *J. Phys. Chem.*, **81**, 2268–70.

Hvoslef, J. (1969). Changes in conformation and bonding of ascorbic acid by ionisation. The crystal structure of sodium ascorbate. *Acta Cryst.*, **B25**, 2214–23.

Iguchi, A. (1958a). Separation of polythionates with anion exchange resins. *Bull. Chem. Soc. Japan*, **31**, 597–600.

Iguchi, A. (1958b). The separation of sulphate, sulphite, thiosulphate and sulphide ions with anion exchange resins. *Bull. Chem. Soc. Japan*, **31**, 600–5.

Ingles, D. L. (1959a). Chemistry of non-enzymic browning. VI. The reaction of aldoses with amine bisulphites. *Aust. J. Chem.*, **12**, 275–9.

Ingles, D. L. (1959b). Chemistry of non-enzymic browning. VIII. Hydrolytic reactions of aldose–bisulphite addition compounds. *Aust. J. Chem.*, **12**, 288–95.

Ingles, D. L. (1959c). Reduction of bisulphite to elemental sulphur by reducing sugars. *Chem. Ind. (London)*, 1045–6.

Ingles, D. L. (1960). Reactions of aldoses with bisulphite and hydrosulphite. The formation of sugar sulphates. *Chem. Ind. (London)*, 1159–60.

Ingles, D. L. (1961a). Chemistry of non-enzymic browning. XII. Bisulphite addition compounds of aldose sugars, osones and some related compounds. *Aust. J. Chem.*, **14**, 302–7.

Ingles, D. L. (1961b). Reactions of reducing sugars with sulphite. The formation of β-sulphopropionic acid. *Chem. Ind. (London)*, 1312–13.

Ingles, D. L. (1962). The formation of sulphonic acids from the reaction of reducing sugars with sulphite. *Aust. J. Chem.*, **15**, 342–9.

Ingram, M. (1947a). An electrometric indicator to replace starch in iodine titrations of sulphurous acid in fruit juices. *J. Soc. Chem. Ind.*, **66**, 50–4.

Ingram, M. (1947b). Investigation of errors arising in iodometric determination of free sulphurous acid by the acetone procedure. *J. Soc. Chem. Ind.*, **66**, 105–15.

Ingram, M. (1948). The germicidal effects of free and combined sulphur dioxide. *J. Soc. Chem. Ind.*, **67**, 18–21.

Inoue, M., Hayatsu, H. and Tanooka, H. (1972). Concentration effect of bisulphite on the inactivation of transforming activity of DNA. *Chem. Biol. Interactions*, **5**, 89–95.

Inouye, B., Morita, K., Ishida, T. and Ogata, M. (1980). Cooperative effect of sulphite and vanadium compounds on lipid peroxidation. *Toxicol. Appl. Pharmacol.*, **53**, 101–7.

Institute of Food Technologists and Committee on Public Information (1976). Sulphites as food additives. *Nutr. Rev.*, **34**, 58–62.

Irreverre, F., Mudd, H., Heizer, W. D. and Laster, L. (1967). Sulphite oxidase deficiency. Studies of a patient with mental retardation, dislocated ocular lenses and abnormal urinary secretion of S-sulpho-L-cysteine, sulphite and thiosulphate. *Biochem. Med.*, **1**, 187–217.

Ishida, M. and Mizushima, S. (1969). Membrane ATPase of *Bacillus megaterium*. I. Properties of membrane ATPase and its solubilised form. *J. Biochem.*, **66**, 33–43.

Issa, I. M., El-Meligy, M. S. and Idriss, K. A. (1974). Oxidation of low valent sulphur compounds with lead(IV) acetate. Potentiometric determination of bisulphite. *Indian. J. Chem.*, **12**, 877–9.

Ivanov, M. A. and Lavrent'ev, S. P. (1967). Mutarotation of glucose. *Izv. Vyssh. Ucheb. Zaved., Les. Zh.*, **10**, 138–41. Through *Chemical Abstracts*, **68**, 96959x.

James, G. T. (1979). Essential arginine residues in human liver arylsulphatase A. *Arch. Biochem. Biophys.*, **197**, 57–62.

Janzen, E. G. (1972). Electron spin resonance study of the SO_3^-. Formation in the thermal decomposition of sodium dithionite, sodium and potassium metabisulphite and sodium hydrogen sulphite. *J. Phys. Chem.*, **76**, 157–62.

Janzen, E. G. and DuBase Jr, C. M. (1965). ESR study of the formation of SO_2^- in the pyrolysis of sodium bisulphite adducts of aldehydes and ketones. *Tetrahedron Lett.*, 2521–4.

Jaroensanti, J. and Panijpan, B. (1981). Kinetics of bisulphite cleavage of three biological phosphorylated derivatives of thiamine. *Int. J. Vit. Nutr. Res.*, **51**, 34–8.

Jaulmes, P. and Hamelle, G. (1961). The determination of sulphite ion iodometrically and of aldehydes by the sulphite method. *Ann. Fals. Expert. Chim.*, **54**, 338–46.

Jay, J. M. (1970). *Modern Food Microbiology*, Van Nostrand, New York.

Jencks, W. P. (1964). Mechanism and catalysis of simple carbonyl group reactions. In *Progress in Physical Organic Chemistry*, Vol. 2 (S. G. Cohen, A. Streiwieser & R. W. Taft, Eds), Interscience Publishers, New York.

Jennings, N., Bunton, N. G., Crosby, N. T. and Alliston, T. G. (1978). A comparison of three methods for the determination of sulphur dioxide in food and drink. *J. Ass. Public Analysts*, **16**, 59–70.

Johansson, L. G., Lindqvist, O. and Vannerberg, N. G. (1980). The structure of caesium hydrogen sulphite. *Acta Cryst.*, **B36**, 2523–6.

Johnstone, H. F. and Leppla, P. W. (1934). The solubility of sulphur dioxide at low partial pressures. The ionisation constant and heat of ionisation of sulphurous acid. *J. Amer. Chem. Soc.*, **56**, 2233–8.

Jones, L. H. and McLaren, E. (1958). Infrared absorption spectrum of SO_2 and CO_2 in aqueous solution. *J. Chem. Phys.*, **28**, 995.

Jorns, M. S. (1980). Studies on the function of FAD in almond oxynitrilase. In *Flavins and Flavoproteins* (K. Yagi & Y. Yamano, Eds), Japan Scientific Societies Press, Tokyo and University Park Press, Baltimore.

Joseph, G. H. and Havighorst, C. R. (1952). Engineering quality pectins. *Food Eng.*, 87–9, 160–2.

Joslyn, M. A. (1955). Determination of free and total sulphur dioxide in white table wines. *J. Agric. Food Chem.*, **3**, 686–95.

Joslyn, M. A. and Braverman, J. B. S. (1954). The chemistry and technology of the pretreatment and preservation of fruit and vegetable products with sulphur dioxide and sulphites. *Adv. Food Res.*, **5**, 97–160.

Jungreis, E. and Anavi, Z. (1969). Determination of sulphite ion (or sulphur dioxide) by atomic absorption spectroscopy. *Anal. Chim. Acta*, **45**, 190–2.

Kalus, W. H. and Filby, W. G. (1976). Formation of monoanion radicals in reactions of vitamin K_3 with sodium sulphite. *Experientia*, **32**, 1366–8.

Kalus, W. H. and Filby, W. G. (1977). The effect of additives on the free radical formation in aqueous solutions of ascorbic acid. *Int. J. Vit. Nutr. Res.*, **47**, 258–64.

Kamogawa, A. and Fukui, T. (1973). Inhibition of α-glucan phosphorylase by bisulphite competition at the phosphate binding site. *Biochim. Biophys. Acta*, **302**, 158–66.

Kaplan, D., McJilton, C. and Luchtel, D. (1975). Bisulphite induced lipid oxidation. *Arch. Environ. Health*, **30**, 507–9.

Karmarkar, K. H. and Guilbault, G. G. (1974). A new design and coatings for piezoelectric crystals in measurement of trace amounts of sulphur dioxide. *Anal. Chim. Acta*, **71**, 419–24.

Kato, H. (1967). Formation of substituted pyrrole-2-aldehydes by reaction of aldoses with alkylamines. *Agric. Biol. Chem.*, **31**, 1086–90.

Kato, H. and Fujimaki, M. (1968). Formation of N-substituted pyrrole-2-aldehydes in the browning reaction between xylose and amino compounds. *J. Food Sci.*, **33**, 445–9.

Kato, H. and Fujimaki, M. (1970). Formation of 1-alkyl-5-(hydroxymethyl)pyrrole-2-aldehydes in the browning reaction between hexoses and alkylamines. *Agric. Biol. Chem.*, **34**, 1071–7.

Kato, H. and Tsuchida, H. (1981). Estimation of melanoidin structure by pyrolysis and oxidation. *Prog. Food Nutr. Sci.*, **5**, 147–56.

Kato, K. (1967). Pyrolysis of cellulose. Part III. Comparative studies of the volatile compounds from pyrolysates of cellulose and its related compounds. *Agric. Biol. Chem.*, **31**, 657–63.

Kaushik, R. L. and Prosad, R. (1969). Periodate as an analytical reagent. Part I. *J. Indian Chem. Soc.*, **46**, 403–10.

Kenyon, J. and Munro, N. (1948). The isolation and some properties of dehydro-L-ascorbic acid. *J. Chem. Soc.*, 158–61.

Kharasch, M. S., Fono, A. and Nudenberg, W. (1951). The chemistry of hydroperoxides VII. The ionic and free radical decomposition of α-phenylethyl hydroperoxide. *J. Org. Chem.*, **16**, 128–30.

Kharasch, M. S., May, E. M. and Mayo, F. R. (1939a). The peroxide effect in the addition of reagents to unsaturated compounds. XVIII. The addition and substitution of bisulphite. *J. Org. Chem.*, **3**, 175–92.

Kharasch, M. S., May, E. M. and Mayo, F. R. (1939b). The peroxide effect in the addition of reagents to unsaturated compounds. XXIII. The reaction of styrene with bisulphite. *J. Amer. Chem. Soc.*, **61**, 3092–8.

Kice, J. L. and Anderson, J. M. (1966). The mechanism of the acid hydrolysis of sodium arylsulphates. *J. Amer. Chem. Soc.*, **88**, 5242–5.

Kice, J. L., Anderson, J. M. and Pawlowski, N. E. (1966). The mechanism of the acid hydrolysis of Bunte salts (*S*-alkyl and *S*-aryl thiosulphates). *J. Amer. Chem. Soc.*, **88**, 5245–50.

Kierkegaard, P., Norrestam, R., Werner, P. E., Csöregh, I., Glehn, M. von, Karlsson, R., Leijonmarck, M., Rönnquist, O., Stensland, B., Tillberg, O. and Torbjörnsson, L. (1971). X-ray structure investigation of flavin derivatives. In *Flavins and Flavoproteins* (H. Kamin, Ed.), University Park Press, Baltimore, and Butterworth & Co. (Publishers) Ltd, London.

King, A. D., Michener, J. D. and Ito, K. A. (1969). Control of *Byssochlamys* and related heat-resistant fungi in grape products. *Appl. Microbiol.*, **18**, 166–73.

King Jr, W. H. (1964). Piezoelectric sorption detector. *Anal. Chem.*, **36**, 1735–9.

Klayman, D. L. and Shine, R. J. (1968). The chemistry of organic thiosulphates. *Quart. Rep. Sulphur Chem.*, **3**, 190–316.

Knowles, M. E. (1971). Inhibition of non-enzymic browning by sulphite: identification of sulphonated products. *Chem. Ind. (London)*, 910–11.

Knowles, M. E. and Eagles, J. (1971). Trimethylsilyl esters of aromatic and aliphatic sulphonic acids. Preparation and mass spectral properties. *Anal. Chem.*, **43**, 1697–8.

Kobayashi, K., Seki, Y. and Ishimoto, M. (1974). Biochemical studies on sulphate-reducing bacteria. XIII. Sulphite reductase from *Desulphovibrio vulgaris*—mechanism of trithionate, thiosulphate and sulphide formation and enzymatic properties. *J. Biochem. (Tokyo)*, **75**, 519–29.

Kobayashi, K., Takahashi, E. and Ishimoto, M. (1972). Biochemical studies on sulphate-reducing bacteria. XI. Purification and some properties of sulphite reductase, desulphoviridin. *J. Biochem. (Tokyo)*, **72**, 879–87.

Koch, R. C. (1960). *Activation Analysis Handbook*, Academic Press, New York.

Kochler, P. E. and Odell, G. U. (1970). Factors affecting the formation of pyrazine compounds in sugar–amine reactions. *J. Agric. Food Chem.*, **18**, 895–8.

Kokesh, F. C. and Hall, R. E. (1975). A re-examination of the equilibrium addition of bisulphite and sulphite ions to benzaldehyde. *J. Org. Chem.*, **40**, 1632–6.

Kolthoff, I. M. and Miller, C. S. (1941). The reduction of sulphurous acid (sulphur dioxide) at the dropping mercury electrode. *J. Amer. Chem. Soc.*, **63**, 2818–21.

Kolthoff, I. M., Belcher, R., Stenger, V. A. and Matsuyama, G. (1957). *Volumetric Analysis. Volume III. Titration methods: oxidation–reduction reactions.* Interscience Publishers, New York and London.

Koppanyi, T., Vivino, A. E. and Veitch Jr, F. P. (1945). A reaction of ascorbic acid with α-amino acids. *Science*, **101**, 541–2.

Kozlov, E. I., L'vova, M. Sh. and Chugunov, V. V. (1979). Stability of water soluble vitamins and coenzymes. IV. Study of kinetics of hydrolysis and stabilisation mechanism of pyridoxal 5'-phosphate by bisulphite ions. *Khim-Farm Zh.*, **13**, 69–73.

Krol, B. and Moerman, P. C. (1959). The effect of sulphite on the colour and bacteriological condition of minced meat. *Conserva*, **8**, 1–22.

Kroneck, P., Lutz, O. and Nolle, A. (1980). NMR investigations of ^{17}O, ^{33}S, ^{95}Mo and ^{97}Mo in thiomolybdates. *Z. Naturforsch.*, **35a**, 226–9.

Küntzel, A. and Schwörzer, L. K. (1957). The preparation of water-soluble leather oiling materials by addition of bisulphite to unsaturated oils. I. Sulphiting of cod liver oil. *Leder*, **8**, 5–13.

Kurata, T. and Fujimaki, M. (1974). Monodehydro-2,2-iminodi-2(2')-deoxy-L-ascorbic acid, a radical product from the reaction of dehydro-L-ascorbic acid with α-amino acid. *Agric. Biol. Chem.*, **38**, 1981–8.

Kurata, T. and Fujimaki, M. (1976). Formation of 3-keto-4-deoxypentosone and 3-hydroxy-2-pyrone by the degradation of dehydro-L-ascorbic acid. *Agric. Biol. Chem.*, **40**, 1287–91.

Kurata, T. and Sakurai, Y. (1967a). Degradation of L-ascorbic acid and mechanism of non-enzymic browning reaction. Part II. Non-oxidative degradation of L-ascorbic acid including the formation of 3-deoxy-L-pentosone. *Agric. Biol. Chem.*, **31**, 170–6.

Kurata, T. and Sakurai, Y. (1967b). Degradation of L-ascorbic acid and mechanism of non-enzymic browning. Part III. Oxidative degradation of L-ascorbic acid (degradation of dehydro-L-ascorbic acid). *Agric. Biol. Chem.*, **31**, 177–84.

Kurata, T., Fujimaki, M. and Sakurai, Y. (1973). Red pigment produced by the reaction of dehydro-L-ascorbic acid with α-amino acid. *Agric. Biol. Chem.*, **37**, 1471–7.

Lafon-Lafourcade, S. and Peynaud, E. (1974). Antibacterial action of the free and combined forms of sulphur dioxide. *Connaiss. Vigne Vin.*, **8**, 187–203.

Lalikainen, T., Joslyn, M. A. and Chichester, C. O. (1958). Mechanism of browning of ascorbic acid–citric acid–glycine system. *J. Agric. Food Chem.*, **6**, 135–9.

Lamaty, G. and Roque, J. P. (1969). Secondary isotope effect of deuterium in the decomposition of 4-*tert*-butyl-cyclohexanone bisulphite addition product. Conformational energy of the sulphite (SO_3^{2-}) group. *Rec. Trav. Chim.*, **88**, 131–8.

Lambeth, D. O. and Lardy, H. A. (1971). Purification and properties of rat-liver-mitochondrial adenosine triphosphatase. *Eur. J. Biochem.*, **22**, 355–63.

Lamikanra, O. (1982). Sulphite-induced oxidation and browning of linolenic acid. *J. Food Sci.*, **47**, 2025–7, 2032.

Lancaster, J. M. and Murray, R. S. (1971). The ferricyanide–sulphite reaction. *J. Chem. Soc. A*, 2755–8.

Landolt, H. (1887). The kinetics of the reaction between iodic and sulphurous acids. *Ber.*, **20**, 745–60.

Lantzke, I. R., Covington, A. K. and Robinson, R. A. (1973). Osmotic and activity coefficients of sodium dithionate and sodium sulphite at 25 °C. *J. Chem. Eng. Data*, **18**, 421–3.

Larson, T. V., Horike, N. R. and Harrison, H. (1978). Oxidation of sulphur dioxide by oxygen and ozone in aqueous solution. A kinetic study with significance to atmospheric rate processes. *Atmospheric Environment*, **12**, 1597–611.

Larsson, L. O. and Kierkegaard, P. (1969). Crystal structure of sodium sulphite. Refinement allowing for the effect of crystal twinning. *Acta Chem. Scand.*, **23**, 2253–60.

Laskin, S., Kuschner, M., Sellakumar, A. and Katz, G. V. (1976). Combined carcinogen–irritant animal inhalation studies. In *Air Pollution and the Lung* (E. F. Aharonson, A. Ben-David & M. A. Klingberg, Eds), John Wiley & Sons, New York.

Lay, A. (1970). Iodometric determination of sulphur dioxide. *Mitt. Rebe. Wein Obstbau Früchteverwert (Klosterneuberg)*, **20**, 85–93.

Lay, A. (1971). New high speed process for determining the total sulphur dioxide content of wines as compared with other methods. *Mitt. Rebe. Wein Obstbau Früchteverwert (Klosterneuberg)*, **21**, 439–43.

Lecher, H. Z. and Hardy, W. B. (1948). Compounds of trialkylamine oxides with sulphur dioxide and trioxide. *J. Amer. Chem. Soc.*, **70**, 3789–92.

Lederer, F. (1978). Sulphite binding to a flavodehydrogenase, cytochrome b_2 from baker's yeast. *Eur. J. Biochem.*, **88**, 425–31.

Lee, K. (1968). [33]S nuclear magnetic resonance in paramagnetic α-MnS. *Phys. Rev.*, **172**, 284–7.

Legault, R. R., Hendel, C. E., Talburt, W. F. and Rasmussen, L. B. (1949). Sulphite disappearance in dehydrated vegetables during storage. *Ind. Eng. Chem.*, **41**, 1447–51.

Leichter, J. and Joslyn, M. A. (1969). Kinetics of thiamine cleavage by sulphite. *Biochem. J.*, **113**, 611–15.

Lem, W. J. and Wayman, M. (1970). The decomposition of aqueous dithionite. III. Stabilisation of dithionite by cations. *Can. J. Chem.*, **48**, 2778–85.

Leonardos, G., Kendall, D. and Barnard, N. (1969). Odour threshold determinations of 53 odourant chemicals. *J. Air Pollut. Contr. Ass.*, **19**, 91–5.

Lidzey, R. G., Sawyer, R. and Stockwell, P. B. (1971). Improvements relating to the automatic determination of alcohol in beers and wines, using a distillation technique. *Lab. Practice*, **20**, 213–16, 219.

Lindberg, B., Tanaka, J. and Theander, O. (1964). Reaction between D-glucose and sulphite. *Acta Chem. Scand.*, **18**, 1164–70.

Lindqvist, I. and Mörtsell, M. (1957). The structure of potassium pyrosulphite and the nature of the pyrosulphite ion. *Acta Cryst.*, **10**, 406–9.

Linke, W. F. (1965). *Solubility of Inorganic and Metal-Organic Compounds*, Vol. II, 4th edn, American Chemical Society, Washington, DC.

Lizada, M. C. C. and Yang, S. F. (1981). Sulphite induced lipid oxidation. *Lipids*, **16**, 189–94.

Lloyd, W. J. W. and Cowle, B. C. (1963). The determination of free sulphur dioxide in soft drinks by a desorption and trapping method. *Analyst (London)*, **88**, 394–8.

Logsdon II, O. J. and Carter, M. J. (1975). Comparison of manual and automated analysis methods for sulphur dioxide in manually impinged ambient air samples. *Env. Sci. Technol.*, **13**, 1172–4.

Lovejoy, R. W., Colwell, J. H., Eggers Jr, D. F. and Halsey Jr, G. D. (1962). Infrared spectrum and thermodynamic properties of gaseous sulphur trioxide. *J. Chem. Phys.*, **36**, 612–17.

Lück, H. and Rickerl, E. (1959). Increased resistance of *Escherichia coli* against preservatives and antibiotics. *Z. Lebensm. Untersuch.-Forsch.*, **109**, 322–9.

Lueck, E. (1980). *Antimicrobial Food Additives*, Springer-Verlag, Berlin, Heidelberg.

Lukes, B., O'Brien, T. J. and Scanlan, R. A. (1980). Residual sulphur dioxide in finished malt: colorimetric determination and relation to *N*-nitrosodimethyl-amine. *J. Am. Soc. Brew. Chem.*, **38**, 146–8.

Lutz, O., Nolle, A. and Schwenk, A. (1973). NMR investigations and nuclear magnetic moment of ^{33}S. *Z. Naturforsch.*, **28a**, 1370–2.

LuValle, J. E., Goldberg, G. M., Davidson, D. L., McNally, A. M., Des Roches, J. and Neary, M. (1958). Technical note on the chemistry of quinone. US Technical Operations Report, No. TO1 58-29.

Lynt Jr, R. K. (1966). Survival and recovery of enterovirus from foods. *Appl. Microbiol.*, **14**, 218–22.

Machell, G. and Richards, G. N. (1960). Mechanism of saccharinic acid formation. Part III. The α-keto-aldehyde intermediate in formation of D-glucometa-saccharinic acid. *J. Chem. Soc.*, 1938–44.

Macris, B. J. (1972). Transport and toxicity of sulphur dioxide in the yeast *Saccharomyces cerevisiae* var. *ellipsoideus*, Ph.D. Thesis, Michigan State University.

Macris, B. J. and Markakis, P. (1974). Transport and toxicity of sulphur dioxide in *Saccharomyces cerevisiae* var *ellipsoideus*. *J. Sci. Food Agric.*, **25**, 21–9.

Mader, P. M. (1958). Kinetics of the hydrogen peroxide–sulphite reaction in alkaline solution. *J. Amer. Chem. Soc.*, **80**, 2634–9.

Maillard, L. C. (1912). Action of amino acids on sugars. Formation of melanoidins in a methodical way. *Compt. Rend.*, **154**, 66–8.

Maillard, L. C. (1916). Synthesis of humus-like substances by the interaction of amino acids and reducing sugars. *Annal. Chim. (Rome)*, **5**, 258–317.

Mallon, R. G. and Rossman, T. G. (1981). Bisulphite (sulphur dioxide) is a comutagen in *E. coli* and in Chinese hamster cells. *Mutation Res.*, **88**, 125–33.

Mangan, J. L. and Doak, B. W. (1949). Behaviour of sulphur dioxide in dehydrated vegetables. *NZ J. Sci. Technol.*, **30B**, 232–40.

Manyai, S. and Varady, Zs. (1958). Selective breakdown of 2,3-diphosphoglyceric acid in red blood cells. *Acta Physiol. Acad. Sci. Hung.*, **14**, 103–14.

Mapson, L. W. (1941). Determination of ascorbic acid in foodstuffs containing SO_2. *J. Soc. Chem. Ind.*, **60**, 802.

Markakis, P. (1982). Stability of anthocyanins in foods. In *Anthocyanins as Food Colours* (P. Markakis, Ed.), Academic Press, London.

Markakis, P. and Embs, R. J. (1966). Effect of sulphite and ascorbic acid on mushroom phenoloxidase. *J. Food Sci.*, **31**, 807–11.

Mascini, M. and Cremisini, C. (1977). A new pH electrode for gas-sensing probes. *Anal. Chim. Acta*, **92**, 277–83.

Mascini, M. and Muratori, T. (1975). Continuous measurement of sulphur dioxide and total oxidant as atmospheric pollutants by means of an I^- membrane electrode. *Annal. Chim. (Rome)*, **65**, 287–300.

Mason, H. M. and Walsh, G. (1928a). Note on the oxidation of sulphites by air. *Analyst (London)*, **53**, 142–4.

Mason, H. M. and Walsh, G. (1928b). Note on the titration of dilute sulphite solutions with standard iodine solutions. *Analyst (London)*, **53**, 144–9.

Mason, L. F. A. (1975). *Photographic Processing Chemistry*, Focal Press, London.

Massey, V., Muller, F., Feldberg, R., Schuman, M., Sullivan, P. A., Howell, L. G., Mayhew, S. G., Matthews, R. G. and Foust, G. P. (1969). The reactivity of flavoproteins with sulphite. *J. Biol. Chem.*, **244**, 3999–4006.

Matsuda, T. (1979). Amperometric determination of sodium hydrogen sulphite with chloramine-T. *Talanta*, **26**, 326–8.

Mauron, J. (1981). The Maillard reaction in food; a critical review from the nutritional standpoint. *Prog. Food Nutr. Sci.*, **5**, 5–35.

Mayer, K., Vetsch, U. and Pause, G. (1975). Inhibition of acid spoilage of biological origin by bound sulphurous acid. *Schweiz. Z. Obst-Weinbau*, **111**, 590–6.

Maylor, R., Gill, J. B. and Goodall, D. C. (1972). Tetra-*n*-alkylammonium bisulphites: a new example of the existence of the bisulphite ion in solid compounds. *J. Chem. Soc., Dalton Trans.*, 2001–3.

Mayo, F. R. and Walling, C. (1940). The peroxide effect in the addition of reagents to unsaturated compounds and in rearrangement reactions. *Chem. Rev.*, **27**, 351–412.

McCord, C. P. and Witheridge, W. N. (1949). *Odours, Physiology and Control*, McGraw-Hill Book Co., New York.

McCord, J. M. and Fridovich, I. (1969a). Superoxide dismutase. An enzymic function for erythrocuprein (haemocuprein). *J. Biol. Chem.*, **244**, 6049–55.

McCord, J. N. and Fridovich, I. (1969b). The utility of superoxide dismutase in studying free radical reactions. I. Radicals generated by the interaction of sulphite, dimethylsulphoxide, and oxygen. *J. Biol. Chem.*, **244**, 6056–63.

McGinnis, R. A. (1971). Juice purification III. In *Sugar Beet Technology* (R. A. McGinnis, Ed.), 2nd edn, Beet Sugar Development Foundation, Colorado.

McKay, H. A. C. (1971). The atmospheric oxidation of sulphur dioxide in water droplets in the presence of ammonia. *Atmospheric Environment*, **5**, 7–14.

McPhee, J. R. (1956). Further studies on the reactions of disulphides with sodium sulphite. *Biochem. J.*, **64**, 22–9.

McWeeny, D. J. (1981). Sulphur dioxide and the Maillard reaction in food. *Prog. Food Nutr. Sci.*, **5**, 395–404.

McWeeny, D. J. and Burton, H. S. (1963). Some possible glucose/glycine browning intermediates and their reactions with sulphite. *J. Sci. Food Agric.*, **14**, 291–302.

McWeeny, D. J., Biltcliffe, D. O., Powell, R. C. T. and Spark, A. A. (1969). The Maillard reaction and its inhibition by sulphite. *J. Food Sci.*, **34**, 641–3.

McWeeny, D. J., Knowles, M. E. and Hearne, J. F. (1974). The chemistry of non-enzymic browning in foods and its control by sulphites. *J. Sci. Food Agric.*, **25**, 735–46.

McWeeny, D. J., Shepherd, M. J. and Bates, M. L. (1980). Physical loss and chemical reactions of SO_2 in strawberry jam production. *J. Food Technol.*, **15**, 613–17.

Merritt, J. and Agazzi, E. J. (1966). Chemical bonding effects on sulphur L emission spectra. *Anal. Chem.*, **38**, 1954.

Meyer, B., Ospina, M. and Peter, L. B. (1980). Raman spectrometric determination of oxysulphur anions in aqueous systems. *Anal. Chim. Acta*, **117**, 301–11.

Meyer, B., Peter, L. and Shaskey-Rosenlund, C. (1979). Raman spectra of isotopic bisulphite and disulphite ions in alkali salts and aqueous solutions. *Spectrochim. Acta*, **35A**, 345–54.

Meyer, B., Peter, L. and Spitzer, K. (1977). Trends in charge distributions in sulphanes, sulphanesulphonic acids, sulphanedisulphonic acids and sulphurous acid. *Inorg. Chem.*, **16**, 27–33.

Meyerhof, O., Ohlmeyer, P. and Mohle, W. (1938). Coupling between oxidation–reduction and phosphorylation in anaerobic carbohydrate splitting. II. Coupling as an equilibrium reaction. *Biochem. Z.*, **297**, 113–33.

Meyers, R. D. and Jorns, M. S. (1981). Identification of the factor responsible for autogenous bleaching of glycollate oxidase. *Biochim. Biophys. Acta*, **657**, 438–47.

Miller, J. M. and Pena, R. de (1972). Contribution of scavenged sulphur dioxide to the sulphate of rain water. *J. Geophys. Res.*, **30**, 5905–16.

Mitchell, S. C. and Waring, R. H. (1978). Detection of inorganic sulphate and other anions on paper and thin-layer chromatograms. *J. Chromatog.*, **166**, 341–3.

Mitsuda, H., Yasumoto, K. and Yokohama, K. (1965). Studies on the free radical in amino–carbonyl reaction. *Agric. Biol. Chem.*, **29**, 751–6.

Mitsuhashi, Y., Hamano, T., Hasegawa, A., Tanaka, K., Matsuki, Y., Adachi, T., Obara, K., Nonogi, H., Fuke, T., Sudo, M., Ikuzawa, M., Fujita, K., Izumi, T., Ogawa, S., Toyoda, M., Ito, Y. and Iwaida, M. (1979). Comparative determination of free and combined sulphite in foods by the modified Rankine method and flame photometric detection gas chromatography. *Z. Lebensm. Untersuch.-Forsch.*, **168**, 299–304.

Miyata, T., Sakumoto, A., Washino, M. and Abe, T. (1975a). Hydrogen isotope effect on radical addition reaction of sodium hydrogen sulphite to allyl alcohol. *Chem. Lett.*, 367–70.

Miyata, T., Sakumoto, A., Washino, M. and Abe, T. (1975*b*). Kinetic study of the radical addition of sodium hydrogen sulphite to allyl alcohol. *Chem. Lett.*, 181–4.

Mochalov, K. N., Groisberg, A. T. and Polovnyak, V. K. (1973). Reduction of hydrosulphite ion by sodium tetrahydroborate. *Izv. Vyssh. Ucheb. Zaved, Khim Tekhnol.*, **16**, 1441–2. Through *Chemical Abstracts*, **80**, 9980d.

Moerck, K. E., McElfresh, P., Wohlman, A. and Hilton, B. (1980). Aflatoxin destruction in corn using sodium bisulphite, sodium hydroxide and aqueous ammonia. *J. Food Prot.*, **43**, 571–4.

Mohan, M. S. and Rechnitz, G. A. (1973). Preparation and properties of the sulphate ion selective membrane electrode. *Anal. Chem.*, **45**, 1323–6.

Monier-Williams, G. W. (1927). Determination of sulphur dioxide. *Analyst (London)*, **52**, 415–16.

Monikowski, E. and Chmielnicka, H. (1967). Role of sulphurous acid in the process of peroxidase oxidation of L-ascorbic acid in the presence of hydrogen peroxide. II. Activity of peroxidases in sulphite-treated edible juices and pulps. *Rev. Ferment. Ind. Aliment.*, **22**, 141–4.

Montgomery, M. W. and Sgarbieri, V. C. (1975). Isoenzymes of banana polyphenol oxidase. *Phytochem.*, **14**, 1245–9.

Mori, Y., Kumano, S., Nango, I. and Kano, M. (1967). Browning reactions between L-ascorbic acid and amino acids. I. Browning reactions of L-ascorbic acid and of dehydro-L-ascorbic acid with glycine. *Eiyo to Shokuryo*, **20**, 211–15.

Morozov, A. I., Sytilin, M. S. and Makolkin, I. A. (1972). Kinetics of the reaction of acrylonitrile with sodium sulphite at constant pH in an unbuffered medium. *Zh. Fiz. Khim.*, **46**, 1295–7.

Moser, W., Chalmer, R. A. and Fogg, A. G. (1965). The Boedeker Reaction I. The interaction of alkali metal sulphite and nitroprussides in aqueous solution. *J. Inorg. Nucl. Chem.*, **27**, 831–40.

Motai, H. (1973). Colour tone of various melanoidins produced from model systems. *Agric. Biol. Chem.*, **37**, 1679–85.

Motai, H. (1975). Viscosity of melanoidin from a glycine xylose system. Support for the equation exhibiting a relationship between colour intensity and molecular weight. *J. Agric. Food Chem.*, **23**, 606–7.

Motai, H. and Inoue, S. (1974). Conversion of colour components of melanoidin produced from the glycine–xylose system. *Agric. Biol. Chem.*, **38**, 233–9.

Moussa, R. S. (1973). The effect of potassium sulphite on the recovery of micro-organisms from dehydrated onion. *J. Sci. Food Agric.*, **24**, 401–6.

Moutonnet, P. (1967). Influence of sulphites on proteolysis: studies *in vitro. Ann. Biol. Anim. Bioch. Biophys.*, **7**, 63–71.

Mukai, F., Hawryluk, I. and Shapiro, R. (1970). The mutagenic specificity of sodium bisulphite. *Biochem. Biophys. Res. Comm.*, **39**, 983–8.

Mukerji, S. K. (1977). Corn leaf phosphoenolpyruvate carboxylases. Inhibition of $^{14}CO_2$ fixation by SO_3^{2-} and activation by glucose-6-phosphate. *Arch. Biochem. Biophys.*, **182**, 360–5.

Muller, F. and Massey, V. (1969). Flavin–sulphite complexes and their structures. *J. Biol. Chem.*, **244**, 4007–16.

Muneta, P. and Wang, H. (1977). Influence of pH and bisulphite on the enzymatic blackening reaction in potatoes. *Amer. Potato J.*, **54**, 73–81.

Murphy, M. J., Siegel, L. M., Kamin, H. and Rosenthal, D. (1973). Reduced nicotinamide adenine dinucleotide phosphate–sulphite reductase of enterobacteria—II. Identification of a new class of haem prosthetic group: an iron–tetrahydroporphyrin (isobacteriochlorin type) with eight carboxylic acid groups. *J. Biol. Chem.*, **248**, 2801–14.

Murray, R. S. (1974). Reinvestigation of the reaction between hexacyanoferrate(III) and sulphite ions. *J. Chem. Soc., Dalton Trans.*, 2381–3.

Murray, R. S. and Stranks, D. R. (1970). Internal oxidation–reduction of *trans*[Co(en)$_2$SO$_3$OH$_2$]$^+$. *Inorg. Chem.*, **9**, 1472–5.

Mushran, S. P. and Agrawal, M. C. (1977). Mechanistic studies on the oxidation of ascorbic acid. *J. Sci. Ind. Res.*, **36**, 274–83.

Nagaraja, K. V. and Manjrekar, S. P. (1971). Microcolorimetric estimation of sulphur dioxide in fruits and fruit products. *J. Ass. Off. Anal. Chem.*, **54**, 1146–9.

Nakamura, H. and Tamura, Z. (1974). Spectrophotometric determination of Bunte salts. *Chem. Pharm. Bull.*, **22**, 2208–10.

Nakamura, H. and Tamura, Z. (1975*a*). Fluorescence reaction of bisulphite with indophenol blue and its application to the determination of bisulphite. *Chem. Pharm. Bull.*, **23**, 1261–70.

Nakamura, H. and Tamura, Z. (1975*b*). Chromatographic detection of Bunte salts and disulphides. *J. Chromatog.*, **104**, 389–98.

Namiki, M. and Hayashi, T. (1973). Free radicals developed in the amino–carbonyl reaction of sugars with amino acids. *Agric. Biol. Chem.*, **37**, 2935–6.

Namiki, M. and Hayashi, T. (1974). Development of novel free radicals during the amino–carbonyl reaction of sugars with amino acids. *J. Agric. Food Chem.*, **23**, 487–91.

Namiki, M. and Hayashi, T. (1981). Formation of novel free radical products in an early stage of Maillard reaction. *Prog. Food Nutr. Sci.*, **5**, 81–91.

Namiki, M., Yano, M. and Hayashi, T. (1974). Free radical product formed by the reaction of dehydroascorbic acid with amino acids. *Chem. Lett.*, 125–8.

Nehring, P. (1961). Determination of sulphurous acid in preserves, jams and similar products. *Deut. Lebensm. Rundschau*, **57**, 145–9.

Nelson, K. E. and Baker, G. A. (1963). Studies on the sulphur dioxide fumigation of table grapes. *Amer. J. Enol. Vitic.*, **14**, 13–22.

Nichols, P. F. and Reed, H. M. (1932). Distillation methods for the determination of sulphur dioxide. *Ind. Eng. Chem. (Anal. Edn)*, **4**, 79–84.

Nickless, G. (1968). The sulphur atom and its nucleus. In *Inorganic Sulphur Chemistry* (G. Nickless, Ed.), Elsevier Publishing Co., Amsterdam.

Nishima, T. and Matsumoto, S. (1969). Effect of food constituents on additive analysis. I. Experimental methods for sulphur dioxide. *Tokyo Toritsu Eisei Kenkyusho Kenkyu Nempo*, **21**, 50–62. Through *Chemical Abstracts*, **76**, 44732p.

Noodleman, L. and Mitchell, K. A. R. (1978). X$_\alpha$ scattered wave calculation of the electronic structures of SO$_2$ and SO$_2$F$_2$. Relationship to π bonding in the cyclic phosphazenes. *Inorg. Chem.*, **17**, 2709–17.

Nord, A. G. (1973). Refinement of the crystal structure of Thenardite, Na$_2$SO$_4$(V). *Acta Chem. Scand.*, **27**, 814–22.

Nord, F. F. and Weiss, S. (1958). Fermentation and respiration. In *The Chemistry and Biology of Yeasts* (A. H. Cook, Ed.), Academic Press, New York.

Norman, R. O. C., Storey, P. M. and West, P. R. (1970). Electron spin resonance studies. Part XXV. Reactions of the sulphate radical anion with organic compounds. *J. Chem. Soc. B*, 1087–95.

Norton, C. J., Seppi, N. F. and Reuter, M. J. (1968). Alkanesulphonate synthesis I. Ion catalysis of sulphite radical-ion addition to olefins. *J. Org. Chem.*, **33**, 4158–65.

Nury, F. S. and Bolin, H. R. (1965). Collaborative study on assay of sulphur dioxide in dried fruit. *J. Ass. Off. Anal. Chem.*, **48**, 796–801.

Nury, F. S., Taylor, D. H. and Brekke, J. E. (1959). Modified direct colorimetric method for determination of sulphur dioxide in dried fruits. *J. Agric. Food Chem.*, **7**, 351–3.

Nutting, M. D., Neumann, H. J. and Wagner, J. R. (1970). Effect of processing variables on the stability of β-carotene and xanthophylls of dehydrated parsley. *J. Sci. Food Agric.*, **21**, 197–202.

O'Brien, T. J., Lukes, B. K. and Scanlan, R. A. (1980). Control of N-nitrosodimethylamine in malt through the use of liquid/gaseous sulphur dioxide. *Tech. Quart., Master Brewers Ass. Am.*, **17**, 196–200.

Ogg Jr, R. A. and Ray, J. D. (1957). Quadrupole relaxation and structures in nitrogen magnetic resonances of ammonia and ammonium salts. *J. Chem. Phys.*, **26**, 1339–40.

Ohmori, M. and Takagi, M. (1978). A facile preparation of dehydro-L-ascorbic acid–methanol solution and its stability. *Agric. Biol. Chem.*, **42**, 173–4.

Okumura, T. and Nishikawa, Y. (1976). Chromatography of inorganic compounds I. Chromatography of inorganic anions using aluminium(III)–morin fluorescent complex as a detection reagent. *Bunseki Kagaku*, **25**, 419–23. Through *Chemical Abstracts*, **86**, 11461c.

Oldfield, J. F. T. (1974). Browning control in beet sugar manufacture. *Chem. Ind.*, 720–1.

Oloman, C. (1970). The preparation of dithionites by electrolytic reduction of sulphur dioxide in water. *J. Electrochem. Soc., Electrochem. Tech.*, **117**, 1604–9.

Olsson, K., Pernemalm, P.-A., Popoff, T. and Theander, O. (1977). Reaction of D-glucose and methylamine in slightly acidic aqueous solutions. *Acta Chem. Scand.*, **B31**, 469–74.

Olsson, K., Penemalm, P.-A. and Theander, O. (1978). Reaction of D-glucose and glycine in slightly acidic aqueous solutions. *Acta Chem. Scand.*, **B32**, 249–56.

Olsson, K., Pernemalm, P.-A. and Theander, O. (1981). Reaction products and mechanisms in some simple model systems. *Prog. Food Nutr. Sci.*, **5**, 47–55.

Osmond, C. B. and Avadhani, P. N. (1970). Inhibition of the β-carboxylation pathway of CO_2 fixation by bisulphite compounds. *Plant Physiol.*, **45**, 228–30.

Otter, G. E. and Taylor, L. (1971). Estimation and occurrence of acetaldehyde in beer. *J. Inst. Brew.*, **77**, 467–72.

Ouyang, J. M., Duan, H., Chang, S. S. and Ho, C. T. (1980). Formation of carbonyl compounds from β-carotene during palm oil deodourisation. *J. Food Sci.*, **45**, 1214–17, 1222.

Owades, J. L. and Dono, J. M. (1967). Simultaneous determination of volatile acids and sulphur dioxide in alcoholic beverages by microdiffusion. *J. Ass. Off. Anal. Chem.*, **50**, 307–11.

Ozawa, T., Sato, M. and Kwan, T. (1972). The reactive characters of the radical anions SO_3^- and SO_4^- with olefinic compounds. *Chem. Lett.*, 591–4.

Ozawa, T., Setaka, M. and Kwan, T. (1971). ESR studies of the sulphite radical anion. *Bull. Chem. Soc. Japan*, **44**, 3473–4.

Pahlich, E. (1972). Are the multiple forms of glutamate dehydrogenase from pea seedlings conformers? *Planta*, **104**, 78–88.

Pahlich, E., Jäger, H. J. and Steubing, L. (1972). Effect of sulphur dioxide on the activities of glutamate dehydrogenase and glutamine synthesase from pea seedlings. *Angew. Bot.*, **46**, 183–97.

Panahi, S. (1982). The separation and partial characterisation of sulphite derived products in dehydrated cabbage. Ph.D. Thesis, University of Leeds.

Parker, A. J. and Kharasch, N. (1959). The scission of the sulphur–sulphur bond. *Chem. Rev.*, **59**, 583–628.

Parker, D. M., Lodola, A. and Holbrook, J. J. (1978). Use of the sulphite adduct of nicotinamide adenine dinucleotide to study ionisations and the kinetics of lactate dehydrogenase and malate dehydrogenase. *Biochem. J.*, **173**, 959–67.

Parker, J. C. (1969). Influence of 2,3-diphosphoglycerate metabolism on sodium–potassium permeability in human red blood cells: studies with bisulphite and other redox agents. *J. Clin. Investigation*, **48**, 117–25.

Parsons, R. (1959). *Handbook of Electrochemical Constants*, Butterworths, London.

Paul, F. (1954). Determination of the aldehyde and sulphurous acid in wine and fruit juices by the apparatus of Lieb and Zacherl. *Mitt. Rebe. Wein Obstbau Früchteverwert* (*Klosterneuberg*), **4A**, 225–34.

Pauling, L. (1952). Interatomic distances and bond character in the oxygen acids and related substances. *J. Phys. Chem.*, **56**, 361–5.

Pavolini, T. (1930). Reactions with sodium nitroprusside. *Boll. Chem. Farm.*, **69**, 713–22.

Pearson, D. and Wong, T. S. A. (1971). Direct titrimetric method for the rapid estimation of sulphur dioxide in fresh pork sausage. *J. Food Technol.*, **6**, 179–85.

Pearson, R. G. (1968). Hard and soft acids and bases. H.S.A.B. Part 1. Fundamental principles. *J. Chem. Educ.*, **45**, 581–7.

Peck Jr, H. D. (1960). Adenosine 5′-phosphosulphate as an intermediate in the oxidation of thiosulphate by *Thiobacillus thioparus*. *Proc. Nat. Acad. Sci.*, **46**, 1053–7.

Peiser, G. D. and Yang, S. F. (1977). Chlorophyll destruction by the bisulphite–oxygen system. *Plant Physiol.*, **60**, 277–81.

Peiser, G. D. and Yang, S. F. (1978). Chlorophyll destruction in the presence of bisulphite and linoleic acid hydroperoxide. *Phytochem.*, **17**, 79–84.

Peiser, G. D. and Yang, S. F. (1979). Sulphite mediated destruction of β-carotene. *J. Agric. Food Chem.*, **27**, 446–9.

Pendergast, W. (1975). Purine studies. Part XV. Addition of hydrogen sulphite ion to purines. *J. Chem. Soc., Perkin Trans. 1*, 2240–3.

Peterson, S. (1953). Anion exchange processes. *Ann. NY Acad. Sci.*, **57**, 144–58.

Pfeilsticker, K., Mark, F. and Bockisch, M. (1975). On the structure of dehydroascorbic acid in aqueous solution. *Carbohydr. Res.*, **45**, 269–74.

Pfleiderer, G. and Hohnholz, E. (1959). Mechanism of action of dehydrogenases; binary and ternary complexes of malic acid dehydrogenase with coenzyme, substrate and sulphite. *Biochem. Z.*, **331**, 245–53.

Pfleiderer, G., Jackel, D. and Wieland, Th. (1956). Action of sulphite on diphosphopyridine nucleotide (DPN) hydrogenating enzymes. *Biochem. Z.*, **328**, 187–94.

Pfleiderer, G., Sann, E. and Stock, A. (1960). The mechanism of action of dehydrogenases. The reactivity of pyridine nucleotides (PN) and PN-models with sulphite as the nucleophilic reagent. *Ber.*, **93**, 3083–99.

Phelps, D. C. and Hatefi, Y. (1981). Inhibition of $D(-)$-β-hydroxybutyrate dehydrogenase by modifiers of disulphides, thiols and vicinal diols. *Biochem.*, **20**, 453–8.

Pierre, M. (1977). Effect of sulphur dioxide on intermediary metabolism II. Effect of sublethal doses of sulphur dioxide on bean leaf enzymes. *Physiol. Vég.*, **15**, 195–205.

Pitman, I. H. and Jain, N. B. (1979). Covalent addition of bisulphite ion to N-alkylated uracils and thiouracils. *Aust. J. Chem.*, **32**, 545–52.

Pitman, I. H. and Sternson, L. A. (1976). Method for estimating the magnitude of equilibrium constants for covalent addition of nucleophilic reagents to heteroaromatic molecules. *J. Amer. Chem. Soc.*, **98**, 5234–8.

Pitman, I. H. and Ziser, M. A. (1970). Thermodynamics and kinetics of covalent addition of bisulphite ion to pyrimidinum ions. *J. Pharm. Sci.*, **59**, 1295–300.

Pivnick, H. (1980). Spices. In *Microbial Ecology of Foods. Vol. 2. Food Commodities* (J. H. Silliker, R. P. Elliot, A. C. Baird-Parker, F. L. Bryan, J. H. B. Christian, D. S. Clark, J. C. Olson Jr & T. A. Roberts, Eds), Academic Press, London.

Pollard, F. H., Nickless, G. and Glover, R. B. (1964). Chromatographic studies on sulphur compounds. Part V. A study to separate thiosulphate, sulphite and the lower polythionates by anion-exchange chromatography. *J. Chromatog.*, **15**, 533–7.

Ponting, J. D. and Johnson, G. (1945). Determination of sulphur dioxide in fruits. *Ind. Eng. Chem. (Anal. Edn)*, **17**, 682–6.

Ponting, J. D. and Joslyn, M. A. (1948). Ascorbic acid oxidation and the browning in apple tissue extracts. *Arch. Biochem.*, **19**, 47–63.

Ponting, J. D., Jackson, R. and Walters, G. (1972). Refrigerated apple slices: preservative effects of ascorbic acid, calcium and sulphites. *J. Food Sci.*, **37**, 434–6.

Postgate, J. R. (1963). The examination of sulphur auxotrophs: a warning. *J. Gen. Microbiol.* **30**, 481–4.

Potter, E. F. (1954). A modification of the iodine titration method for the determination of sulphur dioxide in dehydrated cabbage. *Food Technol.*, **8**, 269–70.

Potter, E. F. and Hendel, C. E. (1951). Determination of sulphite in dehydrated white potatoes by direct titration. *Food Technol.*, **5**, 473–5.

Pourbaix, M. (1966). *Atlas of Electrochemical Equilibria in Aqueous Solutions*, Pergamon Press, Oxford.

Pratella, G. C., Cimino, A. and Foschi, F. (1979). Inhibition of the germination of *Botrytis cinerea* spores by SO_2. *Frutticoltura*, **41**, 67–9.

Prater, A. N., Johnson, C. M., Pool, M. F. and MacKinney, G. (1944). Determination of sulphur dioxide in dehydrated foods. *Ind. Eng. Chem. (Anal. Edn)*, **16**, 153–7.

Prenner, B. M. and Stevens, J. J. (1976). Anaphylaxis after ingestion of sodium bisulphite. *Annals of Allergy*, **37**, 180–3.

Procaccini, R. L., Bürki, K. and Bresnick, E. (1974). Protection of histones isolated from elastase treated mouse epidermis. *Anal. Biochem.*, **62**, 568–72.

Pryor, W. A. (1966). *Free Radicals*, McGraw-Hill Book Company, New York.

Przeszlakowski, S. and Kocjan, R. (1977). Extraction chromatography of common anions in liquid–liquid anion exchange systems. 1. Strong monobasic acids and their sodium salts as eluants. *Chromatographia*, **10**, 358–63.

Pugh, C. E. M. and Raper, H. S. (1927). CLXXX. The action of tyrosinase on phenols. With some observations on the classification of oxidases. *Biochem. J.*, **21**, 1370–83.

Quattrucci, E. (1981). Potential intakes of sulphur dioxide: the European situation. *J. Sci. Food Agric.*, **32**, 1141–2.

Ralph, E. K. and Grunwald, E. (1969). Kinetics of proton exchange in the ionisation and acid dissociation of imidazole in aqueous acid. *J. Amer. Chem. Soc.*, **91**, 2422–5.

Ranganna, S. and Setty, L. (1968). Non enzymatic discolouration in dried cabbage. Ascorbic acid–amino acid interactions. *J. Agric. Food Chem.*, **16**, 529–33.

Ranganna, S. and Setty, L. (1974a). Non enzymatic discolouration in dried cabbage II. Red condensation product of dehydroascorbic acid and glycine ethyl ester. *J. Agric. Food Chem.*, **22**, 719–22.

Ranganna, S. and Setty, L. (1974b). Non enzymatic discolouration in dried cabbage, III. Decomposition products of ascorbic acid and glycine. *J. Agric. Food Chem.*, **22**, 1139–42.

Rankine, B. C. (1962). A new method for determining sulphur dioxide in wine. *Aust. Wine, Brewing and Spirit Review*, 14–15.

Rao, K. B. and Rao, G. G. (1955). Oxidimetric methods for volumetric estimation of sulphite. *Anal. Chim. Acta*, **13**, 312–22.

Ratkowsky, D. A. and McCarthy, J. L. (1962). Spectrophotometric evaluation of activity coefficients in aqueous solutions of sulphur dioxide. *J. Phys. Chem.*, **66**, 516–19.

Reardon, E. J. (1975). Dissociation contents of some monovalent sulphate ion pairs at 25 °C from stoichiometric activity coefficients. *J. Phys. Chem.*, **79**, 422–5.

Rebelein, H. (1973). Rapid method for alcohol, sugar and total sulphur dioxide (by distillation) in wine and fruit juices, and for alcohol in blood. *Chem. Mikrobiol. Technol. Lebensm.*, **2**, 112–21.

Rehm, H. J. and Wittmann, H. (1963). Antimicrobial action of sulphurous acid II. Action of dissociated and undissociated acid on various microorganisms. *Z. Lebensm. Untersuch.-Forsch.*, **120**, 465–78.

Reifer, I. and Mangan, J. L. (1945). A rapid determination of sulphur dioxide in dehydrated foods. *NZ J. Sci. Technol.*, **A27**, 57–64.

Reith, J. F. and Willems, J. J. L. (1958). Determination of sulphurous acid in foods. *Z. Lebensm. Untersuch.-Forsch.*, **108**, 270–80.

Rensburg, N. J. J. van and Swanepoel, O. A. (1967a). Reactions of unsymmetrical disulphides. I. Sulphitolysis of sulphur derivatives of cysteamine and cysteine. *Arch. Biochem. Biophys.*, **118**, 531–5.

Rensburg, N. J. J. van and Swanepoel, O. A. (1967b). Reactions of unsymmetrical disulphides II. Sulphitolysis of disulphide bonds linking proteins to small molecules. *Arch. Biochem. Biophys.*, **121**, 729–31.

Retcofsky, H. L. and Friedel, R. A. (1970). Sulphur-33 magnetic resonance spectrum of thiophene. *Appl. Spectr.*, **24**, 379–80.

Retcofsky, H. L. and Friedel, R. A. (1972). Sulphur-33 magnetic resonance spectra of selected compounds. *J. Amer. Chem. Soc.*, **94**, 6579–84.

Reuger, D. C. (1977). Investigation of the Complexes formed between *E. coli* Sulphite Reductase and Sulphite. PhD Thesis, Duke University, Durham, NC.

Reynolds, T. M. (1963). Chemistry of non enzymic browning I. The reaction between aldoses and amines. *Adv. Food Res.*, **12**, 1–52.

Reynolds, T. M. (1965). Chemistry of non enzymic browning II. *Adv. Food Res.*, **14**, 167–283.

Ribéreau-Gayon, P. and Stonestreet, E. (1965). Determination of anthocyanins in red wine. *Bull. Soc. Chim. France*, 2649—52.

Richards, E. L. (1956). Non enzymic browning: the reaction between D-glucose and glycine in the 'dry' state. *Biochem. J.*, **64**, 639–44.

Richards, L. and Halperin, J. (1976). Isotopic exchange and substitution reactions of sulphitopentamminecobalt(III) and *cis*-disulphitotetramminecobalt(III). Direct evidence for the specific *trans*-labilising influence of the sulphite ligand. *Inorg. Chem.*, **15**, 2571–2.

Ripper, M. (1892). Sulphurous acid in wines and its determination. *J. Prakt. Chem.*, **46**, 428–73.

Ritchie, C. D. and Gandler, J. (1979). Reactivities of alkythiolate ions in aqueous solution. *J. Amer. Chem. Soc.*, **101**, 7318–23.

Ritchie, C. D. and Virtanen, P. O. I. (1973). Cation–anion combination reactions XI. Reactions of cations with amines. *J. Amer. Chem. Soc.*, **95**, 1882–9.

Roberts, A. C. and McWeeny, D. J. (1972). The use of sulphur dioxide in the food industry. A review. *J. Food Technol.*, **7**, 221–38.

Roberts, J. D. and Caserio, M. C. (1965). *Basic Principles of Organic Chemistry*, W. A. Benjamin Inc., New York, Amsterdam.

Robinson, R. A., Wilson, J. M. and Stokes, R. H. (1941). The activity coefficients of lithium, sodium and potassium sulphate and sodium thiosulphate at 25 °C from isopiestic vapour pressure measurements. *J. Amer. Chem. Soc.*, **63**, 1011–13.

Roman, R. and Dunford, H. B. (1973). Studies on horseradish peroxidase. XII. A kinetic study of the oxidation of sulphite and nitrite by compounds I and II. *Can. J. Chem.*, **51**, 588–96.

Ronzio, A. R. and Waugh, T. D. (1944). Glyoxal bisulphite. *Org. Syntheses*, **24**, 61–4.

Rose, S. A. and Boltz, D. F. (1969). The indirect determination of sulphur dioxide by atomic absorption spectrometry after precipitation of lead sulphate. *Anal. Chim. Acta*, **44**, 239–41.

Ross, J. W., Riseman, J. H. and Krueger, J. A. (1973). Potentiometric gas sensing electrodes. *Pure Appl. Chem.*, **36**, 473–87.

Ross, L. R. and Treadway, R. H. (1972). Rapid sulphur dioxide determination in sulphited potato flakes. *Amer. Potato J.*, **49**, 470–3.

Rothenfusser, S. (1929). Detection and determination of sulphur dioxide. *Z. Untersuch. Lebensm.*, **58**, 98–109.

Roy, A. B. (1976). Sulphatases, lysosomes and disease. *Aust. J. Exp. Biol. Med. Sci.*, **54**, 111–35.

Russev, P., Gough, T. A. and Woollam, C. J. (1976). Detection of inorganic gases using a flame ionisation detector. *J. Chromatog.*, **119**, 461–6.

Růžička, J. and Hansen, E. H. (1974). A new potentiometric gas sensor—the air-gap electrode. *Anal. Chim. Acta*, **69**, 129–41.

Saguy, I., Mannheim, C. H. and Passy, N. (1973). The role of sulphur dioxide and nitrate on detinning of canned grapefruit juice. *J. Food Technol.*, **8**, 147–55.

Sahasrabudhe, M. R., Larmond, E. and Nunes, A. C. (1976). Sulphur dioxide in instant mashed potatoes. *Can. Inst. Food Sci. Technol. J.*, **9**, 207–11.

Sakumoto, A., Miyata, T. and Washino, M. (1975). Kinetics of the sulphonation reaction of 1-dodecene with sodium hydrogen sulphite. *Chem. Lett.*, 563–6.

Sakumoto, A., Miyata, T. and Washino, M. (1976). pH effects on the radical addition of hydrogen sulphite ion to olefins. *Bull. Chem. Soc. Japan*, **49**, 3584–8.

Salama, S. B. and Wasif, Saad (1975). Weak complexes of sulphur and selenium. Part III. Effect of solvent on the stability of 1:1 complexes of sulphur dioxide, sulphinyl dichloride and sulphonyl dichloride with halogen ions. *J. Chem. Soc., Dalton Trans.*, 151–3.

Saletan, L. T. and Scharoun, J. (1965). Automated analyses: diastatic power and sulphur dioxide in beer. *Proc. Am. Soc. Brew. Chem.*, 198–207.

Salo, R. J. and Cliver, D. O. (1978). Inactivation of enteroviruses by ascorbic acid and sodium bisulphite. *Appl. Environmental Microbiol.*, **36**, 68–75.

Salomaa, P. (1956). Two volumetric methods for the determination of glyoxal. *Acta Chem. Scand.*, **10**, 306–10.

Samuelson, O. (1963). *Ion Exchange Separations in Analytical Chemistry.* Almqvist and Wiskell, Stockholm. John Wiley & Sons, New York.

Sankaran, L. and Narasimha Rao, P. L. (1977). Sodium bisulphite addition compounds of benzyl isothiocyanates: evidence for their structure as sodium benzylaminothiomethanesulphonates. *Indian J. Chem.*, **15B**, 386–7.

Sauerbrey, G. (1959). The use of quartz oscillators for weighing thin layers and for microweighing. *Z. Physik.*, **155**, 206–22.

Saunders, D. and Pecsok, R. L. (1968). Calculation of distribution coefficients in inorganic gel chromatography. *Anal. Chem.*, **40**, 44–8.

Saunders Jr, W. H. (1961). Some applications of isotopic sulphur. In *Organic Sulphur Compounds*, Vol. 1 (N. Kharasch, Ed.), Symposium Publications Division, Pergamon Press, Oxford.

Scaringelli, F. P., Saltzman, B. E. and Frey, S. A. (1967). Spectrophotometric determination of atmospheric sulphur dioxide. *Anal. Chem.*, **39**, 1709–19.

Schaeffer, K. and Köhler, W. (1918). Optical study of the constitution of hyposulphurous acid, its salt and ester. *Z. Anorg. Allgem. Chem.*, **104**, 212–50.

Schenck, P. W. and Steudel, R. (1968). Oxides of sulphur. In *Inorganic Sulphur Chemistry* (G. Nickless, Ed.), Elsevier Publishing Co., Amsterdam.

Schenk, R. T. E. and Danishefsky, I. (1951). The addition of bisulphite to unsaturated acids and their derivatives. *J. Org. Chem.*, **16**, 1683–9.

Schiff, H. (1866). A new range of organic diamines. *Ann.*, **140**, 92–137.

Schimz, K. L. (1980). The effect of sulphite on the yeast *Saccharomyces cerevisiae*. *Arch. Microbiol.*, **125**, 89–95.

Schimz, K. L. and Holzer, H. (1979). Rapid decrease of ATP content in intact cells of *Saccharomyces cerevisiae* after incubation with low concentrations of sulphite. *Arch. Microbiol.*, **121**, 225–9.

Schmidkunz, H. (1963). Chemilumineszenz der Sulphitoxidation. PhD Thesis, University of Frankfurt.

Schmidt, M. and Wirwoll, B. (1960). Acids of sulphur (XX). The constitution of sulphurous acid. *Z. Anorg. Allgem. Chem.*, **303**, 184–9.

Schönberg, A. and Moubacher, R. (1952). The Strecker degradation of α-amino acids. *Chem. Rev.*, **50**, 261–77.

Schroeter, L. C. (1963). Kinetics of air oxidation of sulphurous acid salts. *J. Pharm. Sci.*, **52**, 559–63.

Schroeter, L. C. (1966). *Sulphur Dioxide. Applications in Foods, Beverages and Pharmaceuticals*, Pergamon Press, London.

Schubert, S. A., Clayton Jr, J. W. and Fernando, Q. (1980). Simultaneous determination of sulphide, elemental sulphur, sulphite and sulphate in solids by time resolved molecular emission spectrometry. *Anal. Chem.*, **52**, 963–7.

Schultz, H. D., Karr, Jr, C. and Vickers, C. D. (1971). Pulsed NMR sulphur-33 investigations with a superconducting magnet. *Appl. Spectr.*, **25**, 363–8.

Scientific Committee for Food (1981). EUR 7421. Reports of the Scientific Committee for Food (Eleventh Series). Commission of the European Communities, Food Science and Techniques, Luxemburg: Office for Official Publications of the European Communities, pp. 47–8.

Scoggins, M. W. (1970). Ultraviolet spectrophotometric determination of sulphur dioxide. *Anal. Chem.*, **42**, 1091–3.

Scott, K. L. (1974). Preparation and characterisation of *cis*- and *trans*-tetra-amminebis(sulphito)cobaltate(III) anions and kinetics of redox decomposition in aqueous acid solution. *J. Chem. Soc., Dalton Trans.*, 1486–9.

Scott, W. D. and Hobbs, P. V. (1967). The formation of sulphate in water droplets. *J. Atmos. Sci.*, **24**, 54–7.

Segel, I. H. and Johnson, M. J. (1963). Synthesis and characterisation of sodium cysteine-*S*-sulphate monohydrate. *Anal. Biochem.*, **5**, 330–7.

Sekerka, I. and Lechner, J. F. (1978). Determination of sulphite and sulphur dioxide by zero-current chromopotentiometry. *Anal. Chim. Acta*, **99**, 99–104.

Semel, A. and Saguy, M. (1974). Effect of electrolytic tinplate and pack variables on the shelf life of canned pure citrus fruit juices. *J. Food Technol.*, **9**, 459–70.

Shapiro, R. (1977). Genetic effects of bisulphite (sulphur dioxide). *Mutation Res.*, **39**, 149–75.

Shapiro, R. and Weisgras, J. M. (1970). Bisulphite-catalysed transamination of cytosine and cytidine. *Biochim. Biophys. Res. Comm.*, **40**, 839–43.

Shapiro, R., DiFate, V. and Welcher, M. (1974). Deamination of cytosine derivatives by bisulphite. Mechanism of the reaction. *J. Amer. Chem. Soc.*, **96**, 906–12.

Shapiro, R., Servis, R. E. and Welcher, M. (1970). Reactions of uracil and cytosine derivatives with sodium bisulphite. A specific deamination method. *J. Amer. Chem. Soc.*, **92**, 422–4.

Sheppard, W. A. and Bourns, A. N. (1954). Sulphur isotope effects in the bisulphite addition reaction of aldehydes and ketones. I. Equilibrium effect and the structure of the addition product. *Can. J. Chem.*, **32**, 4–13.

Shibuya, A. and Horie, S. (1980). Studies on the composition of the mitochondrial sulphite oxidase system. *J. Biochem. (Tokyo)*, **87**, 1773–84.

Shigematsu, H., Kurata, T., Kato, H. and Fujimaki, M. (1971). Formation of 2-(5-hydroxymethyl-2-formylpyrrol-1-yl)alkyl acid lactones on roasting alkyl-α-amino acid with D-glucose. *Agric. Biol. Chem.*, **35**, 2097–105.

Shih, N. T. and Petering, D. H. (1973). Model reactions for the study of the interaction of sulphur dioxide with mammalian organisms. *Biochem. Biophys. Res. Comm.*, **55**, 1319–25.

Shikanova, M. S., Rodichev, A. G. and Krunchak, V. G. (1977). Ion exchange equilibria on ASD-3 and ASD-4 anion-exchange resins in sodium sulphite–sodium hydrogen sulphite solutions. *Z. Fiz. Khim.*, **51**, 2298–300.

Shimasaki, H., Privett, O. S. and Hara, I. (1977). Studies on the fluorescent products of lipid oxidation in aqueous emulsion with glycine on the surface of silica gel. *J. Amer. Oil Chem. Soc.*, **54**, 119–23.

Shinkai, S. (1978). Coenzyme models, 14. Addition reaction of sulphite ion to monomeric and polymer-bound flavins. *Makromol. Chem.*, **179**, 2637–46.

Shinkai, S. and Kunitake, T. (1977). Coenzyme models 6. Addition reaction of sulphite ion to monomeric and polymeric nicotinamides. Correlation between equilibria and reaction rates. *Makromol. Chem.*, **178**, 145–56.

Shipton, J. (1954). Estimation of sulphur dioxide in dried foods. *Food Preserv. Quart.*, **14**, 54–6.

Siegel, L. M. (1975). Biochemistry of the sulphur cycle. In *Metabolic Pathways, Vol. III, Metabolism of Sulphur Compounds* (D. M. Greenberg, Ed.), Academic Press, London.

Sillen, L. G. (1964). Stability constants of metal ion complexes. Section 1. Inorganic ligands, Special Publ. No. 17, The Chemical Society, London.

Simon, A. and Schmidt, W. (1960). Preparative and spectroscopic investigation of the acid sulphites of rubidium and caesium. *Z. Elektrochem.*, **64**, 737–41.

Simon, A., Waldman, K. and Steger, E. (1956). The constitution of acid sulphite V. The structure of disulphite. *Z. Anorg. Allgem. Chem.*, **288**, 131–47.

Singer, T. P., Edmondson, D. E. and Kenney, W. C. (1976). Recent advances, puzzles and paradoxes in the chemistry of covalently bound flavins. In *Flavins and Flavoproteins* (T. Singer, Ed.), Elsevier Scientific Publishing Co., Amsterdam.

Singh, R. and Ahlawat, T. R. (1974). Bisulphite inhibition of wheat (*Triticum aestivum*) tyrosinase. *Biochem. Physiol. Pflanzen.*, **165**, 317–23.

Small, H., Stevens, T. S. and Bauman, W. C. (1975). Novel ion exchange chromatographic method using conductimetric detection. *Anal. Chem.*, **47**, 1801–9.

Smith, R. M. and Martell, A. E. (1976). *Critical Stability Constants. Vol. 4. Inorganic Complexes.* Plenum Press, New York.

Society of Dyers and Colourists (1971). *Colour Index*, Vol. 2, 3rd edn, The Society of Dyers and Colourists, Bradford.

Somers, T. C. and Ziemelis, G. (1980). Gross interference by sulphur dioxide in standard determination of some phenolics. *J. Sci. Food Agric.*, **31**, 600–10.

Sonobe, H., Kato, H. and Fujimaki, M. (1977). Formation of 1-butyl-5-hydroxymethylpyrrole-2-carboxaldehyde in the Maillard reaction between lactose and butylamine. *Agric. Biol. Chem.*, **41**, 609–10.

Spinelly, J. and Koury, B. (1979). Non enzymic formation of dimethylamine in dried fishery products. *J. Agric. Food Chem.*, **27**, 1104–8.

Stachewski, D. (1975). Production of stable carbon, nitrogen, oxygen and sulphur isotopes. *Chem.-Tech. (Heidelberg)*, **4**, 269–80.

Stadtman, E. R., Barker, H. A., Haas, V. and Mrak, E. M. (1946). Storage of dried apricot. Influence of temperature on deterioration of apricots. *Ind. Eng. Chem.*, **38**, 541–3.

Stadtman, F. H., Chichester, C. O. and MacKinney, G. (1952). Carbon dioxide production in the browning reaction. *J. Amer. Chem. Soc.*, **74**, 3194–6.

Stafford, A. E., Bolin, H. R. and Mackey, B. E. (1972). Absorption of aqueous bisulphite by apricots. *J. Food Sci.*, **37**, 941–3.

Stephen, H. and Stephen, D. T. (1963). *Solubilities of Inorganic and Organic Compounds. Vol. I. Binary Systems*, Part 1, Pergamon Press, Oxford.

Stephens, B. G. and Lindstrom, F. (1964). Spectrophotometric analysis of sulphur dioxide suitable for atmospheric analysis. *Anal. Chem.*, **36**, 1308–12.

Stephens, B. G. and Suddeth, H. A. (1970). Spectrophotometric determination of sulphur dioxide by reduction of iron(III) and simultaneous chelation of the iron(II) with 2,4,6-tri(2-pyridyl)($-$)1,3,5-triazine. *Analyst (London)*, **95**, 70–9.

Stevens, D. J. (1966). The reaction of wheat proteins with sulphite II. The accessibility of disulphide and thiol groups in flour. *J. Sci. Food Agric.*, **17**, 202–4.

Stevens, D. J. (1973). Reaction of wheat proteins with sulphite. III. Measurement of labile and reactive disulphide bonds in gliadin and in the protein of aleurone cells. *J. Sci. Food Agric.*, **24**, 279–83.

Stewart, T. D. and Donnally, L. H. (1932a). The aldehyde bisulphite compounds II. The effect of varying hydrogen ion and of varying temperature upon the equilibrium between benzaldehyde and bisulphite ion. *J. Amer. Chem. Soc.*, **54**, 3555–8.

Stewart, T. D. and Donnally, L. H. (1932b). The aldehyde bisulphite compounds I. The rate of dissociation of benzaldehyde sodium bisulphite as measured by its first order reaction with iodine. *J. Amer. Chem. Soc.*, **54**, 2333–40.

Stone, J. and Laschiver, C. (1957). A sensitive colorimetric technique for the determination of traces of sulphur dioxide in beer. *Wallerstein Labs. Comm.*, No. 20, 361–73.

Stricks, W. and Kolthoff, I. M. (1951). Equilibrium constants of the reactions of sulphite with cysteine and with dithiodiglycolic acid. *J. Amer. Chem. Soc.*, **73**, 4569–74.

Stricks, W., Kolthoff, I. M. and Kapoor, R. C. (1955). Equilibrium constants of the reaction between sulphite and oxidised glutathione. *J. Amer. Chem. Soc.*, **77**, 2057–61.

Sutin, N. (1966). The kinetics of inorganic reactions in solution. *Ann. Rev. Phys. Chem.*, **17**, 119–72.

Swinehart, J. H. (1967). The kinetics of the hexacyanoferrate(III)-sulphite reaction. *J. Inorg. Nucl. Chem.*, **29**, 2313–20.

Swoboda, B. E. P. and Massey, V. (1966). On the reaction of the glucose oxidase from *Aspergillus niger* with bisulphite. *J. Biol. Chem.*, **241**, 3409–16.

Sykes, P. (1965). *A Guidebook to Mechanisms in Organic Chemistry*, 2nd edn, Longmans, London.

Tanaka, K., Ishizuka, T. and Sunahara, H. (1979). Elution behaviour of acids in ion exclusion chromatography using a cation-exchange resin. *J. Chromatog.*, **174**, 153–7.

Tanaka, N. and Luker, C. (1978). Effect of sulphite on microorganisms. *Abstracts of the Annual Meeting of the American Society for Microbiology*, **78**, 188.

Taniguchi, K. and Henke, B. L. (1976). Sulphur $L_{II,III}$ emission spectra and molecular orbital studies of sulphur compounds. *J. Chem. Phys.*, **64**, 3021–35.

Tanner, H. (1963). The estimation of total sulphur dioxide in drinks, concentrates and vinegars. *Mitt. Gebiete Lebensm. Hyg.*, **54**, 158–74.

Tanner, H. and Sandoz, M. (1972). Determination of free sulphur dioxide, ascorbic acid and other reductones with the aid of glyoxal. *Schweiz. Z. Obst- Weinbau*, **108**, 331–7.

Tartar, H. V. and Garretson, H. H. (1941). The thermodynamic ionisation constants of sulphurous acid at 25 °C. *J. Amer. Chem. Soc.*, **63**, 808–16.

Tatum, J. H., Shaw, P. E. and Berry, R. E. (1967). Some compounds formed during non enzymic browning of orange powder. *J. Agric. Food Chem.*, **15**, 773–5.

Taylor, D. H., Miller, M. W., Nury, F. S. and Brekke, J. E. (1961). Temperature effect on colorimetric determination of sulphur dioxide in dried fruits. *J. Ass. Off. Anal. Chem.*, **44**, 641–2.

Teixeira Neto, R. O., Karel, M., Saguy, I. and Miztrahi, S. (1981). Oxygen uptake and β-carotene decolouration in a dehydrated food model. *J. Food Sci.*, **46**, 665–9, 676.

Thacker, M. A., Scott, K. L., Simpson, M. E., Murray, R. S. and Higginson, W. C. E. (1974). Redox decomposition of *trans*-tetra-amineaquosulphitocobalt(III) in aqueous solution. *J. Chem. Soc., Dalton Trans.*, 647–51.

Thalacker, R. (1975). Content of fermentatively produced sulphur dioxide in full flavoured and heavy beers. *Mitteilungsbl. GDCh-Fachgruppe Lebensmittelchem. Gerichtl. Chem.*, **29**, 129–32.

Thewlis, B. H. and Wade, P. (1974). An investigation into the fate of sulphites added to hard sweet biscuit doughs. *J. Sci. Food Agric.*, **25**, 99–105.

Til., H. P., Ferron, V. J. and Groot, A. P. de (1972a). The toxicity of sulphite. I. Long-term feeding and multigeneration studies in rats. *Food Cosmet. Toxicol.*, **10**, 291–310.

Til, H. P., Ferron, V. J., Groot, A. P. de and Van der Wal, P. (1972b). The toxicity of sulphite II. Short- and long-term feeding studies in pigs. *Food Cosmet. Toxicol.*, **10**, 463–73.

Timberlake, C. F. (1980). Anthocyanins—occurrence, extraction and chemistry. *Food Chem.*, **5**, 69–80.

Timberlake, C. F. and Bridle, P. (1967). Flavylium salts, anthocyanidins and anthocyanins. II. Reactions with sulphur dioxide. *J. Sci. Food Agric.*, **18**, 479–85.

Timberlake, C. F. and Bridle, P. (1968). Flavylium salts resistant to sulphur dioxide. *Chem. Ind. (London)*, 1489.

Tokuyama, K., Goshima, K., Maezono, N. and Maeda, T. (1971). A novel degradation pathway of L-ascorbic acid. *Tetrahedron Lett.*, 2503–6.

Tomigana, N. (1978). A sulphite-dependent adenosine triphosphatase of *Thiobacillus thiooxidans*. Its partial purification and some properties. *Plant Cell Physiol.*, **19**, 419–28.

Tompkin, R. B., Christiansen, L. N. and Shaparis, A. B. (1980). Antibotulinal efficacy of sulphur dioxide in meat. *Appl. Environ. Microbiol.*, **39**, 1096–9.

Torrence, P. F. and Tieckelmann, H. (1968). Thiamine biosynthesis II. A novel biosynthetic transformation of the *S*-methyl moiety of methionine. *Biochim. Biophys. Acta*, **158**, 183–5.

Trager, W. and Stoloff, L. (1967). Possible reactions for aflatoxin detoxification. *J. Agric. Food Chem.*, **15**, 679–81.

Trommer, W. E. and Glöggler, K. (1979). Solution conformation of lactate dehydrogenase as studied by saturation transfer ESR spectroscopy. *Biochim. Biophys. Acta*, **571**, 186–94.

Trublin, F., Robert, H., Guyot-Hermann, A. M. and Balatre, P. (1976). Determination of sulphite in pharmacopeial solutions, gelatin and wines. *Ann. Fals. Expert. Chim.*, **69**, 907–21.

Tseng, P. K. C. and Gutknecht, W. F. (1976). Direct potentiometric measurement of sulphite ion with mercuric sulphide/mercurous chloride membrane electrode. *Anal. Chem.*, **48**, 1996–8.

Tsuchida, H. and Komoto, M. (1969). Melanoidin produced from the reaction of glucose and ammonia. Part 1. Chromatographic fractionation of the melanoidin on DEAE-cellulose. *Proc. Res. Soc. Japan Sugar Refineries' Technologists*, **21**, 88–93.

Tsuchida, H., Komoto, M., Kato, H. and Fujimaki, M. (1973). Isolation and identification of two *β*-hydroxypyridine derivatives from the non-dialysable melanoidin hydrolysate. *Agric. Biol. Chem.*, **37**, 403–9.

Tsuchida, H., Komoto, M., Kato, H. and Fujimaki, M. (1975). Isolation of deoxyfructosazine and its 6-isomer from the non-dialysable melanoidin hydrolysate. *Agric. Biol. Chem.*, **39**, 1143–8.

Tsuchida, H., Komoto, M., Kato, H. and Fujimaki, M. (1976). Identification of heterocyclic nitrogen compounds produced by pyrolysis of the non-dialysable melanoidin. *Agric. Biol. Chem.*, **40**, 2051–6.

Tuazon, P. T. and Johnson, S. L. (1977). Free radical and ionic reaction of bisulphite with reduced nicotinamide adenine dinucleotide and its analogues. *Biochem.*, **16**, 1183–8.

Uhteg, L. C. and Westley, J. (1979). Purification and steady-state kinetic analysis of yeast thiosulphate reductase. *Arch. Biochem. Biophys.*, **195**, 211–22.

Urch, D. S. (1969). Direct evidence for $3d$-$2p\pi$ bonding in oxyanions. *J. Chem. Soc. A*, 3026–8.

Urch, D. S. (1970). The origin and intensities of low energy satellite lines in X-ray emission spectra: a molecular orbital interpretation. *J. Phys. C.*, **3**, 1275–91.

Urone, P. F. and Boggs, W. E. (1951). Acid-bleached fuchsin in determination of sulphur dioxide in the atmosphere. *Anal. Chem.*, **23**, 1517–19.

Van Ardsel, W. B., Copley, M. J. and Morgan, A. I. (1973). *Food Dehydration.*

Vol. 2. Practices and Applications, 2nd edn, The AVI Publishing Co., Inc., Westport, Connecticut.

Vas, K. (1949). The equilibrium between glucose and sulphurous acid. *J. Soc. Chem. Ind.*, **68**, 340–3.

Veprek-Šiška, J. and Eckschlager, K. (1965). Separation and detection of oxygen-containing sulphur ions by paper electrophoresis. *Coll. Czech. Chem. Comm.*, **30**, 2544–8.

Verhoef, J. C., Kok, W. Th. and Barendrecht, E. (1978). Mechanism and reaction rate of the Karl-Fischer titration reaction. Part III. Rotating ring-disk electrode measurements—comparison with the aqueous system. *J. Electroanal. Chem.*, **86**, 407–15.

Visser, T. J. (1980). Mechanism of inhibition of iodothyronine-5'-deiodinase by thioureylenes and sulphite. *Biochim. Biophys. Acta*, **611**, 371–8.

Vogel, A. I. (1954). *Textbook of Macro and Semimicro Quantitative Inorganic Analysis*, 4th edn, Longmans, London.

Vogel, A. I. (1961). *Textbook of Quantitative Inorganic Analysis Including Elementary Instrumental Analysis*, 3rd edn, Longmans, London.

Vold, R. R., Sparks, S. W. and Vold, R. L. (1978). A determination of the ^{33}S quadrupole coupling constant in CS_2 from linewidth measurements. *J. Magnetic Resonance*, **30**, 497–503.

Volini, M. and Wang, S. F. (1973). The interdependence of substrate and protein transformations in rhodanese catalysis. III. Enzyme changes outside the catalytic cycle. *J. Biol. Chem.*, **248**, 7392–5.

Vonderschmitt, D. J., Vitols, K. S., Huennekens, F. M. and Scrimgeour, K. G. (1967). Addition of bisulphite to folate and dihydrofolate. *Arch. Biochem. Biophys.*, **122**, 488–93.

Vries, J. G. de and Kellogg, R. M. (1980). Reduction of aldehydes and ketones by sodium dithionite. *J. Org. Chem.*, **45**, 4126–9.

Wade, P. (1971). Technology of biscuit manufacture: Investigation of the process for making semi sweet biscuits. *Chem. Ind.*, 1284–93.

Wade, P. (1972). Action of sodium metabisulphite on the properties of hard sweet biscuit dough. *J. Sci. Food Agric.*, **23**, 333–6.

Wade, P. (1974). The use of SO_2 and some of its compounds in the manufacture of semi sweet biscuits. *Chem. Ind.*, 721–2.

Walker, R., Mendoza-Garcia, M. A., Ioannides, C. and Quattrucci, E. (1983a). Acute toxicity of 3-deoxy-4-sulphohexosulose in rats and mice and *in vitro* mutagenicity in the Ames test. *Food Chem. Toxicol.*, **21**, 299–303.

Walker, R., Mendoza-Garcia, M. A., Quattrucci, E. and Zerilli, M. (1983b). Metabolism of 3-deoxy-4-sulphohexosulose, a reaction product of sulphite in foods, by rat and mouse. *Food Chem. Toxicol.*, **21**, 291–7.

Wallnöfer, P. and Rehm, H. J. (1965). Antimicrobial action of sulphurous acid. V. The action of sulphurous acid on the metabolism of respiring and fermenting yeast and *Escherichia coli* cells. *Z. Lebensm. Untersuch.-Forsch.*, **127**, 195–206.

Wang, J. C. and Himmelblau, D. M. (1964). A kinetic study of sulphur dioxide in aqueous solution with radioactive tracers. *A.I.Ch.E. Journal*, **10**, 574–80.

Wang, R. Y. H., Gehrke, C. W. and Ehrlich, M. (1980). Comparison of bisulphite modification of 5-methyldeoxycytidine and deoxycytidine residues. *Nucleic Acid Res.*, **8**, 4777–90.

Weatherill, G. B. (1969). *Sample, Inspection and Quality Control.* Methuen & Co. Ltd, London.

Webber, L. M. and Guilbault, G. G. (1977). The adaptation of coated piezoelectric devices for use in the detection of gases in aqueous solutions. *Anal. Chim. Acta,* **93**, 145–51.

Wedzicha, B. L. (1984). A kinetic model for the sulphite inhibited Maillard reaction. *Food Chem.,* in press.

Wedzicha, B. L. and Bindra, N. K. (1980). A rapid method for the determination of sulphur dioxide in dehydrated vegetables. *J. Sci. Food Agric.,* **31**, 286–8.

Wedzicha, B. L. and Galsworthy, D. (1982). Unpublished results.

Wedzicha, B. L. and Imeson, A. P. (1977). The yield of 3-deoxy-4-sulphopentosulose in the sulphite inhibited non-enzymic browning of ascorbic acid. *J. Sci. Food Agric.,* **28**, 669–72.

Wedzicha, B. L. and Johnson, M. K. (1979). A variation of the Monier Williams distillation technique for the determination of sulphur dioxide in ginger ale. *Analyst (London),* **104**, 694–6.

Wedzicha, B. L. and Lamikanra, O. (1982). Unpublished results.

Wedzicha, B. L. and Lamikanra, O. (1983). Sulphite mediated destruction of β-carotene: the partial characterisation of reaction products. *Food Chem.,* **10**, 275–83.

Wedzicha, B. L. and McWeeny, D. J. (1974a). Non enzymic browning reactions of ascorbic acid and their inhibition. The production of 3-deoxy-4-sulphopentosulose in mixtures of ascorbic acid, glycine and bisulphite ion. *J. Sci. Food Agric.,* **25**, 577–87.

Wedzicha, B. L. and McWeeny, D. J. (1974b). Non-enzymic browning reactions of ascorbic acid and their inhibition. The identification of 3-deoxy-4-sulphopentosulose in dehydrated sulphited cabbage after storage. *J. Sci. Food Agric.,* **25**, 589–92.

Wedzicha, B. L. and McWeeny, D. J. (1975). Concentrations of some sulphonates derived from sulphite in certain foods and preliminary studies on the nature of other sulphite derived products. *J. Sci. Food Agric.,* **26**, 327–35.

Wedzicha, B. L. and Rumbelow, S. J. (1981). The reaction of an azo food dye with hydrogen sulphite ions. *J. Sci. Food Agric.,* **32**, 699–704.

Wei, L. S., Siregar, J. A., Steinberg, M. P. and Nelson, A. I. (1967). Overcoming the bacteriostatic activity of onion in making standard plate counts. *J. Food Sci.,* **32**, 346–9.

Weier, T. E. and Stocking, C. R. (1946). Stability of carotene in dehydrated carrots impregnated with antioxidants. *Science,* **104**, 437–8.

West, P. W. and Gaeke, G. C. (1956). Fixation of sulphur dioxide as disulphitomercurate(II) and subsequent colorimetric estimation. *Anal. Chem.,* **28**, 1816–19.

Weygand, F. (1939). Preparation of *N*-glucosides of aniline and substituted anilines. *Ber.,* **72B**, 1663–7.

Wheaton, R. M. and Bauman, W. C. (1953). Ion exclusion. A unit operation utilising ion exchange materials. *Ind. Eng. Chem.,* **45**, 228–33.

Whitaker, J. R. (1972). *Principles of Enzymology for the Food Sciences.* Vol. 2 of *Food Science,* a series of monographs (O. R. Fennema, Ed.), Marcel Dekker Inc., New York.

Wiberg, K. B., Maltz, H. and Masaya, O. (1968). Mechanism of the ferricyanide oxidation of thiols. *Inorg. Chem.*, **7**, 830–1.

Williams, R. R., Waterman, R. E., Keresztesy, J. C. and Buchman, E. R. (1935). Studies of crystalline vitamin B_1. III. Cleavage of vitamin with sulphite. *J. Amer. Chem. Soc.*, **57**, 536–7.

Wilson, B. J. (1966). *The Radiochemical Manual*, 2nd edn, The Radiochemical Centre, Amersham.

Wilson, C. R. and Andrews, W. H. (1976). Sulphite compounds as neutralisers of spice toxicity for *Salmonella*. *J. Milk Food Technol.*, **39**, 464–6.

Winefordner, J. D., Williams, H. P. and Miller, C. D. (1965). A high sensitivity detector for gas analysis. *Anal. Chem.*, **37**, 161–4.

Winkelmann, D. (1955). Electrochemical measurements of the rate of oxidation of Na_2SO_3 by dissolved oxygen. *Z. Electrochem.*, **59**, 891–5.

Winter, E. R. S. and Briscoe, H. V. A. (1951). Oxygen atom transfer during the oxidation of aqueous sodium sulphite. *J. Amer. Chem. Soc.*, **73**, 496–7.

Wisser, K., Völter, I. and Heimann, W. (1970). About the reaction of sulphurous acid during oxidation. II. Binding of sulphurous acid with oxidation products of ascorbic acid. *Z. Lebensm. Untersuch.-Forsch.*, **42**, 180–5.

Witekowa, S. and Lewicki, A. (1963). Kinetics of the reaction of iodine with sulphur dioxide at the liquid–liquid interface. *Roczniki Chem.*, **37**, 91–107.

Witekowa, S. and Witek, T. (1955). Reaction between iodine and sulphur dioxide. *Zeszyty Nauk. Politech. Lodz No. 5, Chem. Spozywcza*, No. 1, 73–86. Through *Chemical Abstracts*, **50**, 14422g.

Wolfrom, M. L. and Rooney, C. S. (1953). Chemical interactions of amino compounds and sugars. VIII. Influence of water. *J. Amer. Chem. Soc.*, **75**, 5435–6.

Wolfrom, M. L., Kolb, D. K. and Langer Jr, A. W. (1953a). Chemical interactions of amino compounds and sugars. VII. pH dependency. *J. Amer. Chem. Soc.*, **75**, 3471–3.

Wolfrom, M. L., Schlicht, R. C., Langer Jr, A. W. and Rooney, C. S. (1953b). Chemical interactions of amino compounds and sugars. VI. The repeating units in brown polymers. *J. Amer. Chem. Soc.*, **75**, 1013.

Wong, T. C., Luh, B. S. and Whitaker, J. R. (1971). Isolations and characterisation of polyphenol oxidase isoenzymes of clingstone peach. *Plant Physiol.*, **48**, 19–23.

Wood, H. W. (1955). Paper electrophoresis of reducing agents. *Nature*, **175**, 1084–5.

Woodroof, J. G. and Luh, B. S. (1975). *Commercial Fruit Processing*, The AVI Publishing Co., Inc., Westport, Connecticut.

Würdig, G. and Schlotter, H. A. (1971). Occurrence of sulphur dioxide-forming yeasts in a natural grape-must yeast mixture. *Deut. Lebensm. Rundschau*, **67**, 86–91.

Yamazaki, I. and Piette, L. H. (1963). The mechanism of aerobic oxidase reaction catalysed by peroxidase. *Biochim. Biophys. Acta*, **77**, 47–64.

Yang, S. F. (1970). Sulphoxide formation from methionine or its sulphide analogues during aerobic oxidation of sulphite. *Biochem.*, **9**, 5008–14.

Yang, S. F. (1973). Destruction of tryptophan during the aerobic oxidation of sulphite ions. *Environ. Res.*, **6**, 395–402.

Yano, M., Hayashi, T. and Namiki, M. (1974). On the structure of the free radical products formed by the reaction of dehydroascorbic acid with amino acids. *Chem. Lett.*, 1193–6.

Yano, M., Hayashi, T. and Namiki, M. (1976*a*). On the formation of free radical products by the reaction of dehydroascorbic acid with amines. *Agric. Biol. Chem.*, **40**, 1209–15.

Yano, M., Hayashi, T. and Namiki, M. (1976*b*). Formation of free radical products by the reaction of dehydroascorbic acid with amino acids. *J. Agric. Food Chem.*, **24**, 815–19.

Yano, M., Hayashi, T. and Namiki, M. (1978). Formation of a precursor of the free radical species in the reaction of dehydroascorbic acid with amino acids. *Agric. Biol. Chem.*, **42**, 2239–43.

Yllner, S. (1956). Action of neutral sulphite solution on carbohydrates at high temperatures. Part I. A sulphocarboxylic acid derived from xylose. *Acta Chem. Scand.*, **10**, 1251–7.

Yoshida, A. (1981). Determination of dimethylamine in shrimps bleached with sulphite and the formation of nitrosodimethylamine in the presence of nitrite. *Proceedings of the Osaka Prefectural Institute of Public Health, Food Sanitation*, 113–17. Through *Food Science and Technology Abstracts*, 1982, **14**, 9R555.

Yoshida, M., Sone, N., Hirata, H. and Kagawa, Y. (1975). A highly stable adenosine triphosphatase from a thermophilic bacterium. Purification properties and reconstruction. *J. Biol. Chem.*, **250**, 7910–16.

Yoshihiro, Y., Kuroiwa, S. and Nakamura, M. (1961). The role of 5-(hydroxymethyl)-2-furaldehyde in melanoidin formation in heated glucose solution. *Kogyo Kagaku Zasshi*, **64**, 551–5.

Yoshikawa, S. and Orii, Y. (1972*a*). The inhibition mechanism of the cytochrome oxidase reaction. II. Classification of inhibitors based on their modes of action. *J. Biochem. (Tokyo).*, **71**, 859–72.

Yoshikawa, S. and Orii, Y. (1972*b*). The inhibition mechanism of the cytochrome oxidase reaction. III. Combined effects of inhibitors on cytochromeoxidase. *J. Biochem. (Tokyo)*, **71**, 873–6.

Yost, H. T. (1972). *Cellular Physiology*, Prentice-Hall, New York.

Young, P. R. and Jencks, W. P. (1977). Nonenforced catalysis of the bisulphite carbonyl addition reaction by hydrogen bonding. *J. Amer. Chem. Soc.*, **99**, 1206–14.

Young, P. R. and Jencks, W. P. (1979). Separation of polar and resonance substituent effects in the reactions of acetophenones with bisulphite and of benzyl halides with nucleophiles. *J. Amer. Chem. Soc.*, **101**, 3288–94.

Ziegler, I. (1972). The effect of SO_3^{2-} on the activity of ribulose-1,5-diphosphate carboxylase in isolated spinach chloroplasts. *Planta*, **103**, 155–63.

Ziegler, I. (1973). Effect of sulphite on phosphenol pyruvate carboxylase and malate formation in extracts of *Zea mays*. *Phytochem.*, **12**, 1027–30.

Ziegler, I. (1974*a*). Action of sulphite on plant malate dehydrogenase. *Phytochem.*, **13**, 2411–16.

Ziegler, I. (1974*b*). Malate dehydrogenase in *Zea mays*: properties and inhibition by sulphite. *Biochim. Biophys. Acta*, **364**, 28–37.

Ziegler, I., Marewa, A. and Schoepe, E. (1976). Action of sulphite on the substrate kinetics of chloroplastic NADP-dependent glyceraldehyde-3-phosphate dehydrogenase. *Phytochem.*, **15**, 1627–32.

Zoltewicz, J. A. and Kauffman, G. M. (1977). Kinetics and mechanics of the cleavage of thiamine, 2-(1-hydroxyethyl)thiamine and a derivative by bisulphite ion in aqueous solution. Evidence for an intermediate. *J. Amer. Chem. Soc.*, **99**, 3134–42.

Zoltewicz, J. A. and Uray, G. (1981*a*). Nature of the reaction of thiamine in the presence of low concentrations of sulphite ion. Competitive trapping. *J. Org. Chem.*, **46**, 2398–2400.

Zoltewicz, J. A. and Uray, G. (1981*b*). Hydroxide ion as a catalyst for nucleophilic substitution of thiamine by thiolate ions. A rival for sulphite ion. *J. Amer. Chem. Soc.*, **103**, 683–4.

Zoltewicz, J. A., Uray, G. and Kauffman, G. M. (1980). Evidence for an intermediate in nucleophilic substitution of a thiamine anologue. Change from first- to second-order kinetics in sulphite ion. *J. Amer. Chem. Soc.*, **102**, 3653–4.

Zonneveld, H. and Meyer, A. (1960). Determination of sulphurous acid in foods, especially in dried vegetables. *Z. Lebensm. Untersuch.-Forsch.*, **111**, 198–207.

Index

Unless otherwise indicated, sulphur(IV) oxospecies are not distinguished from one another in the index and are represented jointly by S(IV).